About Island Press

Since 1984, the nonprofit Island Press has been stimulating, shaping, and communicating the ideas that are essential for solving environmental problems worldwide. With more than 800 titles in print and some 40 new releases each year, we are the nation's leading publisher on environmental issues. We identify innovative thinkers and emerging trends in the environmental field. We work with world-renowned experts and authors to develop cross-disciplinary solutions to environmental challenges.

Island Press designs and implements coordinated book publication campaigns in order to communicate our critical messages in print, in person, and online using the latest technologies, programs, and the media. Our goal: to reach targeted audiences—scientists, policymakers, environmental advocates, the media, and concerned citizens—who can and will take action to protect the plants and animals that enrich our world, the ecosystems we need to survive, the water we drink, and the air we breathe.

Island Press gratefully acknowledges the support of its work by the Agua Fund, Inc., Annenberg Foundation, The Christensen Fund, The Nathan Cummings Foundation, The Geraldine R. Dodge Foundation, Doris Duke Charitable Foundation, The Educational Foundation of America, Betsy and Jesse Fink Foundation, The William and Flora Hewlett Foundation, The Kendeda Fund, The Forrest and Frances Lattner Foundation, The Andrew W. Mellon Foundation, The Curtis and Edith Munson Foundation, Oak Foundation, The Overbrook Foundation, the David and Lucile Packard Foundation, The Summit Fund of Washington, Trust for Architectural Easements, Wallace Global Fund, The Winslow Foundation, and other generous donors.

The opinions expressed in this book are those of the author(s) and do not necessarily reflect the views of our donors.

About SCOPE

The Scientific Committee on Problems of the Environment (SCOPE) was established by the International Council for Science (ICSU) in 1969. It brings together natural and social scientists to identify emerging or potential environmental issues and to address jointly the nature and solution of environmental problems on a global basis. Operating at an interface between the science and decision-making sectors, SCOPE's interdisciplinary and critical focus on available knowledge provides analytical and practical tools to promote further research and more sustainable management of the Earth's resources. SCOPE's members, national scientific academies and research councils, and international scientific unions, committees and societies, guide and develop its scientific program.

About DIVERSITAS

DIVERSITAS, the international program of biodiversity science, was established, with its current structure, in 2002 under the auspices of ICSU, IUBS, SCOPE and UNESCO. DIVERSITAS has a dual mission: 1) to promote an integrative biodiversity science, linking biological, ecological and social disciplines in an effort to produce socially relevant new knowledge; and 2) to provide the scientific basis for the conservation and sustainable use of biodiversity. The scientific landscape, in relation to global biodiversity issues, is increasingly organizing itself into four interconnected spheres: scientific research, observations, scientific assessments and policymaking. With research as its foundation, DIVERSITAS is located at the intersection of these four scientific spheres. DIVERSITAS is also a founding member of the Earth System Science Partnership (ESSP), a family of the four global change programs (DIVERSITAS, IGBP, IHDP and WCRP) for the integrated study of the Earth System.

SCOPE 69

Biodiversity Change and Human Health

THE SCIENTIFIC COMMITTEE ON PROBLEMS OF THE ENVIRONMENT (SCOPE)

SCOPE SERIES

SCOPE 1–59 in the series were published by John Wiley & Sons, Ltd., U.K. Island Press is the publisher for SCOPE 60 as well as subsequent titles in the series.

SCOPE 60: *Resilience and the Behavior of Large-Scale Systems,* edited by Lance H. Gunderson and Lowell Pritchard Jr.

SCOPE 61: *Interactions of the Major Biogeochemical Cycles: Global Change and Human Impacts,* edited by Jerry M. Melillo, Christopher B. Field, and Bedrich Moldan

SCOPE 62: *The Global Carbon Cycle: Integrating Humans, Climate, and the Natural World,* edited by Christopher B. Field and Michael R. Raupach

SCOPE 63: *Invasive Alien Species: A New Synthesis,* edited by Harold A. Mooney, Richard N. Mack, Jeffrey A. McNeely, Laurie E. Neville, Peter Johan Schei, and Jeffrey K. Waage

SCOPE 64: *Sustaining Biodiversity and Ecosystem Services in Soils and Sediments,* edited by Diana H. Wall

SCOPE 65: *Agriculture and the Nitrogen Cycle: Assessing the Impacts of Fertilizer Use on Food Production and the Environment,* edited by Arvin R. Mosier, J. Keith Syers, and John R. Freney

SCOPE 66: *The Silicon Cycle: Human Perturbations and Impacts on Aquatic Systems,* edited by Venugopalan Ittekkot, Daniela Unger, Christoph Humborg, and Nguyen Tac An

SCOPE 67: *Sustainability Indicators: A Scientific Assessment,* edited by Tomáš Hák Bedřich Moldan, and Arthur Lyon Dahl

SCOPE 68: *Communicating Global Change Science to Society: An Assessment and Case Studies,* edited by Holm Tiessen, Gerhard Breulmann, Michael Brklacich, and Rômulo S. C. Menezes

SCOPE 69: *Biodiversity Change and Human Health: From Ecosystem Services to Spread of Disease,* edited by Osvaldo E. Sala, Laura A. Meyerson, and Camille Parmesan

SCOPE 69

Biodiversity Change and Human Health

From Ecosystem Services to Spread of Disease

Edited by
Osvaldo E. Sala, Laura A. Meyerson,
and Camille Parmesan

A project of SCOPE, the Scientific Committee on
Problems of the Environment, and DIVERSTAS, an interna-
tional programme of biodiversity science, of the
International Council for Science

ISLANDPRESS

Washington • Covelo • London

Library of Congress Cataloging-in-Publication Data

Biodiversity change and human health : from ecosystem services to spread of disease /
edited by Osvaldo E. Sala, Laura A. Meyerson, and Camille Parmesan.
p. cm. — (SCOPE ; 69)
"A project of SCOPE, the Scientific Committee on Problems of the Environment, and
DIVERSITAS, an international programme of biodiversity science, of the International
Council for Science."
Includes bibliographical references.
ISBN-13: 978-1-59726-496-9 (cloth : alk. paper)
ISBN-10: 1-59726-496-2 (cloth : alk. paper)
ISBN-13: 978-1-59726-497-6 (pbk. : alk. paper)
ISBN-10: 1-59726-497-0 (pbk. : alk. paper)
1. Biodiversity. 2. Environmental health. I. Sala, Osvaldo E. II. Meyerson, Laura A.
III. Parmesan, Camille, 1961–
QH541.15.B56B572   2008
304.2—dc22   2008006588

Printed on recycled, acid-free paper ✺

Manufactured in the United States of America
10 9 8 7 6 5 4 3 2 1

We dedicate this book to Hal Mooney who has been a mentor and friend to all of us and a pioneer in studies on the importance of biodiversity to human well-being and the health of our planet.

Contents

Part II: Biodiversity and Quality of Life: Beyond Physical Health

Part III: Decay of Ecosystem Services Following Biodiversity Change, and Consequent Impacts on Human Health

Part IV: Biodiversity Change and the Spread of Infectious Diseases

Part V: Biodiversity as a Resource for Medicine

Acknowledgments

The workshop on Biodiversity, Health and the Environment, held at UNESCO head-quarters in Paris in March 2005, was convened by SCOPE—the Scientific Committee on Problems of the Environment, and DIVERSITAS—the international program of biodiversity science, both ICSU bodies.

It brought together experts from a wide range of disciplines and many countries for an intense week of stimulating discussion. The chapters in this volume draw on and reflect the extraordinary level of intellectual openness, new ideas, and novel approaches generated by the workshop participants. The Editors would like to express their thanks to them, and to the authors and reviewers of the contributions to this volume.

The workshop and this publication benefited greatly from the expertise and experi-ence of the SCOPE and DIVERSITAS Secretariats, critical to the success of the whole project, from the initial planning processes, to the workshop itself, and delivery of the final products. Particular thanks are due to Anne Larigauderie, Anne-Hélène Prieur-Richard, Véronique Plocq Fichelet and Susan Greenwood Etienne.

SCOPE and DIVERSITAS acknowledge with grateful thanks the sponsorship pro-vided by the International Council for Science (ICSU), UNESCO, the French Ministry of Research and Education (Ministère de l'Education Nationale, de l'Enseignement Supérieur et de la Recherche), and the Institut Français de la Biodiversité (IFB). The workshop and this book would not have been possible without their generous support.

The workshop organizers also gratefully acknowledge the following individuals for their participation in the Paris meeting: Salvatore Arico, Catherine Courtet, Jacob Koella, Jean-Claude Lefeuvre, Luiz A. Martinelli, Aditya Purohit, and Jacques Weber.

1

Changes in Biodiversity and Their Consequences for Human Health

Osvaldo E. Sala, Laura A. Meyerson,
and Camille Parmesan

Human actions and their effects on the environment are indisputably the major forces driving changes in biodiversity at both global and local levels. Human-driven climate change, habitat destruction, invasive species introductions, and nitrogen deposition all result in biodiversity losses (Sala et al. 2000). Some of the general impacts of impoverished biodiversity include decreased food production, decreased food security, the loss of resources for indigenous medicine, diminished supplies of raw materials for new pharmaceuticals and biotechnology, and threats to water quality (Grifo and Rosenthal 1997). There is general agreement among the scientific community that maintenance of natural levels of biodiversity is necessary for proper ecosystem functioning and the provision of ecosystem services (e.g., the benefits people receive from nature) to humankind (Chapin et al. 2000; Schulze and Mooney 1994). Research indicates that it is the diversity of biota across the world's ecosystems that underpins the capacity of the Earth's ecosystems to provide most of its goods and services (MEA 2003; Rapport, Costanza, and McMichael 1998; Sala et al. 2000; Schulze and Mooney 1994; WRI 2000).

However, impacts on human health have largely been focused on direct impacts, such as rising human disease exposure due to increased disease vector populations left unchecked by predators. While such blatant consequences of anthropogenic disruption are important, they represent only a fraction of unwanted outcomes. More subtle and more indirect effects have often been suggested, but very few rigorous scientific studies document explicit links between natural biodiversity, its degradation, and human health.

Several authors have addressed the link between biodiversity and ecosystem functioning (Chapin et al. 2000; Folke, Holling, and C. Perrings 1996; Schulze and Mooney

1

1994; Schwartz et al. 2000), but it is still unclear which ecosystem functions are primarily important to sustain our health. Nonetheless, four general types of "human health functions" of ecosystems can be distinguished. First, ecosystems provide us with basic human needs, such as food, clean air, clean water, and clean soils (i.e., ecosystem services). Second, they prevent the spread of disease through biological control. Third, ecosystems provide us with medical and genetic resources that are necessary to prevent or cure diseases. Finally, biodiversity contributes to the maintenance of mental health by providing opportunities for recreation, creative outlets, therapeutic retreats, and cognitive development (de Groot et al. 2002). Thus biodiversity loss may result in compromised ecosystem functions, which, in turn, may negatively influence human health, both directly and indirectly. In this book, we review each of the four general human health functions of ecosystems in turn, highlighting the urgent need for more detailed and comprehensive research on the human health consequences of biodiversity loss.

Previous Efforts

In 1997, Island Press published a volume titled *Biodiversity and Human Health*, edited by Francesca Grifo and Joshua Rosenthal (Grifo and Rosenthal 1997). In this work, contributing authors discussed the benefits and drawbacks of mining biodiversity for herbal and pharmaceutical medicines, and the role of bioprospecting in both exploiting and conserving biodiversity. In the same year, *Nature's Services: Societal Dependence of Natural Ecosystems,* by Gretchen C. Daily (Daily 1997), was also published by the same press. Although neither of these books directly addressed human health issues, they eloquently summarized the critical role that ecosystem services play in society (e.g., decomposition and recycling of wastes, pollination of food crops, and pest control). These research areas have begun to converge, as evidenced both in primary literature (e.g., Daszak, Cunningham, and Hyatt 2000; Dobson et al. 2006) and by recent international efforts, such as the Millennium Ecosystem Assessment publication *Ecosystems and Human Well-Being* (MEA 2003), in which ecosystem change, ecosystem services, and human health are addressed in concert.

Along the way, there have been many other important contributions aimed at increasing awareness and understanding of these relationships by the policy sector and by the general public. Early contributions, such as *The Biophilia Hypothesis,* edited by S. R. Kellert and E. O. Wilson (1993), were very successful in reaching out to a lay audience. This growing panoply continues with *Sustaining Life: How Human Health Depends on Biodiversity,* edited by Eric Chivian and Aaron Bernstein, published in 2008 (Chivian and Bernstein 2008). In this volume, contributors present multiple, and sometimes opposing, views on the benefits, consequences, and trade-offs involved with preserving biological diversity while sustaining human health and well-being.

Other policy-oriented initiatives are also emerging in addition to the Millennium Ecosystem Assessment. In a multi-institutional effort, the U.S. Environmental Protection Agency National Center for Environmental Research (US EPA NCER), the World Conservation Union (IUCN), the Smithsonian Institution, and the Yale Institute for Biospheric Studies organized a multi-institutional, interdisciplinary forum and workshop on biodiversity and human health, which was held in Washington, DC, in September 2006. A specific research solicitation on the topic was issued by the EPA and a proceedings was published (EPA 2006; Pongsiri and Roman 2007). The assembled papers examine such topics as epidemiology and vector ecology, climate change, biodiversity, and health; wildlife trade and the spread of exotics and disease; pharmaceuticals; the role of biodiversity in natural catastrophes; the valuation of biodiversity for public health; and applications of research to the Global Earth Observation System of Systems (GEOSS) (EPA 2006).

The Purpose of This Book

The present volume is the end product of a workshop, Biodiversity, Health, and the Environment (March 14–18, 2005), organized by the Scientific Committee on Problems of the Environment (SCOPE), Diversitas, and the United Nations Educational, Scientific and Cultural Organization (UNESCO). A week of intense work at UNESCO headquarters in Paris resulted in a working draft, followed by extensive editing, incorporation of additional material from outside contributors, and repeated updates as new literature came out. Our aim has been to synthesize the current state of science on the relationships between biological diversity and human health by bringing together information and perspectives from the natural and social sciences as well as the medical community.

The chapters in this book explore the explicit linkages between human-driven alterations of biodiversity and documented impacts of those changes on human health. Our emphasis throughout has been to clearly distinguish the results of rigorous scientific studies from conjecture based on indirect evidence or expert opinion. We also wanted to encompass the broader definition of health used by the World Health Organization as not only physical measures, but overall well-being and quality of life as well. To fulfil this broader mandate, in addition to experts in ecology, evolution, and medicine, our group also included experts from the health-related fields of psychology and sociology. With this diverse group of researchers, we formulated critical assessments of the trade-offs and synergies between human well-being and biodiversity. Finally, we explored potential points of crossover among disciplines, in ways of thinking and in specific methodologies, which we believe could ultimately expand opportunities for humans both to live sustainably and to enjoy a desirable quality of life.

Four Ways in Which Biodiversity Affects Human Health

When biodiversity is considered to be critical to human health and well-being, four major "biodiversity drivers" of human health can be identified: quality of life, medicinal

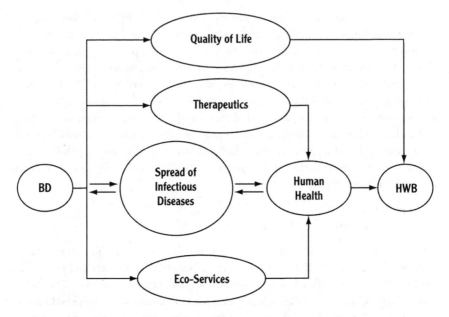

Figure 1.1. The four health functions of ecosystems. Biodiversity and human health can be related through the four "biodiversity drivers" of human health and well-being: quality of life, medicinal and genetic resources, constraints on infectious disease, and ecosystem services. BD = biodiversity; HWB = health and well-being.

and genetic resources, constraints on infectious disease, and ecosystem services (figure 1.1):

- *Quality of life* (including human health) has been defined by the World Health Association as "an individual's perception of their position in life, in the context of the culture and value systems in which they live and in relation to their goals, expectations and concerns (WHOQOL Group 1994; see also this volume, chapter 7). Health includes not only the absence of disease or infirmity, but also physical, mental, and social well-being.
- *Medicinal and genetic resources* are defined as drugs, in pure form or as crude extracts of fauna and flora, used in treatment of illness.
- *Constraints on infectious disease* refers to situations in which higher levels of biodiversity reduce the spread of infectious diseases and in which depleted biodiversity may accelerate the spread of diseases to human populations.
- *Ecosystem services* are defined as those benefits that humans receive from ecosystems (MEA 2003).

Figure 1.1 demonstrates the interrelationships between biological diversity, human health, and these four drivers that are integral to health and well-being and are products of biodiversity. For example, ecosystems can prevent the spread of infectious diseases through biological control of disease vectors and can also provide humans with medicinal and genetic resources. Biodiversity contributes to the maintenance of mental and spiritual health (part of the quality of life) through opportunities for recreation and cognitive development (de Groot, Wilson, and Boumans 2002). Biodiversity loss could therefore negatively affect human health and well-being.

We acknowledge the overlaps among the drivers that we have identified. For example, the provision of medicinal and genetic resources from flora and fauna is also considered an ecosystem service. However, addressing the drivers separately facilitates discussions relating specifically to human health. Identifying the trade-offs and synergies among these biodiversity drivers of human health potentially offers decision makers and resource managers a clearer array of choices when making natural resource decisions that allocate biodiversity among the drivers.

The Structure of This Book

Figure 1.1 is the foundation on which the workshop was based and provides a structure for this book: following this introductory chapter, 11 of the book's 16 chapters focus on one of the biodiversity drivers of human health and well-being. We begin, however, with four cross-cutting chapters resulting from the discussions that occurred during the early spring workshop in Paris. These chapters reflect interdisciplinary work and highlight the connections among biodiversity drivers.

While there is much agreement across disciplines about the importance of biodiversity to human health and well-being, the contrasts in perspectives that arise in the chapters that follow are both enlightening and at times startling. Part I (chapters 2–5) tackles issues that cut across many disciplines, and that require true interdisciplinary research for further progress to be made. Chapter 2 explores the linkages and feedbacks between biodiversity, agriculture, human health, and ecosystem functioning. Chapter 3 argues for comprehensive policies that both protect human health and biodiversity by reducing human stressors and achieving healthy ecosystems. Chapter 4 explores the application of social science methodologies and economic models to decision-making to maximize human health and well-being while simultaneously minimizing negative biodiversity impacts. Chapter 5 explores the theme of sustainability in more detail, discussing the relationships, trade-offs, and synergies that exist between biodiversity, human health, and sustainable development.

Parts II (chapters 6 and 7) and III (chapters 8–10) provide up-to-date syntheses of the different component issues that form the basis of the cross-cutting chapters in part I. Chapters 6 and 7 address the effects of biodiversity on quality of life that have developed over both ecological and evolutionary time, but which in many ways remain cryptic because of the paucity of quantitative approaches to empirically define these relationships. Chapters 8–10 discuss how changes in biodiversity affect the provisioning of ecosystem services, and ultimately human health, by focusing, in turn, on aquatic biodiversity, alterations in the global nitrogen cycle, and microbial biodiversity.

Part IV (chapters 11–14) explores the role of biodiversity on the spread of infectious disease. Chapters 11 and 12 review both the positive and the negative impacts of microbial biodiversity on human health and ecosystem functioning, while chapter 13 reviews mechanisms by which biodiversity can affect the risk and spread of infectious disease. Chapter 14 synthesizes existing evidence that the documented impacts of climate change on the geographic ranges, phenologies, and population dynamics of wild species may have already affected human health, and the implications for future impacts.

Part V (chapter 15) examines biodiversity as a resource for medicinal uses. The part provides two perspectives on the roles of indigenous and modern Western medicine in human health, sustainable biodiversity, and the environment.

Tools for Developing Management Policies for Biodiversity

This book as a whole attempts to synthesize our current understanding of the effects of human-driven alterations of biodiversity on human health and well-being. It is tailored to a scientific audience, and would also serve well as a graduate student text on this topic. We hope that this synthesis contributes novel tools to develop policy at the same time as it highlights research needs. The book emphasizes the fact that issues of biodiversity, human health, development, and conservation are extremely complex. Some chapters stress the multiple benefits of biodiversity for human health through our four drivers, whereas other chapters see biodiversity as a risk associated with the appearance of new diseases for humans and their crops and animals. Consequently, management of biodiversity to benefit human health and well-being will require policies tailored to specific regions. These policies will need to strike a balance between the multiple tensions described throughout the chapters of this book.

References

Chapin, F. S., E. S. Zavaleta, V. T. Eviner, R. L. Naylor, P. M. Vitousek, H. L. Reynolds, D. U. Hooper, et al. 2000. Consequence of changing biodiversity. *Nature* 405:232–42.

Daily, G. C. 1997. *Nature's services: Societal dependence of natural ecosystems.* Washington, DC: Island Press.

Daszak, P., A. A. Cunningham, and A. D. Hyatt. 2000. Emerging infectious diseases of wildlife: Threats to biodiversity and human health science. *Science* 287:443–49.

de Groot, R. S., M. A. Wilson, and R. M. J. Boumans. 2002. A typology for the classification, description and valuation of ecosystem functions, goods and services. *Ecological Economics* 41:393–408.

Dobson, A., I. Cattadori, R. D. Holt, R. S. Ostfeld, F. Keesing, K. Krichbaum, J. R. Rohr, S. E. Perkins, and P. J. Hudson. 2006. Sacred cows and sympathetic squirrels: The importance of biological diversity to human health. *PLoS Medicine* 3:231.

EPA (Environmental Protection Agency). 2006. *Biodiversity and human health: An interdisciplinary approach to examining the links.* Proceedings of an interdisciplinary forum and workshop, September 14–15, 2006, Washington, DC. Washington, DC: EPA, Office of Research and Development, National Center for Environmental Research.

Folke, C., C. S. Holling, and C. Perrings. 1996. Biological diversity, ecosystems, and the human scale. *Ecological Applications* 6:1018–24.

Grifo, F., and J. Rosenthal, eds. 1997. *Biodiversity and human health.* Washington, DC: Island Press.

Kellert, S. R., and E. O. Wilson, eds. 1993. *The biophilia hypothesis.* Washington, DC: Island Press.

(MEA) Millennium Ecosystem Assessment. 2003. *Ecosystems and human well-being: A framework for assessment.* Washington, DC: Island Press.

Pongsiri, M. J., and J. Roman. 2007. Examining the links between biodiversity and human health: An interdisciplinary research initiative of the U.S. Environmental Protection Agency. *EcoHealth* 4 (1): 82–85.

Rapport, D. J., R. Costanza, and A. J. McMichael. 1998. Assessing ecosystem health. *Trends in Ecology & Evolution* 13 (10): 397–402.

Sala, O. E., F. S. Chapin, J. J. Armesto, E. Berlow, J. Bloomfield, R. Dirzo, E. Huber-Sanwald, et al. 2000. Global biodiversity scenarios for the year 2100. *Science* 287:1770–74.

Schulze, E.-D., and H. A. Mooney, eds. 1994. *Biodiversity and ecosystem functioning.* New York: Springer.

Schwartz, M. W., C. A. Brigham, J. D. Hoeksema, K. G. Lyons, M. H. Mills, and P. J. van Mantgem. 2000. Linking biodiversity to ecosystem function: Implications for conservation ecology. *Oecologia* 122:297–305.

WHOQOL (World Health Organisation Quality of Life Assessment) Group. 1994. Development of the WHOQOL: Rationale and current status. *International Journal of Mental Health* 23 (3): 24–56.

WRI (World Resources Institute), United Nations Development Programme, United Nations Environment Programme, and World Bank. 2000. *World resources 2000–2001: People and ecosystems: The fraying web of life.* Washington, DC: WRI.

PART I
Biodiversity and Human Health
Synergisms, Trade-offs, and Road Maps for the Future

As the world's population continues to increase and per capita levels of consumption continue to rise, the concomitant pressures on native biodiversity and natural resources also grow. In many ways, people have become disconnected from the rest of the environment, slipping into the false but comforting assumption that technology and innovation can fulfill human requirements for health and well-being. A more synthetic and realistic assessment is needed in order to ensure that people can continue to thrive and improve their quality of life in the future. Part I of this book utilizes multidisciplinary approaches to understanding and addressing the complex interplays between natural biodiversity and human health and well-being. The entire Paris workshop was engaged in this section of the process, and these chapters thus reflect the views of a wide range of participants. The use of "we" to describe the content of Part I is deliberate and is meant to highlight the truly collaborative nature of the views, concepts, and conclusions embedded in these cross-cutting chapters.

The conceptual framework that guided the original Biodiversity, Health, and the Environment Paris workshop (March 14–18, 2005) out of which this book grew maintains that biodiversity affects human well-being through four parallel paths: effects on the quality of life, provisioning of medicinal and genetic resources, spread of infectious disease, and provisioning of ecosystem services. Sustainable development attempts to raise awareness and to create links between human values, responsibilities, and environmental decisions that in many cases have been decoupled.

In part I (chapters 2–5), we explore the trade-offs between competing uses of biodiversity and highlight synergistic situations in which the conservation of natural biodiversity actually promotes human health and well-being. Using both real and hypothetical examples, we investigate alternative approaches and policies to achieve both biodiversity conservation and human well-being. This goal is frequently seen as unattainable when people assume that there is an inherent conflict between human well-being and biodiversity. We suggest, however, that "win-win" scenarios are often not contemplated because of lack of appropriate expertise in the decision-making process, and that progress in implementation and expansion of truly sustainable development urgently requires cross-disciplinary teams prepared to explore unconventional solutions.

Natural biodiversity supplies food both directly and indirectly by supporting the formation and maintenance of domesticated crops and animals. In chapter 2, Andrew Wilby and his coauthors synthesize a diverse body of literature to review how practices associated with production and harvesting of food affect biodiversity and ecosystem functioning, and how these effects, via both direct and indirect links, can ultimately feed back to impact on food production and human health.

Clearly, the link between food provision and human health is a complex one, involving the combined action of negative and positive relationships mediated by biodiversity. Human societies must maintain a diverse and adequate food supply while ensuring the sustainability and productivity of those ecosystems providing food. The review of literature here, however, suggests that contemporary practices of food provision, while for the most part meeting humanity's food requirements, tend to result in some negative effects on human health, and damage to biodiversity that may threaten the sustainability of the food production systems themselves. Further, global inequalities in food provisions have led to inefficient land use practices that are exacerbating pressures on natural systems. For example, natural land in developing countries is being converted into agricultural land, but formerly agricultural land in developed countries is being abandoned.

Chapter 2 highlights these issues with real-world examples, and begins to tackle the question of how we can harness biological and ecological knowledge to develop technologies that minimize the central conflict between maximizing net output of food production systems and their negative impacts on biodiversity, agricultural sustainability, and human health. Tools such as "precision farming" and biotechnology, among others, are explored in the light of long-term impacts on biodiversity and on human health.

In chapter 3, David J. Rapport and his coauthors broaden the discussion by exploring how multiple anthropogenic stressors jointly impact both biodiversity and human health. Human activities are resulting in global warming, habitat destruction, pollution, overharvesting, and introductions of exotic species. These changes have caused declines in biodiversity at regional and global scales and have had major impacts on human health through direct and indirect impacts on infectious diseases, nutrition, and contaminants. In addition to detailing instances in which an increase in biodiversity has a positive impact on human health, chapter 3 also brings out instances in which an increase in biodiversity has a negative impact. Likewise, while there are cases in which human interventions—particularly with respect to land use changes—shift the balance in favor of human pathogens, there are also instances in which the opposite is true.

Healthy natural ecosystems can buffer society against toxins and pollutants created by development, but degraded ecosystems can amplify the negative impacts of human activities through trophic cascades. In contrast to that scenario, chapter 3 also brings up examples of conflicts, particularly those in which the natural system has a large reservoir of vectors of human pathogens. In those cases, a reduction in biodiversity may be accompanied by an immediate improvement, rather than decline, in human health. However, even in those situations, relationships may not be straightforward. For example, lack of adequate food supply locally has led to increasing encroachment into tropical forests to hunt game animals, which, in turn, has led to the emergence of a series of new zoonoses, most notably human immunodeficiency virus / acquired immune deficiency syndrome (HIV/AIDS) and severe acute respiratory syndrome (SARS). Thus, even when taking a strictly anthropocentric perspective, direct benefit (food) can be strongly offset by an indirect cost (new deadly diseases). Chapter 3 discusses these examples as well as other complex interrelationships between regional and global change, biodiversity, and human health.

The theme of chapter 4 follows directly from the previous two chapters. Camille Parmesan and her coauthors discuss the adaptation of existing tools from a wide range of disciplines and suggest development of new tools to facilitate the implementation of policy shifts that might address the apparent conflicts highlighted in chapters 2 and 3. Here, we accept the inherent trade-offs in short-term versus long-term gains, and advocate the routine use of a suite of tools developed in the social sciences. Methodologies from social psychology allow the quantification of subjective values of biodiversity and of human well-being, which in turn allows such values to be placed into a decision-making framework. Further, we explore economic models that could be adapted to a biodiversity/human health framework, with the goal of developing policies that have the potential to maximize human health and well-being while simultaneously minimizing negative biodiversity impacts.

To end part I, in chapter 5, Laura Meyerson and her coauthors identify four biodiversity drivers of human health, through which they explore the different spatial and temporal scales at which biodiversity affects human health and well-being. These drivers are quality of life, medicinal and genetic resources, constraint on infectious disease, and ecosystem services, each of which responds differently to local and global extinctions. For example, although a local species extirpation may not cause the overall global extinction of a species, there is still a loss of unique population genetic diversity and a consequent loss to the local people who may have depended on that population as a resource.

Using a framework of sustainable development, the authors also discuss the different scales at which trade-offs and synergisms occur in nature when biodiversity is allocated to people. The "green revolution," with its use of fertilizers, has boosted modern food production to unprecedented levels and has supported an ever-growing human world population. However, these technological advances have created conflicts with other local land uses, such as conservation, and have intensified pollution of local waters and soils. Nonetheless, Meyerson and her coauthors envision opportunities that exist to better integrate the needs of people and to achieve conservation of biological diversity at the local and global scales. The chapter therefore ends with a thoughtful discussion of how this long-term ideal compares to sustainable development in practice.

2

Biodiversity, Food Provision, and Human Health

Andrew Wilby, Charles Mitchell, Dana Blumenthal, Peter Daszak, Carolyn S. Friedman, Peter Jutro, Asit Mazumder, Anne-Hélène Prieur-Richard, Marie-Laure Desprez-Loustau, Manju Sharma, and Matthew B. Thomas

The provision of food is a principal link between biodiversity and human health. Human societies derive food by harvesting naturally self-sustaining wild populations or by farming. In its broadest sense, biodiversity is the source of our current food and will be the source of novel foods in the future. There are, however, wider links between biodiversity and the provision of food. All food species, whether hunted or gathered from populations under negligible management, or grown in the most intensive production systems, occur within ecosystems, and their productivity is affected by the activity of other elements of biodiversity within those ecosystems. Understanding how biodiversity, food, and health interact is fundamental to meeting the needs of a growing and largely impoverished world population.

On a global scale, human health has come to depend on the production of vast amounts of food. Greater food provision has been achieved by increasing both the spatial extent of food production and its intensity. From 1980 to 2001, global cereal production increased by about 36% in conjunction with an increase of about 34% in the global land area in permanent crops, and an increase of about 34% in global nitrogen fertilizer use (FAO 2003). At the scale of individual people, food consumption differs dramatically between regions. In developed countries, per capita food availability in 1999–2001 was 3,260 calories per day, while in developing countries, only 2,680 calories per day were available (FAO 2003). Moreover, developing nations relied on cereals for 54% of their caloric nutrition, compared with only 31% in developed nations. While part of this difference is explained by greater sugar consumption in developed nations, the bulk of the difference results from greater consumption of meat, vegetable oil, and milk products in developed countries. Together, these foods make up 33% of caloric nutrition in developed nations, but only 18% in developing nations (FAO 2003).

Key diet-related influences on health have been identified by the World Health Organization as undernutrition, high blood pressure, high cholesterol, obesity (WHO 2002). In impoverished regions of the developing world, energy and protein deficit leaves many people, particularly women and children, underweight, and this deficit is often compounded by deficiencies in micronutrients. Over half of childhood deaths in the developing world are estimated to be due to undernutrition (WHO 2002). By contrast, high blood pressure, high cholesterol levels, and obesity are major causes of ill health and mortality in developed countries and in higher socioeconomic groups in the developing world. Obesity rates in particular have tripled since 1980 in some parts of North America and Europe, accounting for up to 13% of all deaths, as compared with 3% in parts of Africa and Southeast Asia (WHO 2002).

How does biodiversity affect the food-related health of humankind? First, it directly provides the variety of species harvested for food. A diverse diet is important for the provision of essential dietary components such as proteins, fatty acids, and vitamins. In traditional cultures, diets typically include a wide variety of crop plants and animal products (Norton et al. 1984). Unfortunately, the production and utilization of traditional food plants has been declining, as has dietary diversity, in rural and urban communities. Such a decline can result in serious deficiencies in minerals, vitamins, and trace elements. For example, vitamin A deficiency still contributes to high mortality, blindness, stunting, bone deformities, and increased susceptibility to disease—for example, malaria or diarrhea—of millions of people in Africa and Asia (Baker-Henningham et al. 2004; Caulfield, Richard, and Black 2004; Lewallen and Courtright 2001). Another example of human heath impacts resulting from declining dietary diversity is the outbreak of optic and/or peripheral neuropathy that occurred in Cuba in the early 1990s, affecting more than 50,000 people. The disease outbreak occurred following a period when daily per capita caloric availability fell from 3,200 to 1,900 calories (from 1989 to 1993) but was caused primarily by vitamin deficiency (Barnouin et al. 2001). Food diversity also provides some insurance against the vagaries of climate and disease. Ireland's Great Famine of 1845–1849 provides a classical example where lack of food diversity left a human population vulnerable to crop disease. At that time, the potato was the only food affordable for the masses. When late blight epidemics occurred, caused by the introduced organism *Phytophthora infestans*, Ireland's potato crops were totally destroyed. The crop failure led to the death and emigration of millions of people, with Ireland's population falling from 8 to 5 million within only a few years (Bourke 1991).

Biodiversity also interacts with food provision with less obvious impacts on human health. Overharvesting of wild species can lead not only to decline in the production of harvested species, but also to wide-ranging ecosystem impacts, such as occurred with the oyster fishery in Chesapeake Bay in the eastern United States (Burreson, Stokes, and Friedman 2000), discussed later in the chapter. Increasing demand for food and the continuing problems of undernutrition in the developing world have led to widespread extension of total farmed area and intensification of production and harvest. Intensification is generally supported by increased use of synthetic products (e.g., pesticides and fertilizers), which themselves may be of direct harm to human health (e.g., Alavanja, Hoppin, and Kamel 2004). Intensification also has a detrimental impact on biodiversity and may compromise natural ecosystem services (Tilman et al. 2002). For example, wild predator diversity plays an important role in controlling insect pests, and intensification-driven declines in predator populations only increase our reliance on harmful pesticides. Similarly, many crop species require the services of wild pollinators to bear fruit, and this service may be threatened by intensification (Kremen, Williams, and Thorp 2002).

Two components of biodiversity play a crucial role in food production systems: biodiversity that is purposely introduced or managed for production has been termed *planned diversity* (Vandermeer and Perfecto 1995), whereas *unplanned*, or *associated, biodiversity* describes the flora and fauna (soil microorganisms, pollinators, pests, pathogens, natural enemies, etc.) that exist in, or colonize, the agricultural or aquacultural ecosystem. Planned diversity influences and interacts with associated diversity to determine the overall functioning of the production system. In a similar way, the species targeted for hunting, fishing, and gathering, while not necessarily managed or manipulated for production themselves, are dependent on other elements of biodiversity.

In this chapter we explore the direct and indirect linkages between biodiversity, food provision, and human health. We discuss how practices associated with production and harvesting of food affect biodiversity and ecosystem functioning, and how these effects, via both direct and indirect links, can ultimately feed back to impact on food production and human health. The framework for discussion of these links is a conceptual model presented in figure 2.1. In this model, biodiversity provides the food species in harvested and farmed production systems. Through this provision of food, and its quality and quantity, biodiversity has a direct link with human health. However, these food provision activities do not occur in isolation; they feed back to influence biodiversity, especially the associated biodiversity that interacts with planned or target diversity to determine productivity. Harvesting and farming also interact in that together they fulfill human food requirements; an increase in one will allow a decrease in the other if food requirements are constant. In addition to providing food, harvesting and farming may have a direct influence on human health via the cultural value of these activities, and biodiversity may have a direct influence on human health via use and nonuse values unassociated with food provision. In sum, biodiversity is the source of the food products on which human health and well-being depend. Further, the effects of farming and harvesting food on human health are mediated in part via changes in biodiversity and/or the goods and services it provides. In this chapter, we characterize the complex links between food, biodiversity, and human health across food production systems, from the harvest of wild species to intensive cropping. Our

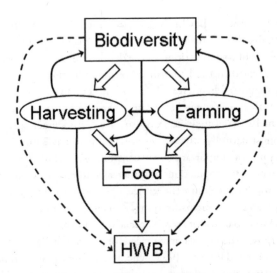

Figure 2.1. A conceptual model linking biodiversity, food provision, and human health. The core process is shown by the open arrows: Biodiversity is farmed or harvested (unmanaged populations), providing food, the quality and quantity of which affects human health. However, other links are also evident (solid arrows). Farming and harvesting activities are not independent, as they combine to meet human food requirements. Farming and harvesting activities also feed back on biodiversity, which can affect the efficiency and sustainability of food production, and can have positive and negative direct effects on human health. Other links, which are not exclusive to food provision systems, connect biodiversity directly to human health (dashed arrows).

emphasis is not restricted to the nature of relationships and interactions in the present but also includes future trends—in particular, how growing demands for food can be met while also increasing the sustainability of production and safeguarding human health.

Harvesting Wild Populations

Fishing, hunting, and gathering are important activities of food provision in most societies. In some, they may be the main source of food, whereas in others they may be a source of recreation or luxury food. As a traditional means of obtaining food, hunting and fishing have high cultural and recreational value in many societies. However, the direct benefits of food provision and the recreational/cultural value of these activities should be weighed against their potential negative impact on human health via biodiversity impacts, changes in ecosystem function, and the emergence of zoonotic diseases.

The effects of terrestrial hunting on biodiversity are well known. Hunting has directly resulted in the extinction or near extinction of many species, including high-profile examples such as the passenger pigeon (*Ectopistes migratorius*) (Blockstein and Tordoff

1985) and the American bison (*Bison bison*), as well as the many island species exploited by sailors in the 17th–19th centuries (Ehrlich and Ehrlich 1981). Currently, overhunting of wildlife for meat is thought to be causing numerous local and global extinctions in the humid tropics. In Vietnam, for example, 12 large mammal species are thought to have been lost due to hunting over the past 40 years (Bennett and Rao 2002). Although there is a long history of hunting by human societies, consumption of wild meat has increased markedly in the humid tropics over recent decades, and the problem is particularly acute in tropical forests, where productivity for meat is an order of magnitude less than that in more open savannah habitats (Milner-Gulland and Bennett 2003). Increased human population size and access, improved hunting technologies, and loss of traditional hunting controls all contribute to increased harvest of wild meat (Bennett and Rao 2002). Within regions, large species tend to disappear first, and as they become scarce, people typically switch to hunting smaller species, and the resulting local extinctions impact on forest dynamics and on local people (Milner-Gulland and Bennett 2003). In addition, habitat destruction, human encroachment, and hunting have also reduced human health via an increase in zoonotic disease expression (e.g., Ebola virus; Walsh et al. 2003).

Harvesting of wild populations is still a common method of seafood provision across the developed and developing worlds. With improvement of trawling technology and fish-targeting technologies, harvesting efficiency in commercial fisheries has increased substantially during the last 50 years. Larger fish, which have greater economic value but slower replacement rates, exhibit greater decline under intensive fishing pressure. As the larger and more economically important species decline, commercial fishers have been exploiting smaller pelagic and benthic species of lower commercial value. Overexploitation of wild oceanic fish stocks has been linked to reductions in the abundance of larger fish on a global scale. There are also many local examples where overfishing has led to declines in fish and invertebrate fisheries (Goldburg and Naylor 2005; Karpov et al. 2000; R. Myers and Worm 2005; R. Myers et al. 2004).

The Chesapeake Bay estuary provides an example of how harvest, habitat alteration, disease, and nutrient input have seriously impacted biodiversity, ecosystem function, and human health. Anthropogenic habitat changes, overfishing, and disease introductions (Burreson, Stokes, and Friedman 2000) have led to a decline in oyster recruitment. Oyster landings declined from 14.5 million kilograms in 1959 to less than 0.5 million kilograms in 1990 (Leonard 1992) and are currently at less than 1% of historic levels (Newell et al. 1999). Prior to intensive harvest and disease introduction, the Chesapeake Bay oyster biomass was able to filter the entire bay volume in three to four days; currently it is estimated that the remaining oysters require approximately one year to perform this function. In addition, baywide biodiversity has been altered by eutrophication, caused largely by fertilizer runoff from terrestrial systems, and by a loss of oyster reef habitat, which provides a home for a variety of marine and estuarine species (Harding and Mann 1999). Concern over the water quality in this embayment has led to considering introducing an exotic oyster species, *Crassostrea ariakensis*.

Impacts of harvesting wild populations in aquatic and terrestrial systems are not always independent. Recent studies in West Africa highlight the potential for indirect linkages between increased intensification of fishing, conservation, and health effects. Communities in coastal West Africa have traditionally utilized both fish and bushmeat for protein. However, over the last 20 years, intensive commercial fishing by European boats off the coast of West Africa has depleted fish stocks substantially. The reduced availability of fish has led to increased exploitation of bushmeat to meet the demand for protein. Not only has the increased demand for bushmeat resulted in population reductions and local extinctions of a range of large mammals (Brashares et al. 2004), it has also increased risk of disease through exposure to new disease agents (see this volume, chapter 3).

Clearly, many societies in the developing world are dependent on wild species for food, and the collection of wild food has important social and cultural value to many societies, irrespective of their wealth. Negative impacts on human health take place where overharvesting is such that harvest becomes unsustainable. Severe degradation of the ecosystems providing food can occur to the extent that continued food provision is threatened, along with the social and cultural services associated with hunting, fishing, and gathering activities.

Grazing Animal Systems

The inability of wild populations to supply the needs of growing human populations was the driving force behind domestication of food species, the first step in the intensification of food provision. Pastoralism, in which domesticated or semidomesticated animals are managed and utilized to harvest wild plant communities, allows increased meat production compared with harvesting wild animal populations. As with wild animal harvests, the productivity of pastoral systems relies directly on biological diversity, and is vulnerable to overexploitation.

The influence of livestock grazing on plant species diversity appears to be highly dependent on grazing intensity, the evolutionary history of grazing within the ecosystem, and resource availability, particularly precipitation. Heavy grazing can greatly reduce plant cover, leading to decreased infiltration, increased runoff and erosion (Belsky and Blumenthal 1997; Fleischner 1994), and, in dry climates, desertification (Schlesinger et al. 1990); both processes can lead to dramatic reductions in food production. In contrast, moderate grazing within historically grazed ecosystems can be an important factor in maintaining plant diversity (Milchunas et al. 1988) and may be necessary management for some traditionally grazed ecosystems. Where grazing is a novel disturbance for an ecosystem, even moderate grazing can reduce diversity by altering plant species composition (Mack and Thompson 1982), and particularly by facilitating invasion of non-native species (Hobbs 2001).

Livestock grazing also influences biodiversity indirectly, a prominent example being hunting of predators of livestock. Concern for the effect of predators on livestock has led to the hunting and extirpation of predators in many regions, especially the regional extinction of large mammals. Lions, for example, ranged over much of Asia, Africa, Europe, and North America in prehistoric times but are now restricted to a small fraction of the range they occupied only a century ago (Patterson et al. 2004). Similar range reduction has occurred in many wolf populations (Sidorovich et al. 2003). Birds of prey have also been hunted to protect livestock. The red kite (*Milvus milvus*), for example, was driven almost to extinction by overzealous farmers and gamekeepers in the United Kingdom, even though it preys mainly on small mammals and carrion (Wotton et al. 2002). Decimation of predator populations may also pose a risk to human health if those predators regulated prey populations that serve as reservoirs of zoonotic pathogens that are transmitted to humans (Ostfeld and Holt 2004).

The level of biological diversity within grazed ecosystems can contribute to productivity, and therefore human health, in several ways. Experimental studies suggest that diversity can increase the amount and decrease the variability of plant production (Loreau et al. 2001; Naeem et al. 1994; Tilman, Wedin, and Knops 1996), and therefore livestock forage. Plant community productivity increases with increasing species diversity, both through complementary resource use by different species and through increasing the probability of the occurrence of strong competitors (Golluscio, Sala, and Lauenroth 1998; Hector et al. 1999; Tilman et al. 2001). Increases in productivity with increased plant diversity in the tallgrass prairie have been shown to be largest during drought years (Tilman and Downing 1994), suggesting that plant diversity may be particularly important in buffering people from the effects of environmental variation (Loreau et al. 2001).

The importance of both plant and animal diversity in maintaining productivity in the face of environmental variation can be seen most clearly in pastoral systems. For example, the Ngisonyoka Turkana of northwest Kenya use a diverse assemblage of livestock species, feeding on a diverse plant community, to obtain a consistent food supply in the face of a highly variable climate (Coughenour et al. 1985). Herbaceous vegetation, which accounts for the majority of plant productivity but is present only following the wet season, is harvested by cattle, while woody plants provide more consistent camel production. Both cattle and camels are used primarily for milk. Sheep and goats have a mixed diet and are used for meat when drought limits milk production.

Plant diversity may also contribute to forage production by inhibiting pest and weed populations. Although observational and experimental studies have found relatively high invasion rates in diverse plant communities (Robinson, Quinn, and Stanton 1995; Stohlgren, Barnett, and Kartesz 2003), experimental studies often demonstrate that diversity reduces invasion of non-native species (Diemer and Schmid 2001; Joshi, Matthies, and Schmid 2000; Kennedy et al. 2002; Levine 2000; Levine and D'Antonio 1999; Tilman 1997; see also Levine, Adler, and Yelenik 2004 and Prieur-Richard and

Lavorel 2000b for reviews). The apparent contradiction can be explained by confounding factors such as spatial scale and/or resource availability driving differences in both diversity and invasion among communities (Davis, Grime, and Thompson 2000; Dukes 2001; Hobbs and Huenneke 1992; Shea and Chesson 2002). In addition to richness per se, other components of diversity can reduce invasion, including species identity (Crawley et al. 1999) and functional group richness and identity (Hector et al. 2001; Prieur-Richard and Lavorel 2000a; Symstad 2000). For example, native late-successional species, which are easily lost from disturbed plant communities, may provide the strongest competition with invaders under local climatic and edaphic conditions (Blumenthal, Jordan, and Svenson 2003). Finally, more diverse plant communities will support a more diverse herbivore community (Koricheva et al. 2000; Siemann 1998), which in turn can inhibit invasive species (Prieur-Richard et al. 2002).

Aquaculture

Although aquaculture has been practiced for millennia, it has only become a commercial industry over the past few decades. It is now the fastest-growing segment of agriculture, with an average annual increase in production of 10%–11% globally since 1980 (FAO 1996, 2004). Although this industry meets a large global demand for seafood and has the potential to relieve pressure caused by the harvest of wild aquatic populations, there are growing concerns over the negative impacts of aquaculture on biodiversity and human health. Reported negative impacts include organic enrichment of ecosystems, escape of farmed fish into wild communities, depletion of wild populations for farm stocks, transfer of disease and parasites from farmed fish to wild stocks, and impacts on water quality (Goldburg, Elliot, and Naylor 2001). As with terrestrial farming and other methods of food provision, the degree of influence on biodiversity and the wider environment is largely dependent on the intensity and scale of the operation.

The extent of habitat conversion for aquaculture varies from small-scale impacts, such as the placement of anchors for suspended culture of marine bivalves, to large-scale impacts associated with deforestation or the conversion of land for pond or raceway culture. Intertidal bivalve culture may influence local biodiversity both positively and negatively through the placement of nets, tubes, racks, and buoys, and through alterations to substrata. These structural changes may exclude some local species, but their addition of complexity to the habitat may also serve to increase biodiversity while in place (Dumbauld 2002). Culture of species such as marine algae and bivalves and some herbivorous fishes are often cited as having neutral or positive influences on biodiversity and environmental sustainability while providing food and economic benefits to local communities (Goldburg, Elliot, and Naylor 2001). However, other systems cause considerable ecological damage. For example, milkfish (*Chanos chanos*) and shrimp aquaculture have a marked impact on associated biodiversity through the conversion of mangrove ecosystems. Equatorial mangrove forests supply many products and ecosystem functions, including timber, firewood, and food for local communities. They also generate soil, cycle nutrients, improve water quality, produce oxygen, are sinks for CO_2, stabilize watersheds, protect coastal areas, provide habitat, act as nurseries for invertebrates and fish, and influence associated marine ecosystems, such as seagrass beds and coral reefs (Ronnback 1999; Ruitenbeek 1992).

Although farming in former mangrove forests dates back over 500 years, conversion of mangrove ecosystems for shrimp culture is thought to be a major current threat to mangrove ecosystems due to the magnitude of, and the methods employed by, this growing industry (Ronnback 1999). The value of ecological functioning of mangrove ecosystems is estimated to exceed that of harvested products, such as wood. Ruitenbeek (1992) suggested that ecological functions netted US$4,800–$16,750/ha versus US$3,600/ha for timber, and intact mangrove ecosystems at US$10 million in services and US$25 million in fishery protection annually. Ironically, the shrimp culture industry is reliant on the ecosystem functioning of intact mangrove forests for its success and sustainability (McKinnon et al. 2002; Ronnback 1999). Many nearshore marine species (including the shrimp species being cultured) use mangrove forests as a nursery or habitat for other life stages (Mumby et al. 2004). Reliance on the capture of wild brood stock, while reducing concerns about species introduction and the genetic integrity of wild species, raises concern over the impact on mangrove and coastal biodiversity (Mumby et al. 2004; Sarkar and Bhattacharya 2003).

In addition to the concerns caused by habitat destruction, other potential negative impacts of aquaculture on biodiversity and human health may include pesticide and drug use and species introductions. A wide range of contaminants can be introduced into ecosystems, including both intentionally used chemicals (e.g., pesticides, antimicrobials) and chemicals that are unintentionally introduced as contaminants of feed, which have also been found in wild fish (Howgate 1998). There is concern over therapeutic drug use and the potential exposure of humans to drug residues. For example, chloramphenicol is banned in the United States due to the development of aplastic anemia in sensitive people (Rich, Ritterhoff, and Hoffmann 1950). Concern also exists over the development of antibiotic resistance in pathogenic bacteria and the potential for transfer of drug resistance to human-pathogenic bacteria (Alderman and Hastings 1998). Recently developed alternatives to drug use include the use of probiotic immunostimulants, and specific microbial assemblages to competitively exclude potential pathogens (Elston et al. 2004; Sambasivam, Chandran, and Khan 2003). Aquaculture has also been associated with the introduction of non-native diseases (Burreson, Stokes, and Friedman 2000), which can be detrimental to both planned and unplanned diversity associated with aquaculture systems. Wider negative impacts on biodiversity are also recorded. The increasing demand for fish feed, for example, has the potential to result in overexploitation of forage fish like anchovy (*Engraulis* spp.) and herring (*Clupea harengus*) (Goldburg, Elliot, and Naylor 2001), species commonly used in terrestrial agricultural feeds and increasingly in aquacultural feeds (FAO 2004).

Terrestrial Crop Production

Humankind is dependent on terrestrial agriculture for the bulk of its food, and the provisioning of food is clearly the principal benefit of agriculture to human health. However, farming provides many less obvious cultural and social benefits. While cultural ecosystem services may be under threat due to the loss of biodiversity caused by food production and harvest (MEA 2003), traditional farming practices may often themselves have strong cultural significance and thus a positive relationship with human health. In modern farming systems, efficient production may be a source of social value in farming communities (Burton 2004), and the move toward more sustainable, less productive farming systems can incur significant social costs in addition to economic ones. In contrast, some regions with long histories of agricultural production, such as Europe, have developed strong social values of rural landscapes, and regulation protects these farming systems against change driven by economic rationalization (Vanclay 2003).

In addition to these benefits, agriculture also has a range of detrimental effects on human health, particularly in its high-intensity modern form. The use of external inputs such as fertilizers, biocides, drugs, and other therapeutic substances has greatly increased production in many agricultural systems but has also resulted in unintended side effects on human health. Chronic exposure to a range of insecticides is known to increase the risk of neurotoxicity and cancer (Alavanja, Hoppin, and Kamel 2004), and there is growing evidence that occupational exposure to pesticides is associated with increased likelihood of fetal death due to congenital abnormalities (Bell, Hertz-Picciotto, and Beaumont 2001). In the developing world, pesticide exposure continues to be problematic despite increasing regulation. In Sri Lanka, for example, there was an increase of 50% in the numbers of hospital admissions due to pesticide poisoning between 1986 and 2000, however, regulation of some highly toxic pesticide compounds has resulted in reduced fatality due to poisoning over this period (Roberts et al. 2003).

In addition to its direct impact on human health, crop production also has a dramatic effect on biodiversity. Conversion of unmanaged ecosystems to agricultural systems almost invariably involves large structural changes and is considered to be the greatest threat to global biological diversity, with estimates of the percentage of total land area that is transformed ranging from 39% to 50% (Vitousek et al. 1997). Much higher levels of land transformation are found within particular areas, such as Europe and parts of Southeast Asia, with many regions containing little or no untransformed land (Hannah et al. 1994).

The impact of agriculture on biodiversity is not confined simply to the effects of habitat conversion. Modern agricultural practices further reduce the biodiversity of agricultural ecosystems. Taking pest control as an example, it has been estimated that 2.5×10^9 kg of pesticides are applied globally to agricultural systems, but only an estimated 0.1% of this total is actually delivered to the intended target (Naylor and Ehrlich 1997). As a consequence of such inefficiencies, agricultural systems tend to harbor lower biodiversity than the natural systems they replace, and, significantly, functional composition may be altered to the extent that effects of ecosystem functioning are broader than those caused by diversity reduction per se. The impact of agricultural intensification also propagates through food webs with the result that fertilization and biocide inputs affect higher trophic levels (de Kraker et al. 2000; Hites et al. 2004). Pesticides have been implicated in the demise of a suite of once common farmland birds in the United Kingdom due to general declines in the availability of invertebrate food resources (Benton et al. 2002). The effects of agricultural practices are not confined to aboveground biodiversity; low plant diversity, soil tilling, pesticides, irrigation, and fertilization can affect the diversity of soil organisms and the functioning of important ecosystem processes, such as nutrient cycling. In general, agricultural systems have lower soil diversity than the natural systems that they replace, and intensification of agriculture further reduces diversity (Decaens and Jimenez 2002; Mikola, Bardgett, and Hedlund 2002; Oehl et al. 2004).

Although the impact of agriculture on biodiversity is in itself cause for concern, there is also increasing evidence that aspects of the functionality of agricultural ecosystems may be compromised by declining biodiversity via impacts on production-supporting ecosystem services. The first aspect of biodiversity that influences agricultural production is the "planned" (directly managed) component, encompassing a continuum of scales from genetic/intraspecific diversity (i.e., within crop plant species), through species diversity (both within field and between fields), to habitat or landscape diversity. Diversity can be actively manipulated (i.e., planned) at any one of these levels through farm management practices. Such management may affect crop pests, weeds, or diseases directly, or indirectly through interactions with associated biodiversity.

The most obvious and direct element of planned agrobiodiversity is the crop plant. For many agricultural species, we now rely heavily on a few "modern" varieties that tend to be very uniform, containing less genetic diversity than traditional varieties. As indicated by Matthew B. Thomas and his coauthors (see this volume, chapter 13), reduced genetic diversity creates risks; if a pest or disease is able to exploit the one dominant variety, it has almost unlimited potential to spread throughout the field and landscape. Genetic diversity within a crop species can be increased through conventional breeding or new biotechnological methods. Planting multiple genetic varieties of a crop species produces a varietal mixture that often reduces spread of plant pathogens and resulting losses of yield compared with monoculture (Mundt 2002). Use of varietal mixtures can also buffer against unpredictable abiotic variables, leading to increased stability of yield over different environments relative to monocultures (Wolfe 2000).

Species diversity can be directly managed by combining different plant species as intercrops or polycultures. This practice effectively reduces risk by stabilizing yields over the long term, promotes diet diversity, and maximizes returns under low levels of technology and limited resources (Richards 1985). Many studies report reduced insect pest populations (Andow 1991), reduced viral incidence (Power and Flecker 1996), and lower weed abundance (Olasantan, Lucas, and Ezumah 1994) in intercropping or polyculture systems compared with monocultures. However, the diversity of crop species in an agro-ecosystem has a much less predictable effect on directly transmitted microbial pathogens, such as most fungi (Matson et al. 1997). Fungal diseases can be less severe in polycultures than in monocultures (Boudreau 1993), but the opposite effect is also seen (Boudreau and Mundt 1997). Generalizations are difficult because the effects of intercropping on disease dynamics depend on a variety of factors, including microclimate effects and the spatial scale of pathogen dispersal (Boudreau and Mundt 1997).

At the farm and landscape level, planned diversity can be manipulated through spatial and temporal patterns of planting; changes in field size, shape, and position (relative to other crop or noncrop habitats); and active management of landscape features such as hedgerows, shelterbelts, watercourses, and so forth. Numerous studies demonstrate increased abundance of beneficial insects where wild vegetation is maintained at field edges and in association with crops (see Gurr, Wratten, and Altieri 2004). These non-crop habitats can provide overwintering sites and alternative food sources for predators and parasites of pest species, which often, but not always (Weibull, Östman, and Granqvist 2003), lead to increase in pest control function. Crop rotation is also used across farming landscapes to control pests and pathogens. Crop rotation is particularly effective for host-specialized pathogens with limited dispersal ability, such as the soilborne pathogens *Gaeumannomyces graminis* (cause of take-all of wheat) and the soybean cyst nematode *Heterodera glycines* (Boudreau and Mundt 1997). The benefits of crop rotation for weed control are also well established. Rotating crops changes not only the primary competitor, but also the management system, leading to variable environments among years, limiting the suitability of particular weed species to an agroecosystem (Doucet et al. 1999) and reducing the ability of weeds to evolve resistance to management (Beckie et al. 2004; Gorddard, Pannell, and Hertzler 1996).

Many of the advantages of manipulating planned biodiversity in agricultural ecosystems derive from the secondary effects of unplanned or associated biodiversity on nutrient cycling, pest control, pollination, feed provision, water quality maintenance, and soil stabilization. In soil communities, microbial diversity generally has been shown to be positively related to microbial activity in the soil and the rate of nutrient processing (Hedlund and Sjögren Öhrn 2000). Other studies have concluded that soil functioning is independent of diversity, or that the relationship varies, depending on the particular ecosystem process under question (Griffiths et al. 2001). A general consensus is that soil diversity, at higher taxonomic levels, at least, is functionally significant, and there is growing evidence that soils in less intensively managed agricultural systems may benefit from greater functional stability and increased ability to resist and recover from stresses compared with those in intensively managed systems (Mikola, Bardgett, and Hedlund 2002).

The implications of biodiversity for natural pest control and pollination are, perhaps, better understood. Studies of a range of agricultural systems suggest that associated biodiversity is key to the suppression of potential pest species. For example, there is strong evidence that much of the devastating damage to Asian rice in the 1970s and 1980s by the brown planthopper (*Nilaparvata lugens*) was the result of overuse of insecticides through nontarget impacts on generalist predator species (Schoenly et al. 1994; Settle et al. 1996). Several studies of crops as diverse as carrots, tomatoes, and apples have shown that higher diversity in low-input or organic production systems results in more efficient pest control than in high-input systems (Berry et al. 1996; Bogya and Marko 1999; Drinkwater et al. 1995). Similar effects of biodiversity loss occur on pollination services derived from crop-associated biodiversity. Many farmed vegetables, legumes, and fruits, providing approximately a third of human diets in total (N. Myers 1996), are dependent to some extent on pollination by wild animal pollinators. Evidence is mounting that the production of these crops may be limited by a lack of pollination in some intensified systems (e.g., Kevan 1999; Kremen, Williams, and Thorp 2002; see also this volume, chapter 3).

Intensified Animal Production Systems

In parallel with the intensification of terrestrial cropping systems, we have seen continuing intensification of animal production systems, from the pastoral grazing systems described above to industrial-scale "factory farms." This intensification has resulted in a novel array of detrimental impacts on human health. For example, some animal production systems can act as reservoirs for pathogenic organisms. Intensification of livestock production can increase the likelihood that wildlife pathogens will invade the production system and ultimately gain entry to the human population. Where intensive, non-biosecure livestock production occurs in areas of high wildlife biodiversity (e.g., the tropics), it provides a catchment zone for the emergence of novel zoonotic pathogens. Here livestock populations allow wildlife pathogens to become amplified or genetically modified, making them able to move into the human population. For example, in 1999 a new paramyxovirus, Nipah virus, emerged in pig farmers and abbattoir workers in Malaysia and Singapore, where it killed 103 of 263 people infected, a case fatality rate of 39% (Chua et al. 2000). The virus originates in fruit bats but was passed to pigs via urinary, fecal, or salival contamination of pigsties. Recent modeling work shows that intensification of pig farming, which involves high turnover of animals, allows the virus to persist on the farm and increases the number of human infections as pigs are sold to other farms (Daszak et al., 2006).

Less intensive production of pigs in Southeast Asia has been associated with repeated, annual outbreaks of influenza in humans. Here the close association of pigs with poultry allows for the presence of avian and mammalian strains of influenza virus within pig hosts (Alexander and Brown 2000). The ability of influenza virus genes to reassort (e.g., to switch over whole genes from avian to mammalian strains) allows rapid shifts in phenotype and the sudden emergence of new, virulent strains in the human population. Thus pig farming is interacting with poultry production, and a diversity of avian hosts and viruses, to produce a series of major disease outbreaks in humans. More recently, a series of small-scale outbreaks has been reported caused by avian influenza virus strains directly infecting humans (Li et al. 2004). These outbreaks represent an important threat to global health in that the avian strains are novel to the human population, and the viruses are particularly lethal.

Management strategies for intensifying livestock production can have serious consequences for human health. During the 1980s, a new disease of cattle, bovine spongiform encephalopathy (BSE, or mad cow disease), emerged in the United Kingdom. This disease, caused by a prion protein, results in severe pathological changes to the brain of cattle and is similar in etiology to Creutzfeldt-Jakob disease (CJD) in humans. By January 1998, over 172,000 confirmed cases of BSE had been reported in British cattle, and infected cattle had been found in various other European countries and Japan (Ferguson et al. 1999). More important, clear evidence had emerged that the agent had jumped species again into the human population, where a series of cases of a "new variant" form of CJD (vCJD) had been found in British people who had eaten BSE-infected meat (Bruce et al. 1997). The emergence of BSE appears to have its origins in a change in the meat rendering process whereby meat from infected cattle carcasses was turned into protein feed for cattle.

In addition to providing a reservoir for diseases and promoting host shifts by pathogenic organisms, animal production systems have also been linked with development of drug resistance. To help prevent disease associated with high density of animals in production facilities, livestock are often given subtherapeutic doses of the same antibiotics used in human medicine. Antibiotic-resistant *Salmonella* or *Escherichia coli* strains pathogenic to humans are increasingly common in poultry or beef produced in large-scale facilities (Tilman et al. 2002).

Inputs of nutrients and pesticides to terrestrial agricultural systems can negatively impact human health and biodiversity both within those systems and in distant systems linked by aquatic or atmospheric transport of animal waste and synthetic fertilizers. Nitrate contamination of groundwater (Agrawal et al. 1999; Oenema et al. 1998) can have a range of direct consequences for human health, including reproductive problems, methemoglobinemia ("blue baby" syndrome), and several types of cancer. Indirect effects of fertilization also are possible through increased abundance of disease vectors, such as mosquitoes, and restricted food supplies due to eutrophication of important fishing zones (Townsend et al. 2003). Economic pressure to consolidate animal feeding operations in small areas requires these nutrient inputs to be increasingly from highly distant source ecosystems (Mallin and Cahoon 2003). Nitrogen from these high-density animal production systems and from crop fertilizer is volatilized into the atmosphere and deposited in other ecosystems tens, hundreds, or thousands of kilometers away (Galloway and Cowling 2002; Matson, Lohse, and Hall 2002). Such increased atmospheric nitrogen deposition rates can decrease the biodiversity of grassland ecosystems by allowing nitrophilic species to dominate (Matson, Lohse, and Hall 2002). Together, aqueous transport and atmospheric deposition of nitrogen into estuaries contribute to decreased biodiversity by increasing harmful algal blooms such as *Pfiesteria piscicida* (Glasgow and Burkholder

2000) and other species, which in turn can directly cause fish kills and have negative impacts on human health. At an even larger scale, agricultural fertilizer inputs produce an annual hypoxic, or "dead," zone in the Gulf of Mexico of around 20,000 square kilometers in which few aerobic species can exist—a large area with literally very low diversity and a very low abundance of fish, algae, or microbes. This condition results from the transport of nutrients from cropping systems in the midwestern United States down to the Mississippi River. Plankton biomass produced in the surface water sinks and is decomposed in lower layers, depleting the oxygen supply in the water column and creating hypoxic conditions in which almost nothing survives (Rabelais, Turner, and Scavia 2002).

Food Provision, Biodiversity, and Health into the Future

Clearly, the link between food provision and human health is a complex one involving the combined action of negative and positive relationships mediated by biodiversity. Human societies must maintain a diverse and adequate food supply while at the same time ensuring the sustainability and productivity of those ecosystems providing food. Contemporary practices of food provision, although for the most part meeting humanity's food requirements, tend to result in some negative effects on human health, and damage to biodiversity that may threaten the sustainability of the food production systems themselves. Moreover, increasing demand for food in the future is likely to exacerbate these pressures.

This chapter has outlined abundant evidence of negative relationships or trade-offs between the intensity or level of production/harvest and biodiversity, and, by extension, between intensification of food production and the key production-supporting ecosystem services that biodiversity provides. These trade-offs are summarized in figure 2.2. The exact nature of these negative relationships is the key determinant of the impact of food provision on biodiversity and the implications for future food production. Increased productivity is an important outcome of intensification, which is required to meet growing demand for food, but such productivity is often achieved by substituting external inputs for essential natural ecosystem services such as pest control, pollination, and nutrient cycling. Unfortunately, reliance on external inputs can further reduce biological diversity, and also be of direct harm to human health. The challenge for future food provision is to maintain and even increase productivity without severe loss of diversity and the essential services it provides.

Given current technology and economic incentives, substantial reliance on low-diversity intensive agriculture is likely to remain necessary to produce sufficient food (Tilman et al. 2002). There are, however, examples of diverse, yet productive, agroecosystems, such as well-managed fisheries (top right of figure 2.2a). The Marine Stewardship Council (MSC), an independent global nonprofit organization that has developed standards for well-managed and sustainable fisheries, evaluates major fisheries on a rotating five-year basis and has deemed numerous fisheries as being well managed. The South West

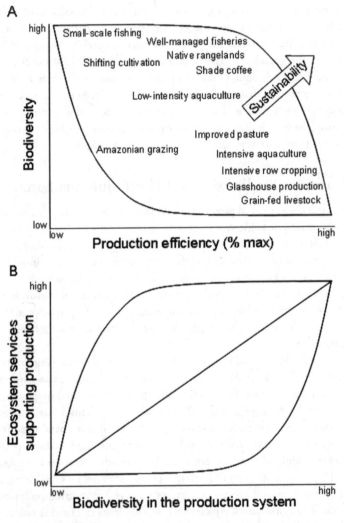

A

high

Small-scale fishing

Well-managed fisheries

Native rangelands

Shifting cultivation

Shade coffee

Low-intensity aquaculture

Sustainability

Improved pasture

Amazonian grazing

Intensive aquaculture

Intensive row cropping

Glasshouse production

Grain-fed livestock

Biodiversity

low

low   **Production efficiency (% max)**   high

B

high

Ecosystem services supporting production

low

low   **Biodiversity in the production system**   high

Figure 2.2. *a,* Intensification of food production systems aimed at increasing the efficiency of production inevitably results in loss of biodiversity, though the shape of this trade-off is of fundamental importance to human health. Current food production systems vary considerably in their position with respect to this trade-off. One goal of research should be to develop technologies that allow efficient production while minimizing the impact on biodiversity. Such technologies are a route to sustainable food provision. *b,* Nevertheless, intensification is likely to continue to have a negative impact on biodiversity, at least in some production systems. An important goal of research is to understand the impact of this biodiversity loss on the key ecosystem services (e.g., pest control, pollination, nutrient cycling) that support food production. Maintenance of these services is necessary for sustainability of future food provision.

mackerel (*Scomber scombrus*) handline fishery and the Alaska pollock (*Theragra chalco-gramm*) fisheries of the Gulf of Alaska, Bering Sea and Aleutian Islands, are examples of the 26 fisheries that are considered sustainable (MSC 2008). Hilborn, Orensanz, and Parma (2005) reported that although many fisheries are overexploited, many are biologically and economically sustainable. The primary driving forces in a successful fishery include the governing institutional structure and provision of incentives to participants in the fishery. Most (unsuccessful) fisheries are based on a "race for fish" concept in which participants compete with one another to reach a quota or attempt to catch the fish/shellfish before they are caught by others (Hilborn et al. 2003). Such management is ineffective, as illustrated by the many overexploited fisheries (e.g., California abalone *Haliotis* spp., New England groundfish, Argentinian hake *Merluccius hubbsi*), and leads to high fishing pressure resulting in poor-quality products and short-term gains. Fisheries with effective institutional structure and incentive programs, and which do not try and reach the maximum biological yields, have the potential to be sustainable (Hilborn et al. 2003). Several of the fisheries deemed sustainable by the MSC are also stated as successes by Hilborn, Orensanz, and Parma (2005), including the Alaska pollock (*Theragra chalco-gramma*) cooperative and the West Australian rock lobster (*Panulirus cygnus*) fishery. Priority should be given to preserving and understanding these systems. Such systems point to the possibility of developing more effective and sustainable management strategies that minimize the productivity/diversity trade-off. There is now growing evidence that biodiversity can be incorporated into productive, intensive agricultural systems, at both local and landscape scales (Mundt 2002; Wolfe 2000; Zhu et al. 2000). By making incorporation of biodiversity into agriculture a research priority, we have the potential to change the shape of the relationship between productivity and diversity for other ecosystems by essentially shifting productive ecosystems toward the top right in figure 2.2a. This shift, should, in our view, be a central goal of future agroecological research.

However, it remains that intensification will tend to impact biodiversity, and what becomes key (at least from a food provision perspective) is to understand what this loss of biodiversity means in terms of change in ecosystem functioning and the provision of services essential to the production system (figure 2.2b). In this regard, quantifying the functional elements of diversity and the relative influence of mechanisms such as species complementarity, redundancy, and composition in determining the shape of the biodiversity–ecosystem functioning relationship is a central challenge. The relationship between biodiversity and ecosystem functioning has been the subject of much research effort in ecology over recent years; however, there has been relatively little application of this research to food production systems (but see Cardinale et al. 2003; Ives, Cardinale, and Snyder 2005; Thomas 1999; Wilby and Thomas 2002a, 2002b).

In the future, tools such as biotechnology and precision farming, which allow better targeting of inputs, could provide new opportunities to allow intensification of food production while minimizing the impact on biodiversity. Biotechnology has opened up new potential for the breeding of crop plants by allowing the use of a wider gene pool.

Improvements might be achieved in production capacity, lower costs to environment, and direct health benefits through improved nutritional quality (Bouis, Chassy, and Ochanda 2003), and several examples are already available. The gene producing the natural Bt insecticide originating from bacteria (*Bacillus thuringiensis*) has been introduced into several crop species (maize *Zea mays*, cotton *Gossypium* sp., potato *Solanum tuberosum*), making them resistant to some insect pests (Qaim and Zilberman 2003). Another strategy involving transformation of plants with viral genes has produced plants that are protected from a variety of viral diseases, including diseases for which there is no known disease resistance, such as papaya ringspot virus, a major pathogen of the squash family (Cucurbitaceae) (Gonsalves 1998). Plant engineering has also been used to increase crop nutritional quality, thereby improving the direct impact on human health. Transgenic cassava (*Manihot esculenta*) plants have been developed that contain 35%–45% more protein than conventional varieties (Zhang et al. 2003). Rice has been genetically engineered to contain provitamin A (beta carotene) and iron (Egana 2003).

If agricultural biotechnology is to result in long-term benefits to human health, however, these benefits need to be carefully weighed against the environmental, health, and social risks associated with the technology. For example, although pest-resistant crops may eventually reduce pesticide use, to date, farmers have adopted herbicide-tolerant (HT) varieties at a much faster pace than insect-resistant varieties. Adoption of HT varieties has been associated with a significant increase in glyphosate use (i.e., the target molecule) and a decline in other herbicides (Benbrook 2001; Bonny and Sausse 2004). More generally, the use of genetically engineered (GE) crops results in changes in the entire system of production, inducing highly unpredictable ecological impacts (Bohan et al. 2005; White, Harmon, and Andow 2004). In addition, the potential effects of crops on health, including allergies and antibiotic resistance, remain a concern (Royal Society 2002). Perhaps the greatest controversy with GE crops has been about the multinational corporate control of the seed supply. Farmers who purchase and grow GE crops already sign agreements not to replant the next year's crop from saved seeds; instead they must purchase new seed from dealers each year, and tools ("Terminator Technology") are already available that could force them to do so. Such practices could be highly detrimental to small farmers, who depend on seed-saving practices and cannot afford to purchase new seeds every year. Finally, there are already examples of introduced genes spreading from engineered crops into related species (Marvier and Van Acker 2005), leading to the possibility of unintended consequences for biological diversity, particularly with the use of ecologically relevant genes.

Global patterns of future trade in agricultural products are also likely to have global impacts on biodiversity. Currently, national agriculture is controlled by the availability of food stocks, the availability of arable land and labor, nutritional needs, and the economics of food production and distribution. As crop production methods have achieved higher yields, we have seen a social migration from populations engaged in agriculture and a parallel decrease in the amount of land devoted to agriculture in developed countries, as fewer people and less land can produce more food with greater efficiency. In fact, since World War II, national agricultural self-sufficiency has become a political goal. At odds with this goal has been the evolution of the concept of globalization. First the world saw a gradual decrease in tariffs designed to protect domestic agricultural products. Then, as these tariffs fell, governments began to subsidize the cost of production to sustain domestic agricultural production. More recently, international legal instruments have deemed governmental subsidies to be illegal. As subsidies fall, agriculture is joining the international trend of moving production to locations that are able to produce goods at the lowest conventional economic cost. As a result, natural land in developing countries is being converted into agricultural land, but formerly agricultural land in developed countries is being abandoned. Both have begun to demonstrate enormous effects on local and regional biodiversity.

What are the implications of the complex interdependencies between food provision, biodiversity, and human health for the future development of agricultural research and policy? A recently revived debate concerns the relative merits of low-input extensive agriculture and intensive production for biodiversity conservation. For conservation of any species inhabiting agricultural ecosystems, the most appropriate course of agricultural development depends on the shape of the response between intensification and population viability. If small increases in the intensity of production cause a proportionally large impact on viability, then it is most appropriate to concentrate production on the smallest amount of land possible. If, however, small increments of intensification have a proportionately minor impact, then spreading production across larger areas is the least damaging option (Green et al. 2005). This framework can be extended to include impacts on biodiversity in a wider sense and, by extension, to impacts on human health. The key question to address is the shape of the human health–intensification relationship. Is human health best served by extensive, perhaps traditional, agricultural practices over relatively large areas or, alternatively, by land-sparing policies that aim to produce food on sacrificial, relatively small areas with high-intensity production? A further key issue concerns the location of agricultural production on national and global scales: how do we ensure that agriculture is practiced where there will be minimal negative impact on biodiversity and human health? Finally, and perhaps most important, how can we harness biological and ecological knowledge to develop technologies that minimize the central conflict between the productivity of food production systems and their negative impacts on biodiversity, agricultural sustainability, and human health?

References

Agrawal, G. D., S. K. Lunkad, and T. Malkhed. 1999. Diffuse agricultural nitrate pollution of groundwaters in India. *Water Science and Technology* 39:67–75.

Alavanja, M. C. R., J. A. Hoppin, and F. Kamel. 2004. Health effects of chronic pesticide exposure: Cancer and neurotoxicity. *Annual Review of Public Health* 25:155–97.

Alderman, D. J., and T. S. Hastings. 1998. Antibiotic use in aquaculture: Development of antibiotic resistance: Potential for consumer health risks. *International Journal of Food Science and Technology* 33:139–55.

Alexander, D. J., and I. H. Brown. 2000. Recent zoonoses caused by influenza A viruses. *Scientific and Technical Review OIE* 19:197–225.

Andow, D. A. 1991. Vegetational diversity and arthropod population response. *Annual Review of Entomology* 36:561–86.

Baker-Henningham, H., and S. Grantham-McGregor. 2004. Nutrition and child development. *Public Health Nutrition* 8:247–63.

Barnouin, J., T. Verdura Barrios, M. Chassagne, R. Perez Cristia, J. Arnaud, P. Fleites Mestre, M. E. Montoya, and A. Favier. 2001. Nutritional and food protection against epidemic emerging neuropathy: Epidemiological findings in the unique disease-free urban area of Cuba. *International Journal for Vitamin and Nutrition Research* 71:274–85.

Beckie, H. J., L. M. Hall, S. Meers, J. J. Laslo, and F. C. Stevenson. 2004. Management practices influencing herbicide resistance in wild oat. *Weed Technology* 18:853–59.

Bell, E., I. Hertz-Picciotto, and J. Beaumont. 2001. A case study of pesticides and fetal death due to congenital abnormalities. *Epidemiology* 12:148–56.

Belsky, A. J., and D. M. Blumenthal. 1997. Effects of livestock grazing on stand dynamics and soils in upland forests of the interior West. *Conservation Biology* 11:315–27.

Benbrook, C. 2001. Do GM crops mean less pesticide use? *Pesticide Outlook* 12:204–7.

Bennett, E., and M. Rao. 2002. Wild meat consumption in Asian tropical forest countries: Is this a glimpse of the future for Africa? In *Links between biodiversity, conservation, livelihoods and food security: The sustainable use of wild species for meat,* ed. S. Mainka and M. Trivedi, 39–44. Cambridge, UK: IUCN.

Benton, T., D. Bryant, L. Cole, and Q. Crick. 2002. Linking agricultural practice to insect and bird populations: A historical study over three decades. *Journal of Applied Ecology* 39:673–87.

Berry, N. A., S. D. Wratten, A. McErlich, and C. Frampton. 1996. Abundance and diversity of beneficial arthropods in conventional and "organic" carrot crops in New Zealand. *New Zealand Journal of Crop and Horticultural Science* 24:307–13.

Blockstein D., and H. Tordoff. 1985. Gone forever: A contemporary look at the extinction of the passenger pigeon. *American Birds* 39:845–51.

Blumenthal, D. M., N. R. Jordan, and E. L. Svenson. 2003. Weed control as a rationale for restoration: The example of tallgrass prairie. *Conservation Ecology* 7 (1): 6.

Bogya, S., and V. Marko. 1999. Effect of pest management systems on ground-dwelling spider assemblages in an apple orchard in Hungary. *Agriculture Ecosystems and Environment* 73:7–18.

Bohan, D. A., C. W. H. Boffey, D. R. Brooks, S. J. Clark, A. M. Dewar, L. G. Firbank, A. J. Haughton, et al. 2005. Effects on weed and invertebrate abundance and diversity of herbicide management in genetically modified herbicide-tolerant winter-sown oilseed rape. *Proceedings of the Royal Society of London* B 272:463–74.

Bonny, S., and C. Sausse. 2004. Does the transgenic crop make it possible to reduce the use of pesticides? The case of glyphosate-tolerant soybeans. *OCL (Oléagineux, Corps gras, Lipides)* 11:85–91.

Boudreau, M. A. 1993. Effects of intercropping beans with maize on angular leaf spot of bean in Kenya. *Plant Pathology* 42:16–25.

Boudreau, M. A., and C. C. Mundt. 1997. Ecological approaches to disease control. In

Environmentally safe approaches to crop disease control, ed. J. Rechcigl and N. Rechcigl. Boca Raton, FL: CRC/Lewis Press.

Bouis, H. E., B. M. Chassy, and J. O. Ochanda. 2003. Genetically modified food crops and their contribution to human nutrition and food quality. *Trends in Food Science and Technology* 14:191–209.

Bourke, A. 1991. Potato blight in Europe in 1845: The scientific controversy. In *Phytophthora*, ed. J. A. Lucas, R. C. Shuttock, D. S. Shaw, and L. R. Cooke, 12–24. Cambridge: Cambridge University Press.

Brashares, J. S, P. Arcese, M. K. Sam, P. B. Coppolillo, A. R. E. Sinclair, and A. Balmford. 2004. Bushmeat hunting, wildlife declines, and fish supply in West Africa. *Science* 306:1180–83.

Bruce, M. E., R. G. Will, J. W. Ironside, I. McConnell, D. Drummond, A. Suttie, L. McCardle, et al. 1997. Transmissions to mice indicate that "new variant" CJD is caused by the BSE agent. *Nature* 389:498.

Burreson, E. M., N. A. Stokes, and C. S. Friedman. 2000. Increased virulence in an introduced pathogen: *Haplosporidium nelsoni* (MSX) in the eastern oyster *Crassostrea virginica*. *Journal of Aquatic Animal Health* 12:1–8.

Burton, R. J. F. 2004. Seeing through the "good farmer's" eyes: Towards developing an understanding of the social symbolic value of "productivist" behaviour. *Sociologia Ruralis* 44:195–99.

Cardinale, B. J., C. T. Harvey, K. Gross, and A. R. Ives. 2003. Biodiversity and biocontrol: Emergent impacts of a multi-enemy assemblage on pest suppression and crop yield in an agroecosystems. *Ecology Letters* 6:857–65.

Caulfield, L. E., S. A. Richard, and R. E. Black. 2004. Undernutrition as an underlying cause of malaria morbidity and mortality in children less than five years old. *American Journal of Tropical Medicine and Hygiene* 71 (2): 55–63.

Chua, K. B., W. J. Bellini, P. A. Rota, B. H. Harcourt, A. Tamin, S. K. Lam, T. G. Ksiazek, et al. 2000. Nipah virus: A recently emergent deadly Paramyxovirus. *Science* 288:1432–35.

Coughenour, M. B., J. E. Ellis, D. M. Swift, D. L. Coppock, K. Galvin, J. T. McCabe, and T. C. Hart. 1985. Energy extraction and use in a nomadic pastoral ecosystem. *Science* 230:619–25.

Crawley, M. J., S. L. Brown, M. S. Heard, and G. R. Edwards. 1999. Invasion-resistance in experimental grassland communities: Species richness or species identity? *Ecology Letters* 2:140–48.

Daszak, P., R. Plowright, J. H. Epstein, J. Pulliam, S. Abdul Rahman, H. E. Field, C. S. Smith, et al. 2006. The emergence of Nipah and Hendra virus: pathogen dynamics across a wildlife-livestock-human continuum. In *Disease ecology: Community structure and pathogen dynamics*, ed. S. K. Collinge and C. Ray. Cary, NC: Oxford University Press.

Davis, M. A., J. P. Grime, and K. Thompson. 2000. Fluctuating resources in plant communities: A general theory of invisibility. *Journal of Ecology* 88:528–34.

Decaens, T., and J. J. Jimenez. 2002. Earthworm communities under an agricultural intensification gradient in Colombia. *Plant and Soil* 240:133–43.

de Kraker J., R. Rabbinge, A. van Huis, J. C. van Lenteren, and K. L. Heong. 2000. Impact of nitrogenous-fertilization on the population dynamics and natural control of rice leaffolders (Lep.: Pyralidae). *International Journal of Pest Management* 46:219–24.

Diemer, M., and B. Schmid. 2001. Effects of biodiversity loss and disturbance on the survival and performance of two *Ranunculus* species with differing clonal architectures. *Ecography* 24:59–67.

Doucet, C., S. E. Weaver, A. S. Hamill, and J. H. Zhang. 1999. Separating the effects of crop rotation from weed management on weed density and diversity. *Weed Science* 47:729–35.

Drinkwater, L. E., D. K. Letourneau, F. Workneh, A. H. C. Vanbruggen, and C. Shennan. 1995. Fundamental differences between conventional and organic tomato agroecosystems in California. *Ecological Applications* 5:1098–1112.

Dukes, J. S. 2001. Biodiversity and invasibility in grassland microcosms. *Oecologia* 126:563–68.

Dumbauld, B. R. 2002. The role of oyster aquaculture as habitat in the (USA) West Coast estuaries: A review. *Proceedings of the 2001 Puget Sound Research Conference.* Feb. 12–14, 2001, Bellevue WA, Puget Sound Action Team, Olympia, WA.

Egana, N. E. 2003. Vitamin A deficiency and golden rice: A literature review. *Journal of Nutritional and Environmental Medicine* 13:169–84.

Ehrlich, P., and A. Ehrlich. 1981. *Extinction: The causes and consequences of the disappearance of species.* New York: Random House.

Elston, R., K. Humphrey, A. Gee, D. Cheney, and J. David. 2004. Progress in the development of effective probiotic bacteria for bivalve shellfish hatcheries and nurseries. *Journal of Shellfish Research* 23:288–89.

FAO (Food and Agriculture Organization). 1996. *The state of the world fisheries and aquaculture (SOFIA).* Rome: Food and Agricultural Organization of the United Nations.

———. 2003. *Summary of food and agricultural statistics.* Rome: Food and Agriculture Organization of the United Nations.

———. 2004. *The state of the world fisheries and aquaculture (SOFIA).* Rome: Food and Agricultural Organization of the United Nations.

Ferguson, N. M., C. A. Donnelly, M. E. J. Woolhouse, and R. M. Anderson. 1999. Estimation of the basic reproduction number of BSE: The intensity of transmission in British cattle. *Proceedings of the Royal Society of London* B 266:23–32.

Fleischner, T. L. 1994. Ecological costs of livestock grazing in western North America. *Conservation Biology* 8:629–44.

Galloway, J. N., and E. B. Cowling. 2002. Reactive nitrogen and the world: 200 years of change. *Ambio* 31:64–71.

Glasgow, H. B., and J. M. Burkholder. 2000. Water quality trends and management implications from a five-year study of a eutrophic estuary. *Ecological Applications* 10:1024–46.

Goldburg, R., M. S. Elliot, and R. L. Naylor. 2001. *Marine aquaculture in the United States: Environmental impacts and policy options.* Arlington, VA: Pew Oceans Commission.

Goldburg, R., and R. L. Naylor. 2005. Future seascapes, fishing, and fish farming. *Frontiers in Ecology and the Environment* 3:21–28.

Golluscio, R. A., O. E. Sala, and W. K. Lauenroth. 1998. Differential use of large summer rainfall events by shrubs and grasses: A manipulative experiment in the Patagonian steppe. *Oecologia* 115:17–25.

Gonsalves, D. 1998. Control of papaya ringspot virus in papaya: A case study. *Annual Review of Phytopathology* 36:415–37.

Gorddard, R. J., D. J. Pannell, and G. Hertzler. 1996. Economic evaluation of strategies for management of herbicide resistance. *Agricultural Systems* 51:281–98.

Green, R. E., S. J. Cornell, J. P. W. Scharlemann, and A. Balmford. 2005. Farming and the fate of wild nature. *Science* 307:550–55.

Griffiths, B. S., K. Ritz, R. Wheatley, H. L. Kuan, B. Boag, S. Christensen, F. Ekelund, S. J. Sorensen, S. Muller, and J. Bloem. 2001. An examination of the biodiversity-ecosystem function relationship in arable soil microbial communities. *Soil Biology and Biochemistry* 33:1713–22.

Gurr, M., S. D. Wratten, and M. A. Altieri. 2004. *Ecological engineering for pest management: Advances in habitat manipulation for arthropods.* Wallingford, UK: CAB International.

Hannah, L., D. Lohse, C. Hutchinson, J. L. Carr, and A. Lankerani. 1994. A preliminary inventory of human disturbance of world ecosystems. *Ambio* 23:246–50.

Harding, J. M., and R. Mann. 1999. Fish species richness in relation to restored oyster reefs, Piankatank River, Virginia. *Bulletin of Marine Science* 65:289–300.

Hector, A., J. Joshi, S. P. Lawler, E. M. Spehn, and A. Wilby. 2001. Conservation implications of the link between biodiversity and ecosystem functioning. *Oecologia* 129:624–28.

Hector, A., B. Schmid, C. Beierkuhnlein, M. C. Caldeira, M. Diemer, P. G. Dimitrakopoulos, J. A. Finn, et al. 1999. Plant diversity and productivity experiments in European grasslands. *Science* 286:1123–27.

Hedlund, K., and M. Sjögren Öhrn. 2000. Tritrophic interactions in a soil community enhance decomposition rates. *Oikos* 88:585–91.

Hilborn, R., T. A. Branch, B. Ernst, A. Magnusson, C. V. Minte-Vera, M.D. Scheuerell, and J. L. Valero. 2003. State of the world's fisheries. *Annual Review of Environment and Resources* 28:559–99.

Hilborn, R., J. M. Orensanz, and A. M. Parma. 2005. Institutions, incentives and the future of fisheries. *Philosophical Transactions of the Royal Society* B 360:47–57.

Hites, R. A., J. A. Foran, D. O. Carpenter, M. C. Hamilton, B. A. Knuth, and S. J. Schwager. 2004. Global assessment of organic contaminants in farmed salmon. *Science* 303:226–29.

Hobbs, R. J. 2001. Synergisms among habitat fragmentation, livestock grazing, and biotic invasions in southwestern Australia. *Conservation Biology* 15:1522–28.

Hobbs, R. J., and L. F. Huenneke. 1992. Disturbance, diversity, and invasion: Implications for conservation. *Conservation Biology* 6:324–37.

Howgate, P. 1998. Review of the public health safety of products from aquaculture. *International Journal of Food Science and Technology* 33:99–125.

Ives, A. R., B. J. Cardinale, and W. E. Snyder. 2005. A synthesis of subdisciplines: Predator-prey interactions, and biodiversity and ecosystem functioning. *Ecology Letters* 8:102–16.

Joshi, J., D. Matthies, and B. Schmid. 2000. Root hemiparasites and plant diversity in experimental grassland communities. *Journal of Ecology* 88:634–44.

Karpov, K. A., P. L. Haaker, I. K. Taniguchi, and L. Rogers-Bennett. 2000. Serial depletion and the collapse of the California abalone (*Haliotis* spp.) fishery. In *Workshop on rebuilding abalone stocks in British Columbia*, ed. A. Campbell, 11–34. Canadian Special Publication on Fisheries and Aquatic Sciences 130. Ottawa, ON: NRC Research Press.

Kennedy, T. A., S. Naeem, K. M. Howe, J. M. H. Knops, D. Tilman, and P. Reich. 2002. Biodiversity as a barrier to ecological invasion. *Nature* 417:636–38.

Kevan P. 1999. Pollinators as bioindicators of the state of the environment: Species, activity and diversity. *Agriculture, Ecosystems and Environment* 74:373–93.

Koricheva, J., C. P. H. Mulder, B. Schmid, J. Joshi, and K. Huss-Danell. 2000. Numerical responses of different trophic groups of invertebrates to manipulations of plant diversity in grasslands. *Oecologia* 125:271–82.

Kremen, C., N. M. Williams, and R. W. Thorp. 2002. Crop pollination from native bees at risk from agricultural intensification. *Proceedings of the National Academy of Science* 99:16812–16.

Leonard, D. L. 1992. Workshop on water quality and alternate species in the Canadian mollusk culture industry. *Bulletin of the Aquaculture Association of Canada* 92 (4): 45–58.

Levine, J. M. 2000. Species diversity and biological invasions: Relating local process to community pattern. *Science* 288:852–54.

Levine, J. M., P. B. Adler, and S. G. Yelenik. 2004. A meta-analysis of biotic resistance to exotic plant invasions. *Ecology Letters* 7:975–89.

Levine, J. M., and C. M. D'Antonio. 1999. Elton revisited: A review of evidence linking diversity and invisibility. *Oikos* 87:15–26.

Lewallen, S., and P. Courtright. 2001. Blindness in Africa: Present situation and future needs. *British Journal of Ophthalmology* 85:897–903.

Li, K. S., Y. Guan, J. Wang, G. J. D. Smith, K. M. Xu, L. Duan, A. P. Rahardjo, et al. 2004. Genesis of a highly pathogenic and potentially pandemic H5N1 influenza virus in eastern Asia. *Nature* 430:209–13.

Loreau, M., S. Naeem, P. Inchausti, J. Bengtsson, J. P. Grime, A. Hector, D. U. Hooper, et al. 2001. Biodiversity and ecosystem functioning: Current knowledge and future challenges. *Science* 294:804–8.

Mack, R. N., and J. N. Thompson. 1982. Evolution in steppe with few large, hoofed mammals. *American Naturalist* 119:757–73.

Mallin, M. A., and L. B. Cahoon. 2003. Industrialized animal production: A major source of nutrient and microbial pollution to aquatic ecosystems. *Population and Environment* 24:369–85.

Marvier, M., and R. C. Van Acker. 2005. Can crop transgenes be kept on a leash? *Frontiers in Ecology and the Environment* 3:99–105.

Matson, P. A., K. A. Lohse, and S. J. Hall. 2002. The globalization of nitrogen deposition: Consequences for terrestrial ecosystems. *Ambio* 31:113–19.

Matson, P. A., W. J. Parton, A. G. Power, and M. J. Swift. 1997. Agricultural intensification and ecosystem properties. *Science* 277: 504–9.

McKinnon, A. D., L. A. Trott, D. M. Davidson, and A. Davidson. 2002. Water column production and nutrient characteristics in mangrove creeks receiving farm effluent. *Aquaculture Research* 33:55–73.

MEA (Millennium Ecosystem Assessment). 2003. *Ecosystems and human well-being.* Washington, DC: Island Press.

Mikola, J., R. D. Bardgett, and K. Hedlund. 2002. Biodiversity, ecosystem functioning and soil decomposer food webs. In *Biodiversity and ecosystem functioning: Synthesis and perspectives*, ed. M. Loreau, P. Inchausti, and S. Naeem, 169–80. Oxford: Oxford University Press.

Milchunas, D. G., O. E. Sala, and W. K. Lauenroth. 1988. A generalized model of the effects of grazing by large herbivores on grassland community structure. *American Naturalist* 132:87–106.

Milner-Gulland, E. J., and E. L. Bennett. 2003. Wild meat: The bigger picture. *Trends in Ecology and Evolution* 18:351–57.

Mumby, P. J., A. J. Edwards, J. E. Arias-Gonzalez, K. C. Lindeman, P. G. Blackwell, A.

Gall, M. I. Gorczynska, et al. 2004. Mangroves enhance the biomass of coral reef fish communities in the Caribbean. *Nature* 527:533–36.

Mundt, C. C. 2002. Use of multiline cultivars and cultivar mixtures for disease management. *Annual Review of Phytopathology* 40:381–410.

Myers, N. 1996. Environmental services of biodiversity. *Proceedings of the National Academy of Sciences USA* 93:2764–69.

Myers, R. A., S. A. Levin, R. Lande, F. C. James, W. W. Murdoch, and R. T. Paine. 2004. Hatcheries and endangered salmon. *Science* 303:1980.

Myers, R. A., and B. Worm. 2005. Extinction, survival or recovery of large predatory fishes. *Philosophical Transactions of the Royal Society* B 360:13–20.

Naeem, S., L. J. Thompson, S. P. Lawler, J. H. Lawton, and R. M. Woodfin. 1994. Declining biodiversity can alter the performance of ecosystems. *Nature* 368:734–37.

Naylor, R. L., and P. R. Ehrlich. 1997. Natural pest control services and agriculture. In *Nature's services*, ed. G. C. Daily, 151–74. Washington, DC: Island Press.

Newell, R. I. E., J. C. Cornwell, M. Owens, and J. Tuttle. 1999. Role of oysters in maintaining estuarine water quality. *Journal of Shellfish Research* 18:300–301.

Norton, H. H., E. S. Hunn, C. S. Martinsen, and P. B. Keely. 1984. Vegetable food products of the foraging economies of the Pacific Northwest. *Ecology of Food and Nutrition* 14:219–28.

Oehl, F., E. Sieverding, P. Mader, D. Dubois, K. Ineichen, T. Boller, and A. Wiemken. 2004. Impact of long-term conventional and organic farming on the diversity of arbuscular mycorrhizal fungi. *Oecologia* 138:574–83.

Oenema O., P. C. M. Boers, M. M. van Eerdt, B. Fraters, H. G. van der Meer, C. W. J. Roest, J. J. Schroder, and W. J. Willems. 1998. Leaching of nitrate from agriculture to groundwater: The effect of policies and measures in the Netherlands. *Environmental Pollution* 102:471–78.

Olasantan, F. O., E. O. Lucas, and H. C. Ezumah. 1994. Effects of intercropping and fertilizer application on weed control and performance of cassava and maize. *Field Crops Research* 39:63–69.

Ostfeld, R. S., and R. D. Holt. 2004. Are predators good for your health? Evaluating evidence for top-down regulation of zoonotic disease reservoirs. *Frontiers in Ecology and the Environment* 2:13–20.

Patterson, B. D., S. M. Kasiki, E. Selempo, and R. W. Kays. 2004. Livestock predation by lions (*Panthera leo*) and other carnivores on ranches neighboring Tsavo National Parks, Kenya. *Biological Conservation* 119:507–16.

Power, A. G., and A. S. Flecker. 1996. The role of biodiversity in tropical managed ecosystems. In *Biodiversity and ecosystem processes in tropical forests*, ed. G. H. Orians, R. Dirzo, and J. H. Cushman, 173–94. New York: Springer-Verlag.

Prieur-Richard, A. H., and S. Lavorel. 2000a. Do more diverse plant communities show greater resistance to invasions? *Revue d'Écologie—La Terre et la Vie* 37–51.

———. 2000b. Invasions: The perspective of diverse plant communities. *Australian Ecology* 25:1–7.

Prieur-Richard, A. H., S. Lavorel, Y. B. Linhart, and A. Dos Santos. 2002. Plant diversity, herbivory and resistance of a plant community to invasion in Mediterranean annual communities. *Oecologia* 130:96–104.

Qaim, M., and D. Zilberman. 2003. Yield effects of genetically modified crops in developing countries. *Science* 299:900–902.

Rabelais, N. N., R. E. Turner, and D. Scavia. 2002. Beyond science into policy: Gulf of Mexico hypoxia and the Mississippi River. *BioScience* 52:129–42.

Rich, M. L., R. J. Ritterhoff, and R. J. Hoffmann. 1950. A fatal case of aplastic anaemia following chloramphenicol (chloromycetin) therapy. *Annals of Internal Medicine* 33:1459–67.

Richards, P. 1985. *Indigenous agricultural revolution.* Boulder, CO: Westview Press.

Roberts, D. M., A. Karunarathna, N. A. Buckley, G. Manuweera, M. H. R. Sheriff, and M. Eddleston. 2003. Influence of pesticide regulation on acute poisoning deaths in Sri Lanka. *Bulletin of the World Health Organisation* 81:789–98.

Robinson, G. R., J. F. Quinn, and M. L. Stanton. 1995. Invasibility of experimental habitat islands in a California winter annual grassland. *Ecology* 76:786–94.

Ronnback, P. 1999. The ecological basis for economic value of seafood production supported by mangrove ecosystems. *Ecological Economics* 29:235–52.

Royal Society. 2002. *Genetically modified plants for food use and human health: An update.* London: Royal Society.

Ruitenbeek, H. 1992. *Mangrove management: An economic analysis of management options with a focus on Bintuni Bay, Irian Jaya, Indonesia.* EMDI Environmental Report No. 8. Jakarta, Indonesia: Environmental Management in Indonesia Project (EMDI).

Sambasivam, S., R. Chandran, and S. A. Khan. 2003. Role of probiotics on the environment of shrimp pond. *Journal of Environmental Biology* 24:103–6.

Sarkar, S. K., and A. K. Bhattacharya. 2003. Conservation of biodiversity of the coastal resources of Sundarbans, Northeast India: An integrated approach through environmental education. *Marine Pollution Bulletin* 47:260–64.

Schlesinger, W. H., J. F. Reynolds, G. L. Cunningham, L. F. Huenneke, W. M. Jarrell, R. A. Virginia, and W. G. Whitford. 1990. Biological feedbacks in global desertification. *Science* 247:1043–48.

Schoenly, K., J. Cohen, K. L. Heong, G. Arida, A. Barrion, and J. A. Litsinger. 1994. Quantifying the impact of insecticides on food web structure of rice-arthropod populations in a Philippine farmer's irrigated field. In *Food webs: Integration of patterns and dynamics.* New York: Chapman and Hall.

Settle, W. H., H. Ariawan, E. T. Astruti, W. Cahyana, A. L. Hakim, D. Hindayana, A. S. Lestari, and P. Sartanto. 1996. Managing tropical pests through conservation of generalist natural enemies and alternative prey. *Ecology* 77:1975–88.

Shea, K., and P. Chesson. 2002. Community ecology theory as a framework for biological invasions. *Trends in Ecology and Evolution* 17:170–76.

Sidorovich, V. E., L. L. Tikhomirova, and B. Jedrzejewska. 2003. Wolf *Canis lupus* numbers, diet and damage to livestock in relation to hunting and ungulate abundance in northeastern Belarus during 1990–2000. *Wildlife Biology* 9:103–11.

Siemann, E. 1998. Experimental tests of effects of plant productivity and diversity on grassland arthropod diversity. *Ecology* 79:2057–70.

Stohlgren, T. J., D. T. Barnett, and J. T. Kartesz. 2003. The rich get richer: Patterns of plant invasions in the United States. *Frontiers in Ecology and the Environment* 1:11–14.

Symstad, A. J. 2000. A test of the effects of functional group richness and composition on grassland invisibility. *Ecology* 81:99–109.

Thomas, M. B. 1999. Ecological approaches and development of "truly integrated" pest management. *Proceedings of the National Academy of Sciences USA* 96:5944–51.

Tilman, D. 1997. Community invasibility, recruitment limitation, and grassland biodiversity. *Ecology* 78:81–92.

Tilman, D., K. Cassman, P. Matson, R. Naylor, and S. Polasky. 2002. Agricultural sustainability and intensive production practices. *Nature* 418:671–77.

Tilman, D., and J. A. Downing. 1994. Biodiversity and stability in grasslands. *Nature* 367:363–65.

Tilman, D., P. B. Reich, J. Knops, D. Wedin, T. Mielke, and C. Lehman. 2001. Diversity and productivity in a long-term grassland experiment. *Science* 294:843–45.

Tilman, D., D. Wedin, and J. Knops. 1996. Productivity and sustainability influenced by biodiversity in grassland ecosystems. *Nature* 379:718–20.

Townsend A. R., R. W. Howarth, F. A. Bazzaz, M. S. Booth, C. C. Cleveland, S. K. Collinge, A. P. Dobson, et al. 2003. Human health effects of a changing global nitrogen cycle. *Frontiers in Ecology and the Environment* 1:240–46.

Vanclay, F. 2003. The impacts of deregulation and agricultural restructuring for rural Australia. *Australian Journal of Social Issues* 38:81–94.

Vandermeer, J., and I. Perfecto. 1995. *Breakfast of biodiversity: The truth about rainforest destruction*. Oakland, CA: Food First Books.

Vitousek, P. M., H. A. Mooney, J. Lubchenco, and J. M. Melillo. 1997. Human domination of Earth's ecosystems. *Science* 277:494–99.

Walsh, P. D., K. A. Abernethy, M. Bermejo, R. Beyers, and P. De Wachter. 2003. Catastrophic ape decline in western equatorial Africa. *Nature* 422:611–14.

Weibull, A. C., O. Östman, and A. Granqvist. 2003. Species richness in agroecosystems: The effect of landscape, habitat and farm management. *Biodiversity and Conservation* 12:1335–55.

White, J. A., J. P. Harmon, and D. A. Andow. 2004. Ecological context for examining the effects of transgenic crops in production systems. *Journal of Crop Improvement* 12 (1/2): 457–89.

WHO (World Health Organisation). 2002. *The world health report: Reducing risks, promoting healthy life*. Geneva: World Health Organisation.

Wilby, A., and M. B. Thomas. 2002a. Are the ecological concepts of assembly and function of biodiversity useful frameworks for understanding natural pest control? *Agricultural and Forest Entomology* 4:237–43.

———. 2002b. Natural enemy diversity and natural pest control: Patterns of pest emergence with agricultural intensification. *Ecology Letters* 5:353–60.

Wolfe, M. S. 2000. Crop strength through diversity. *Nature* 406:681–82.

Wotton, S. R., I. Carter, A. V. Cross, B. Etheridge, N. Snell, K. Duffy, R. Thorpe, and R. D. Gregory. 2002. Breeding status of the red kite *Milvus milvus* in Britain in 2000. *Bird Study* 49:278–86.

Zhang, P., J. M. Jaynes, I. Potrykus, W. Gruissem, and J. Puonti-Kaerlas. 2003. Transfer and expression of an artificial storage protein (ASP1) gene in cassava (*Manihot esculenta* Crantz). *Transgenic Research* 12:243–50.

Zhu, Y., H. Chen, J. Fan, Y. Wang, Y. Li, J. Chen, J. Fan, S. Yang, L. Hu, H. Leung, T. W. Mew, P. S. Teng, Z. Wang, and C. C. Mundt. 2000. Genetic diversity and disease control in rice. *Nature* 406:718–22.

3

The Impact of Anthropogenic Stress at Global and Regional Scales on Biodiversity and Human Health

David J. Rapport, Peter Daszak, Alain Froment, Jean-François Guégan, Kevin D. Lafferty, Anne Larigauderie, Asit Mazumder, and Anne Winding

Levels and layers of complexity are the rule rather than the exception in describing the relationships between humans and the environment, especially when it comes to understanding the relationships between biodiversity, the environment, and human health. Since humans are very much a part of the environment and ecosystems, and since most ecosystems may be described as human dominated, the socioeconomic and cultural dimensions are as critical to our understanding of the dynamics of ecosystems as their biological/ecological dimensions (Vitousek et al. 1997). In effect, we are confronted with an ecocultural system—a system in which each component impacts and is impacted by all the others. The links between local, regional, and global anthropogenic stress and human health are well known, particularly with reference to emerging infectious diseases (Morens et al. 2004; Patz et al. 2004; Smolinski, Hamburg, and Lederberg 2003; Weiss and McMichael 2004). However, the role of biodiversity in human health is less understood and is only just becoming the focus of research (Daszak, Cunningham, and Hyatt 2000; Dobson 1995).

Figure 3.1 portrays the complex web of connectivity between human activity, global and regional environments, biodiversity, and human health. Clearly, every element of this ecocultural system affects, and is affected by, the others. Anthropogenic drivers are human activities that can compromise the functions of ecosystems and thus have a major role in placing both biodiversity and human health at risk.

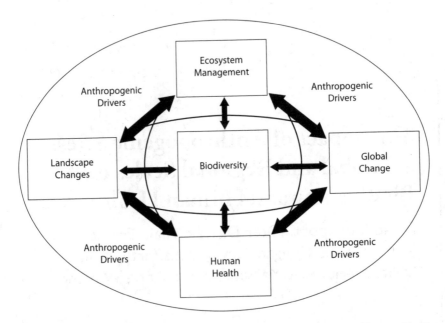

Figure 3.1. Interactions between environmental change, biodiversity, and human health. This figure portrays the complex web of connectivity between human activity, global and regional environments, biodiversity, and human health. Clearly, every element of this ecocultural system impacts, and is impacted by, the others. Anthropogenic drivers are human activities that, through their effect on the functions of ecosystems, play a major role in impacting biodiversity and human health.

The Complex Linkages between Health, Environment, and Biodiversity

In reality, the linkages between health, the environment, and biodiversity are highly complex, and few generalizations can be made. Rather, each linkage can, depending on the context, be either positive or negative, and its effects may be nonlinear as well. In many situations, an increase in biodiversity has a positive impact on human health—for example, increases in the diversity of available food supplies can have a positive impact on nutritional health. However, in some cases, particularly those in which the natural system has a large reservoir of vectors of human pathogens as a natural component, a reduction in biodiversity may also be accompanied by an improvement in human health (see this volume, chapter 12).

Drainage of the swamps in the Lower Laurentian Great Lake Basin, while reducing biodiversity (i.e., biota that thrived in swamps), improved human health through elimi-

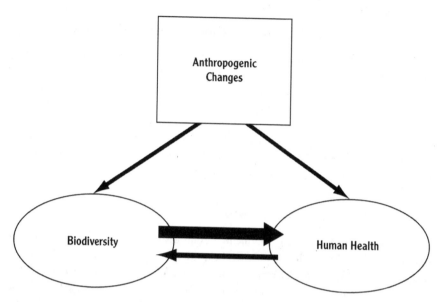

Figure 3.2. Ways in which anthropogenic changes in global and regional environments impact biodiversity and human health. These changes affect human health directly, and also indirectly through changes in biodiversity.

nation of malaria-carrying mosquitoes (*Anopheles* spp.). Similarly, in many parts of southern Europe, swamp drainage has eliminated the vector for malaria to the benefit of human health and well-being. Removal of wildlife reservoirs of zoonotic or livestock diseases is another reduction in biodiversity that has benefited humans. Elimination of vampire bats (*Desmodus rotundus*), which spread rabies to cattle in Latin America, for example, increased cattle productivity and thus improved human health (via better nutrition) and well-being (Mayen 2003).

There are also examples in which negative impacts on human health and well-being have had a positive impact on biodiversity. In Africa, the presence of tsetse flies (*Glossina* spp.) and their host species over large regions has prevented the colonization of some grazing areas by people and cattle, thus preserving biodiversity in those areas. Therefore, in this complex web of connections, changes to the environment that benefit our economy or health also can be viewed as trade-offs against biodiversity. The term *trade-offs*, however, has to be understood not in policy terms, where environment is traded off for short-term economic gains, but rather in scientific terms, in which one recognizes that gains in one domain are often at the expense of losses in another. Figure 3.2 illustrates the interactions between anthropogenic environmental change, biodiversity, and human health.

Impacts of Human Activity on Biodiversity and Health

Environmental change is increasingly driven by human activities (Friend and Rapport 1991; Rapport 2007a, 2007b; Rapport and Singh 2006; Rapport and Ullsten 2006; Vitousek et al. 1997). Human activities are resulting in global warming, habitat destruction, pollution, overharvesting, and introduction of exotic species. These changes have caused declines in biodiversity at regional and global scales and have had major impacts on human health through direct and indirect impacts on infectious diseases, nutrition, and contaminants (figure 3.3) (Epstein and Rapport 1996; Rapport, Costanza, and McMichael 1998; Rapport and Mergler 2004).

Infectious Diseases and Biodiversity

Lyme disease is one of the best-documented examples of how environmental changes may alter biodiversity and human health (Schmidt and Ostfeld 2001). Lyme disease is caused by a tick-transmitted bacterial (spirochete) pathogen (*Borrelia burgdorferi*) that is carried by a range of mammal, bird, and reptile reservoirs. Its presence in multiple hosts, and its vector-borne life cycle, mean that its transmission is susceptible to changes in biodiversity, land use, and climate (see this volume, chapter 13). In the northeastern United States, Lyme disease has emerged as a significant infection in humans, and although it causes little mortality, it is a significant national burden through morbidity and health care costs (Barbour and Fish 1993). Its emergence is an example of how recent shifts in land use, biodiversity, and human demography can combine to produce a new disease risk.

In the northeastern United States, in the 18th and 19th centuries, agricultural and industrial development resulted in rapid deforestation of the region, and, in the process, eliminated the deer populations that are essential for the Lyme disease cycle. In the early 20th century, the locus of agricultural activity shifted to the Midwest, and thus the northeastern United States became reforested, resulting in ideal habitat for supporting a rebound of deer populations and a consequent increase in the incidence of Lyme disease (Barbour and Fish 1993). Further, the almost complete removal of top predators contributed to population explosion in deer in the region. Thus biodiversity changes, both positive (reestablishment of forests in the northeast) and negative (elimination of top predators), in this case combined to promote overpopulation of deer to enhance the potential for the spread of Lyme disease.

Recent work has also shown that the fragmented habitat characteristic of human-modified landscapes in the northeastern United States provides an improved environment relative to dense forest for one key Lyme disease reservoir, the white-footed deer mouse (*Peromyscus maniculatus*) (LoGiudice et al. 2003). This reservoir supports relatively high densities of larval ixodid ticks and an abundant supply of the Lyme disease spirochete and is very common in human-dominated habitats. In more pristine areas,

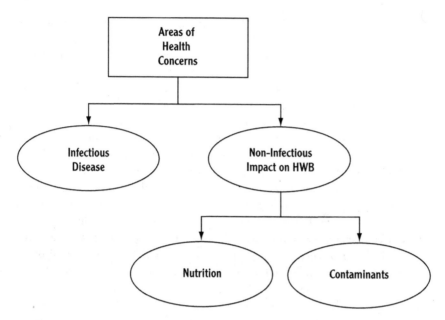

Figure 3.3. Major routes through which local, regional, and global anthropogenic changes can affect health and human well-being (HWB).

other, less competent reservoirs (e.g., squirrels) lower the relative abundance of deer mice and dilute their influence on disease risk for humans (the dilution effect). Thus, in this case, change in species composition, and possibly biodiversity loss, increases disease risk for humans. Ultimately, Lyme disease may become a more significant human pathogen as climate change affects the ecology of the disease-host interactions, and thus transmission dynamics (see this volume, chapter 13).

In the southwestern United States, lizard reservoir hosts sharply reduce the infection rate of ixodid tick vectors, potentially protecting human health. In southern and Pacific coastal states, the host of tick larvae and nymphs is frequently a lizard. Certain lizards, such as the western fence lizard (*Sceloporus occidentalis*), are not susceptible to infection with the Lyme disease spirochete. Owing to this feeding preference, in the southwestern part of the United States, most tick larvae and nymphs do not acquire the pathogen responsible for Lyme disease. Thus the actual infection rate may be 50% (or more) in adult ticks in hyperendemic areas of the Northeast, while only 1%–2% in most areas of the South and West. As in so many vector-borne diseases, it is not entirely the local species number per se that is a predictor of human health vulnerability, but rather the species composition within the ecosystem.

In this case, the effect of biodiversity on Lyme disease hinges on the incompatibility of hosts fed on by ticks. This life cycle mode does not apply to all infectious diseases.

Many infectious disease agents have high within-stage host specificity and multihost life cycles. For this reason, communities with high species heterogeneity (relative species turnover) should support high parasite heterogeneity (Hechinger and Lafferty 2005). The diversity of infectious diseases affecting human populations roughly corresponds to regions of high biodiversity due to independent effects of climate, biogeography, and evolutionary history (Guernier, Hochberg, and Guégan 2004). For example, ascariasis prevalence is higher in the diverse tropical forests of Cameroon than in the arid regions in the northern part of the country. Local-scale anthropogenic changes may result in positive or negative associations between biodiversity and infectious disease for reasons that can be idiosyncratic, and therefore difficult to predict. Regions of high biodiversity are generally correlated with regions of high cultural diversity at the global scale—thus adding another layer of complexity in the relationship between disease and diversity (both biological and cultural) (Maffi 2001). However, predictive approaches have been developed, and recent work on global patterns of zoonotic disease emergence demonstrates a strong relationship between the risk of new zoonoses from wildlife, high wildlife diversity and dense human populations (Jones et al. 2008).

Zoonoses

Encroachment into tropical forests also brings humans into direct contact with a high diversity of wildlife reservoirs that carry potentially new zoonotic diseases (diseases transmitted between humans and other animals). Such encroachment has led to the emergence of a series of new zoonoses over the last few decades. For example, the use of wildlife as a food source (bushmeat or wildmeat) has occurred throughout human history, but recent demographic and land use changes in some regions (e.g., West Africa, Indonesia, China) have greatly increased consumption rates and the efficiency of harvesting (Fa, Peres, and Meeuwig 2002). Wildlife harvesting has resulted in a significant loss of biodiversity (the empty forest syndrome) and has become a major conservation threat to some groups of mammals. Because wildlife are often the reservoirs of zoonotic diseases, including a number of highly lethal viruses, increased bushmeat consumption often leads to an increase in contact with these viruses and to their emergence in the human population (Peeters et al. 2002). Two notable examples are human immunodeficiency virus/acquired immune deficiency syndrome (HIV/AIDS) and severe acute respiratory syndrome (SARS).

Analysis of gene sequences of the two viruses responsible for HIV/AIDS (HIV-1 and HIV-2) demonstrates their recent origin in the human population, as simian immunodeficiency viruses (SIVs) transmitted from chimpanzees (*Pan troglodytes*) and sooty mangabeys (*Cercocebus atys*), respectively (Hahn et al. 2000). Tracing back the geographical and temporal signal to viral sequences suggests that the HIV-1 epidemic developed from an initial spillover event in the early part of the 20th century in west equatorial

Africa (Hahn et al. 2000). It is likely that multiple introductions of SIVs occurred historically, but that HIV-1 and HIV-2 have only spread regionally and globally following large-scale changes to demography, trade, and travel in Africa and the rest of the world since the 1950s (Hooper 1999). HIV/AIDS is now probably the most significant single infectious disease in current times, with conservative estimates of over 40 million people infected, and projections of 45 million new infections by 2010 if the virus continues to be transmitted at the current rate (Stover et al. 2002).

Apart from the obvious need for cheaper therapeutics and for the development of an HIV vaccine, the emergence of this disease provides a sharp indication of trade-offs between economic development and human health. In Africa, road building, increased volume of bushmeat consumption, increased temporary migration of the workforce to urban sites, and human behavioral changes all played a role in the spread of HIV/AIDS across the region. Increasing volume and reach of air travel to new destinations facilitated the regional epidemic in becoming a pandemic. Finally, changes to human sexual and drug use behavior led to the establishment of this disease across the globe as a major cause of mortality, morbidity, and sociopolitical change.

Preventing the negative consequences of encroachment on biodiversity to human health is a complex policy challenge. Strategies include assessing the relative risk of viral spillover from wildlife to humans due to local-scale factors (bushmeat hunting, deforestation, animal husbandry practices) and regional development plans (road building and demographic changes). These situations and other similar circumstances require knowledge of the *potentially* zoonotic pathogens present in wildlife populations where demographic changes bring humans and wildlife into closer contact. New techniques to rapidly survey for novel viruses (e.g., microarray technology) have been proposed as solutions. Some groups have begun to examine the risk of new zoonotic disease emergence from bushmeat in areas where people are rapidly encroaching on wildlife habitat: e.g., recent reports of simian viruses in bushmeat hunters in Cameroon (Switzer et al. 2005; Wolfe et al. 2004). However, such approaches are only just being developed, and policy decisions are currently not based on a valid examination of trade-offs. Increasing our knowledge of the risk of zoonotic transfer due to hunting, deforestation, and encroachment may benefit biodiversity by providing further evidence for the benefits of taking actions to reduce hunting, deforestation, and encroachment pressure.

In 2002, a novel coronavirus (SARS CoV) emerged that is lethal to humans (case fatality rate of around 10%) and spread as a pandemic outbreak to affect people in the United States, Europe, and across Asia in 2003. The available data suggest that SARS CoV originated from a group of viruses found in *Rhinolophus* spp. bats (Li et al. 2005), and initially spilled over to humans within wild animal markets ("wetmarkets") in China (Guan et al. 2003). The virus was responsible for the severe acute respiratory syndrome (SARS) that rapidly spread to population centers around the world, with globally more than 5,000 cases and approximately 750 deaths in 2002–2003.

The SARS outbreak had a significant impact on the global economy, virtually closing down important air transport hubs in southeast Asia during the height of the outbreak. It is estimated that SARS cost the global economy between US$30 and $100 billion in this single outbreak (Fedson 2003). Despite this huge economic cost, and the potential for future outbreaks from the wildlife markets of the world, policy for SARS and future zoonoses tends to be reactive rather than predictive and proactive. There are no nationwide systematic studies of the risk of future viral emergence from wildmeat markets in Asia, for example. However, mathematical modelers have started to produce convincing models of disease propagation in a globalized world that clearly show the importance of tackling the disease at its source (Hufnagel, Brockmann, and Geisel 2004).

The SARS epidemic illustrates how the control of new zoonotic viruses from wildlife can threaten biodiversity. Control measures for many zoonotic pathogens (e.g., rabies virus, hantavirus and SARS coronavirus) involve local or even national plans to remove their wildlife host reservoirs. Initially, the Chinese authorities responded to SARS by closing the markets and blocking the sale of these species. Later, authorities planned (but did not execute) the eradication of wild civets (*Paguma larvata*). Clearly, an ecosystem-centered approach to human health provides an impetus to shift attention to the potential for effective "upstream" interventions through ecosystem management to reduce "downstream" human health vulnerabilities (Epstein and Rapport 1996; Rapport and Lee 2003; Rapport and Mergler 2004).

Species introductions provide new hosts and parasites to native biological communities. Although only a fraction of the parasites and diseases of an introduced animal or plant coinvade (Torchin and Mitchell 2004), those that do can have dramatic impacts on naive host populations (Lafferty and Gerber 2002). In cases where new introduced hosts are vectors or reservoirs of human infectious diseases, impacts on human health can result. In addition to promoting disease, invasions can impact local native biodiversity.

West Nile virus (WNV) is endemic to Europe and Africa, where it is transmitted between birds by mosquitoes and spills over to humans, horses, and other mammals. It was first observed in the New World in 1999, where it caused 67 human deaths and mass mortalities in wild birds. It has since spread rapidly across North America and into Central and South America (CDC 2008). Between 1999 and 2004, approximately 6,700 cases of neuroinvasive disease (meningitis, encephalitis, or acute flaccid paralysis) have occurred, with 650 fatalities. Serosurveys indicate that symptoms develop in only 20% of persons with WNV and neuroinvasive disease in only 1 in 140 (L. Petersen and Hayes 2004). We might infer from these results that approximately 940,000 persons have been infected with WNV in the United States, and that 190,000 of them have become ill (L. Petersen and Hayes 2004). West Nile virus is now, therefore, the dominant vector-borne pathogen in the United States (L. Petersen and Hayes 2004).

West Nile virus is an unusual zoonotic disease in that it causes mortality in humans, other mammals, and a diverse range of bird species. In the New World, WNV causes high mortality in some bird species and has the capacity to reduce biodiversity at local spatial scales (Marra et al. 2004). Corvid species are particularly susceptible, and mortality rates of 100% have been observed in experimentally infected American crows (*Corvus brachyrhynchos*). In the wild, estimating WNV's impact is more difficult due to problems of defining populations for highly mobile species. The virus is known to have killed tens of thousands of birds and likely has been responsible for 10 to 100 times more (Marra et al. 2004). One study showed that crow populations in Manhattan decreased by 90% following WNV emergence, although crow populations in other parts of New York State varied in their response (Hochachka et al. 2004). Endangered species with small populations and reduced geographic ranges may be at risk from WNV-induced extinctions. In the western United States, a WNV-induced die-off of the endangered greater sage-grouse (*Centrocercus urophasianus*) and high mortality of birds of prey, including owls, suggest that these may undergo substantial declines now that WNV is endemic in the United States. A recent paper analyzed U.S. bird census data for species predicted to be at risk from WNV and showed that significant declines in these species correlated with the likely date of spread of the virus into that region in the United States (Ladeau et al. 2007).

The introduction of WNV to the New World provides an example of global anthropogenic influence on disease spread. The genetic sequence of WNV from New York is almost identical to those from an outbreak in the Middle East in 1998–1999, suggesting very recent introduction into the New World (Lanciotti et al. 1999). The most likely explanation is that the virus was introduced via international air travel. This theory is supported by the emergence of the virus in New York, a large international port of entry into the United States, and by the relative ease by which infected mosquitoes, or birds, would be able to travel in airplanes from Europe or the Middle East relative to other pathways (via ships or migration). West Nile virus also illustrates how changes in land use can increase the risk of disease for people. Recent research shows that in some parts of the United States, West Nile virus transmission is higher in urban regions than in rural or pristine environments for some avian taxonomic groups (Ezenwa et al. 2006), suggesting that there may be an effect on risk of infection for people.

A number of stressors can reduce biodiversity. Infectious diseases have complex interactions with environmental stressors. Sometimes stressors have opposing effects on the population dynamics of disease. For example, a stress might increase the susceptibility of individual hosts to infectious disease but ultimately reduce the spread of disease through a host population if the disease increases parasite or host death rates, or reduces parasite or host birthrates. For this reason, predictions from individual hosts do not necessarily apply to host populations. Theory suggests that host-specific infectious diseases should decline under stress (if that stress also impacts host abundance) while non-host-specific infectious diseases should generally increase under stress (Lafferty and Holt 2003). It is questionable whether these theoretical predictions extend directly to humans. The ability of humans to modify their environment, transport resources, and apply medical care may reduce the extent to which environmental changes and disease interact through host population density. In fact, stressors to human populations often correspond to dense populations, such as in refugee camps or developing urban areas.

Habitat Modification, Biodiversity, and Human Health

Sometimes stressors and infectious diseases come hand in hand. Agriculture, livestock farming, residential development, industries, and recreation tend to decrease water quality and thus impair human health by exposure to pathogens and chemicals. For example, the ability of Canada's urban and rural communities to sustain healthy drinking water is seriously threatened because of the increased incidences of pathogens in source water (Davies and Mazumder 2003). Fecal contamination with coliforms, *Vibrio cholerae*, and other pathogens like *Cryptosporidium* and *Giardia* cause major outbreaks of enteric illnesses and deaths of millions of humans per year.

How important are the integrity, complexity, and biodiversity of ecosystems in controlling epidemic outbreaks of pathogens? Riparian cover and diversity can reduce the loading of pathogens from land used for livestock farming to source water (see this volume, chapter 8). Intact consumer communities should reduce the impacts of nutrient loading on algal and bacterial biomass and therefore may help to regulate disease outbreaks (Mazumder, Lean, and Taylor 1992; Mazumder et al. 1988). These interactions underscore how biodiversity can buffer exposure to disease.

For example, leishmaniasis is an important public health problem in many tropical and subtropical regions. Phlebotomine sand flies are vectors of this parasitic disease, and rodents and canines serve as animal reservoirs. The increased incidence of leishmaniasis reported in many Mediterranean countries, such as Tunisia, is due to several reasons: (1) flux of nonimmune humans into natural sites of transmission; (2) change in the ecology of vectors and reservoir hosts; (3) reduction in the use of insecticides for malaria control; and (4) improvements in the diagnosis and reporting of disease.

In Tunisia and Iran, leishmaniasis outbreaks started after an increase of both reservoir and vector abundance (density of sand flies) in the regions as a consequence of planting native shrubs to provide natural barriers against soil erosion and desert progression. Thus, while this initiative restored biodiversity in some Mediterranean countries, it also resulted in the creation of new habitats favoring the survival of disease vectors and hosts. Removal of herbs and shrubs and planting of larger trees has brought about a significant reduction in the incidence of cutaneous leishmaniasis among the local communities in those countries (Kamhawi, Adwan, and Rida 1993).

The case study of schistosomiasis further underscores how habitat modification can facilitate disease. Schistosomiasis constitutes one of the most debilitating tropical infectious diseases, infecting hundreds of millions of people and killing tens of thousands. These snail-transmitted trematodes (*Schistosoma* spp.) are reemerging in different African and East Asian countries, despite the undisputable improvement in sanitary and socioeconomic conditions in those areas. Larval worms (cercariae) leave snails (*Biomphalaria* spp.) and penetrate humans in contact with the worms' freshwater habitat. Adult worms live in the circulatory system of humans, consuming blood. Their eggs lodge in various tissues of their human final hosts, leading to failure of the heart, liver, or other organs.

Conditions that benefit snails often correspond to a reduction in biodiversity. Most snail species that host human schistosomes thrive in human-created or modified habitats. Deforestation reduces acidic leaf litter and increases algal growth in ponds and streams, creating conditions suitable for snails (Southgate 1997). Snails also benefit from rice-field cultures, dams, and aquaculture production. These habitats may lack snail predators and have nutrient inputs favorable for snail population growth. For instance, large impoundments throughout Africa (e.g., Paperna 1969)—notably, construction of the Aswan High Dam, which created Lake Nasser (Heyneman 1979)—have substantially increased schistosome transmission, resulting in increased human morbidity and mortality (Gryseels et al. 1994). Copper-based molluscicides can help reduce snail populations, but may add to further losses in biodiversity.

A community ecology perspective provides additional predictions for the interactions between environmental changes, human health, and biodiversity. For example, population biology predicts that the reduction of a predator (such as by harvesting) would reduce the predator's infectious diseases. In turn, community ecology predicts that reducing predators could increase prey density and make lower trophic levels more susceptible to disease (Lafferty and Kuris 2005). On the other hand, harvesting may increase the severity of wildlife disease if it results in more susceptible newborn organisms within the ecosystem (Choisy and Rohani 2006). In some cases, trophic cascades can have complicated effects on disease and biodiversity. For instance, visit a fish market in Paris and you can take home a cut of fresh Nile perch (*Lates niloticus*). This export from Africa to Europe illustrates how economic benefits can have indirect impacts on human health and biodiversity. A biodiverse assemblage of cichlid fishes evolved in Lake Victoria and a few highly specialized species adapted to feed on mollusks, causing reduced snail density, which hampered the transmission of schistosomes to humans. Fisheries biologists introduced the Nile perch in an attempt to stimulate the local economy. The introduction unexpectedly resulted in a trophic cascade that favored snail population increase while reducing native biodiversity. Nile perch drastically reduced the abundance of native cichlids. The subsequent explosion of mollusk populations on lakeshores, plus an increase in human settlement to benefit from the new fish economy in the area, created foci for schistosomiasis transmission. Thus the Nile perch fishery, while creating direct economic profits for local populations, also resulted in the loss of biological diversity of the endemic cichlid fauna and consequently increased the prevalence of schistosomiasis in people.

Improving conditions for vectors can lead to wide-scale impacts on human health and biodiversity. In Africa, Asia, and Latin America, deforestation and increased movement of human populations have coincided with an upsurge of malaria and/or its vectors (Patz et al. 2004). For example, the building of the Panama Canal was associated with the production of new habitat for mosquitoes (temporal ponds and ditches), thus paving the way for the introduction and spread of malaria and yellow fever vectors (anopheline mosquitoes) with ultimately a very high mortality due to yellow fever and malaria. Currently, an increase in malaria in Amazonia has resulted from deforestation for agriculture and road building.

Increasing settlement of the Amazon Basin has uncalculated impacts for biodiversity as well as emergence of disease in indigenous peoples. The mosquito *Anopheles darlingi* normally lives in the rainforest canopy but can readily adapt to human-made ecosystems. Immigrants infected with *Plasmodium* protozoans have brought malaria into those areas where mosquitoes were newly available for malaria transmission. Such combinations of circumstances have led to increases in the distribution of malaria in other regions where deforestation corresponds with human immigration. These interactions represent how land use change can drive the introduction of mosquito vectors into biodiverse regions (Vittor et al. 2006).

Disease Control

Sometimes efforts to address impacts on human health significantly affect biodiversity. With failures in the development of vaccines and evolved resistance of malarial parasites to chemotherapy, vector control remains one of the best options for reducing human health impacts. Insecticides have nontarget effects, whose impact on tropical arthropod faunas could be substantial when broadly applied. However, limited topical application to interior walls and bed nets has proved to be an effective malaria control strategy. More environmentally damaging has been the management of water sources for breeding mosquitoes, through drainage and controlled water levels. This oldest of malaria control strategies has led to the reduction and elimination of wetland habitats worldwide, with suspected losses to biodiversity in these sensitive areas. The benefit of this approach to human health has been dramatic. However, wetland destruction has effectively eliminated malaria *and* wetland biodiversity from many regions, such as the southern United States, the North American Lower Great Lakes, and Israel/Palestine (Kitron 1987; see also chapter 4 in this book).

Noninfectious Contaminant/Pollution Issues

Contaminants can increase human mortality/morbidity and decrease birthrates (generally decreasing human well-being). They may also increase susceptibility to infectious disease through impacts on immune systems. Contaminants similarly impact biodiversity because they have a net impact on most nonhuman species. The net effects of contaminants on human populations must take into account the fact that they are often by-products of human activities that otherwise increase human well-being (mining, pesticides, transportation, industry) (Rapport et al. 2003). For example, a farmworker and local wildlife might each be exposed to the toxic effects of pesticides and herbicides, while at the regional level, the impacts of pesticides on biodiversity correspond with benefits resulting from increased food production. However, when considering the costs and benefits of contaminants, it is important to consider that the beneficiaries of contaminant production and sale (e.g., a mining company) are often different individuals than those that suffer health impacts (e.g., consumers of mercury-laden fish).

The human activities of burning fossil fuels for heating and transportation have resulted in contamination of many parts of the environment with oil and tar components, especially in the industrialized parts of the world. Around major cities in Europe, Canada, and the United States, numerous sites are affected by coal tar, which is a waste product of gas production. Coal tar contains a high amount of polycyclic aromatic hydrocarbons (PAHs), many of which are carcinogenic. The presence of coal tar has rendered the soil toxic and a threat to human health. Contamination by oil and derivatives like gasoline and fuel oil have resulted from point sources such as gas stations and accidental spills, or from deposition along roads. These compounds are a threat to human health as many of them are carcinogenic and are readily consumed by children playing in soil, or by inhalation of contaminated dust particles.

Depending on the geology and organic matter content of the soil at the site, the biomass and biodiversity of bacteria are usually decreased. If the right conditions are present (sufficient nutrients and aerobic conditions), bacteria and fungi present in the soil will degrade contaminants. Initially, the volatile compounds will evaporate or quickly be degraded by microorganisms, followed by the straight-chain aliphatic hydrocarbons. Such processes will increase the number and diversity of microorganisms able to degrade these contaminants. Later, the more recalcitrant PAHs will be degraded by other microorganisms, leading to a new change in biodiversity (Andreoni et al. 2004), and to a less toxic soil with less negative impact on human health. The specific bacteria and fungi responsible for the degradation of contaminants will differ depending on the soil type, the physical and chemical conditions in the soil environment, and the actual contaminant. Efforts to clean up contaminated soils include introducing contaminant-degrading microorganisms, soil aeration, and nutrient addition (Johnsen, Wick, and Harms 2005). In many cases, however, the native microbial community diversity completely metabolize the contaminants. These mechanisms testify to the fact that the large and partly unknown microorganisms present in the environment can be beneficial for human health by degrading contaminants to CO_2 and H_2O.

Microorganisms can degrade many other toxic organic compounds (e.g., polychlorinated biphenyls [PCBs], flame retardants, some chemical pesticides, additives to washing powders). The degradation of organic compounds by the microbial community occurs in urban soils, as well as in agricultural soils and aquatic systems. Bioremediation can only be accomplished by having a viable microbial community in the environment with sufficient diversity of genes and functions to degrade organic contaminants, through often very complex degradation pathways and perhaps involving several different species.

Since World War II, there has been an enormous increase in the application of chemical pesticides to agricultural fields to enhance plant production. Safety evaluations assume eventual microbial degradation of the pesticide. However, in numerous cases, pesticides persist in groundwater, in aquatic systems and, through bioaccumulation, in the food chain. The persistence of pesticides has a negative effect on human health, leading to a search for alternative approaches. Presently, available alternatives include replacing pesticides with (1) microbial pest control agents (MPCAs); (2) organic farming without input of chemical pesticides; (3) development of disease-suppressive microorganisms in soils; or (4) development of disease-resistant crops. These approaches reduce contaminants in food and soil but can lead to lower yield. Generally, organic farming leads to higher biodiversity within and above ground (Aude, Tybirk, and M. B. Pedersen 2003). The preservation of microbial diversity can, in itself, result in fewer and less severe pathogen attacks because of the development of disease-suppressive soils (Weller, Raaijmakers, and Gardener 2002).

Significant results have been published during the last decade on the development and mechanism of disease-suppressive soils (De la Fuente, Landa, and Weller 2006; Glandorf et al. 2001). Several methods of obtaining disease-suppressive soils and mechanisms of disease suppression within the soil have been described (Whipps 2002). In disease-suppressive soils, the biodiversity is changed to a diversity better suited for agricultural purposes and hence having a positive effect on human health.

Noninfectious Nutrition Issues

In developing countries, changes in food resources can alter human population growth through mortality and birth. Improved food production increases child growth rate and decreases diseases associated with malnutrition. However, as discussed in chapter 2, the interactions between food production, human well-being, and biodiversity are complex. Food production tends to reduce biodiversity directly, to compete with it for space (through landscape conversion to agriculture), and to release contaminants to protect dense monocultures, suggesting that increased food production should lead to a negative and indirect association between human well-being and biodiversity. For example, the reduction in diet diversity and the shift toward lower-quality food sometimes associated with large-scale terrestrial agriculture can lead to declines in both human well-being and biodiversity.

Treated sewage sludge provides an inexpensive and readily available means to fertilize plant crops. Adding sewage sludge alters the microbial community or agricultural soils. Depending on the efficacy of the sewage treatment and later treatment of the sludge, many anaerobic microorganisms and contaminants may be present. Feces and urine from livestock are also used as plant fertilizers and will likewise contain a wealth of different microorganisms, medicine residues (e.g., antibiotics), and growth promoters (e.g., copper) in addition to the nitrogen and organic matter beneficial for plant growth.

The added microorganisms will affect the diversity of the indigenous microbial community and fauna by competition for food and space and by acting as food for the bacteria-eating fauna and fungi. The contaminants will, in some cases, be fed on by microorganisms and microfauna, or may be toxic. The fertilizers will increase microbial activity and diversity because microorganisms will mineralize the organic matter and make it available to the plants (A. Winding and S. Petersen, pers. comm.).

The addition of these kinds of wastes to agricultural soils changes biodiversity in this environment and benefits human health by disposing of waste and increasing crop production. It can, however, also threaten human health by increasing the risk of transfer of human pathogens and antibiotic-resistant genes to food crops or livestock, and by increasing the risk of bioaccumulation of heavy metals. If these risks are evaluated and handled properly, this use of wastes is a benefit for human health (S. Petersen et al. 2003). There may be a better case to be made for use of the precautionary principle, given that growth promoters and antibiotics released to the environment are a potential risk to public health.

Anthropogenic changes such as agriculture, harvesting, and industrial and residential developments have caused major changes in the loading of nutrients (nitrogen and phosphorus) in freshwater and marine ecosystems worldwide. These nutrient inputs increase algal biomass, toxic algal blooms, anoxic waters, and fish kills (Epstein and Rapport 1996; Lefeuvre et al. 2004; Smith 2003). While nitrogen and phosphorus are known to be the two most important nutrients regulating productivity and diversity of all trophic levels, an excessive concentration leads to reduced biodiversity of algae, with significant consequences for human, animal, and ecosystem health (see this volume, chapter 8). Often there is a shift from small algae to large filamentous blue-green algae (also known as cyanobacteria), with a significant reduction of biodiversity. There are about 50 different species of blue-green algae in marine, brackish, and fresh waters, with about 12 of them capable of producing toxins.

There are three major classes of algal toxins responsible for a variety of health effects, such as skin irritation, respiratory ailments, and neurotoxical and carcinogenic effects. Some of the organic compounds are peptides (nodularin and microcystin, causing skin irritation). Anatoxins and saxitoxins are neurotoxic alkaloids, which affect the nervous system, skin, liver, and gastrointestinal tract. Anatoxins in high doses can cause death due to respiratory failure, while the saxitoxins are responsible for paralytic shellfish poisoning in humans consuming contaminated shellfish.

Several notable outbreaks have been associated with algal blooms and toxic blue-green algae in different parts of the world (Australia, Europe, Asia, North America, and South

America). In most of these cases, there have been significant economic and health losses due to contaminated drinking water, ranging from declines in tourism and fisheries, to extensive gastrointestinal sickness, to deaths of humans. Among these cases, some of the most serious algal bloom–related sicknesses and fatalities have been in Brazil, where, on two occasions, 82 and 60 people died, thousands developed enteric illnesses, and 150 developed severe illness and kidney failure (Reynolds 2004). In many of these outbreaks, children and immunocompromised individuals were the most susceptible to exposure to algal toxins. Expensive coagulation and filtration can remove the toxic algae, but these are not 100% effective in removing dissolved toxins.

Thus the main management strategy to reduce health risks from toxic algal blooms should be focused on reducing point- and nonpoint-source loading of nitrogen and phosphorus. While high loading of nutrients from agricultural and sewage sources and the associated decline in nitrogen-to-phosphorus ratio cause blue-green algal blooms, it is still not possible to predict the timing and composition of blooms. It is also unclear what factors trigger toxin production by blue-green and other toxin-producing algae.

Ecosystem Approaches to Human Health and Biodiversity

The interrelationships between regional and global change, biodiversity, and human health are indeed complex. When it comes to directionality, few generalizations can be made. There are instances in which an increase in biodiversity has a positive impact on human health, and instances in which it has a negative impact. There are instances in which human interventions—particularly in land use change—shift the balance in favor of human pathogens; there are instances in which the opposite is true. Further, when it comes to human health, many anthropogenic environmental changes unleash a diversity of consequences, some positive—as, for example, when increased agricultural productivity enhances nutrition—and some negative—as, for example, when nutrient runoff from agricultural fields favors the abundance of human pathogens.

One might then ask if there are some overarching policy recommendations that emerge with respect to the interrelationship of biodiversity and human health, taking all these complexities into account. One is clearly that improving human health outcomes has much to do with maintaining healthy ecosystems—that is, with harmonizing human activity within an ecosystem context so as to minimally compromise ecological and cultural evolution. This implies that we look for solutions that simultaneously improve human health while maintaining biodiversity. Thus an overarching policy for human health and biodiversity must be based on achieving healthy ecosystems. In most parts of the world, this requires policies that pull back anthropogenic environmental changes rather than add to the total burden. While no doubt this is easier said than done, if there is to be a viable human future, "designing with nature" for healthy ecosystems becomes paramount.

References

Andreoni, V., L. Cavalca, M. A. Rao, G. Nocerino, S. Bernasconi, E. Dell'Amico, M. Colombo, and L. Gianfreda. 2004. Bacterial communities and enzyme activities of PAHs polluted soils. *Chemosphere* 57:401–12.

Aude, E., K. Tybirk, and M. B. Pedersen. 2003. Vegetation diversity of conventional and organic hedgerows in Denmark. *Agriculture Ecosystems and Environments* 99:135–47.

Barbour, A. G., and D. Fish. 1993. The biological and social phenomenon of Lyme disease. *Science* 260:1610–16.

CDC (Centers for Disease Control and Prevention). 2008. West Nile virus: Statistics, surveillance, and control. http://www.cdc.gov/ncidod/dvbid/westnile/surv&control.htm.

Choisy, M., and P. Rohani. 2006. Harvesting can increase severity of wildlife disease epidemics. *Proceedings of the Royal Society of London* B 273:2025–34.

Daszak, P., A. A. Cunningham, and A. D. Hyatt. 2000. Emerging infectious diseases of wildlife: Threats to biodiversity and human health. *Science* 287:443–49.

Davies, J.-M., and A. Mazumder. 2003. Health and environmental policy issues in Canada: The role of watershed management in sustaining clean drinking water quality at surface sources. *Journal of Environmental Management* 68:276–86.

De La Fuente, L., B. B. Landa, and D. M. Weller. 2006. Host crop affects rhizosphere colonization and competitiveness of 2,4-diacetylphloroglucinol-producing *Pseudomonas fluorescens*. *Phytopathology* 96:751–62.

Dobson, A. 1995. Biodiversity and human health. *Trends in Ecology and Evolution* 10:390–91.

Epstein, P., and D. J. Rapport. 1996. Changing coastal marine environments and human health. *Ecosystem Health* 2 (3): 166–76.

Ezenwa, V. O., M. S. Godsey, R. J. King, and S. C. Guptill. 2006. Avian diversity and West Nile virus: Testing associations between biodiversity and infectious disease risk. *Proceedings of the Royal Society* B 273:109–17.

Fa, J. E., C. A. Peres, and J. Meeuwig. 2002. Bushmeat exploitation in tropical forests: An intercontinental comparison. *Conservation Biology* 16:232–37.

Fedson, D. S. 2003. Vaccination for pandemic influenza and severe acute respiratory syndrome: Common issues and concerns. *Clinical Infectious Diseases* 36:1562–63.

Friend, A. M., and D. J. Rapport. 1991. Evolution of macro-information systems for sustainable development. *Ecological Economics* 3:59–76.

Glandorf, D. C. M., P. Verheggen, T. Jansen, J.-W. Jorritsma, E. Smit, P. Leeflang, K. Wernars, et al. 2001. Effect of the genetically modified *Pseudomonas putida* WCS358r on the fungal rhizosphere microflora of field-grown wheat. *Applied and Environmental Microbiology* 67:3371–78.

Gryseels, B., F. Stelma, I. Talla, G. Van Dam, S. Polman, M. Sow, R. Diaw, et al. 1994. Epidemiology, immunology and chemotherapy of *Schistosoma mansoni* infections in a recently exposed community in Senegal. *Tropical and Geographical Medicine* 46:209–19.

Guan, Y., B. J. Zheng, Y. Q. He, X. L. Liu, Z. X. Zhuang, C. L. Cheung, S. W. Luo, et al. 2003. Isolation and characterization of viruses related to the SARS coronavirus from animals in southern China. *Science* 302:276–78.

Guernier, V., M. E. Hochberg, and J. F. Guégan. 2004. Ecology drives the worldwide distribution of human infectious diseases. *PLoS Biology* 2:740–46.

Hahn, B. H., G. M. Shaw, K. M. de Cock, and P. M. Sharp. 2000. AIDS as a zoonosis: Scientific and public health implications. *Science* 287:607–14.

Hechinger, R. F., and K. D. Lafferty. 2005 Host diversity begets parasite diversity: Bird final hosts and trematodes in snail intermediate hosts. *Proceedings of the Royal Society of London* B 272:1059–66.

Heyneman, D. 1979. Dams and disease. *Human Nature* 2:50–57.

Hochachka, W. M., A. A. Dhondt, K. J. McGowan, and L. D. Kramer. 2004. Impact of West Nile virus on American crows in the northeastern United States, and its relevance to existing monitoring programs. *EcoHealth* 1:60–68.

Hooper, E. 1999. *The river.* Boston: Little, Brown.

Hufnagel, L., D. Brockmann, and T. Geisel. 2004. Forecast and control of epidemics in a globalized world. *Proceedings of the National Academy of Sciences* 101:15124–29.

Johnsen, A. R., L. Y. Wick, H. Harms. 2005. Principles of microbial PAH-degradation in soil. *Environmental Pollution* 133:71–84.

Jones, K. E., N.G. Patel, M. A. Levy, A. Storeygard, D. Balk, J. L. Gittleman, and P. Daszak. 2008. Global trends in emerging infectious diseases. *Nature* 451: 990–93.

Kamhawi, S., A. Arbagi, S. Adwan, and M. Rida. 1993. Environmental manipulation in the control of a zoonotic cutaneous leishmaniasis focus. *Archives d'Institut Pasteur de Tunis* 70 (3–4): 383–90.

Kitron, U. 1987. Malaria, agriculture and development: Lessons from past campaigns. *International Journal of Health Services* 17 (2): 295–326.

Ladeau S. L., A. M. Kilpatrick, and P. P. Marra. 2007. West Nile Virus emergence and large-scale declines of North American bird populations. *Nature* 447:710–13.

Lafferty, K. D., and L. Gerber. 2002. Good medicine for conservation biology: The intersection of epidemiology and conservation theory. *Conservation Biology* 16 (3): 593–604.

Lafferty, K. D., and R. D. Holt. 2003. How should environmental stress affect the population dynamics of disease? *Ecology Letters* 6 (7): 797–802.

Lafferty K. D., and A. M. Kuris. 2005. Parasitism and environmental disturbances. In *Parasitism and ecosystems,* ed. F. Thomas, J. F. Guégan, and F. Renaud, 113–23. Oxford: Oxford University Press.

Lanciotti, R. S., J. T. Roehrig, V. Deubel, J. Smith, M. Parker, K. Steele, B. Crise, et al. 1999. Origin of the West Nile virus responsible for an outbreak of encephalitis in the northeastern United States. *Science* 286:2333–37.

Li, W., Z. Shi, M. Yu, W. Ren, C. Smith, J. H. Epstein, H. Wang, et al. 2005. Bats are natural reservoirs of SARS-like coronaviruses. *Science* 310:676–79.

LoGiudice, K., R. S. Ostfeld, K. A. Schmidt, and F. Keesing. 2003. The ecology of infectious disease: Effects of host diversity and community composition on Lyme disease risk. *Proceedings of the National Academy of Sciences of the United States of America* 100:567–71.

Maffi, L. 2001. *On biocultural diversity: Linking language, knowledge, and the environment.* Washington, DC: Smithsonian Institution Press.

Marra, P. P., S. Griffing, C. Caffrey, A. M. Kilpatrick, R. McLean, C. Brand, E. Saito, A. P. Dupuis, L. D. Kramer, and R. Novak. 2004. West Nile virus and wildlife. *Bioscience* 54:393–402.

Mayen, F. 2003. Haematophagous bats in Brazil, their role in rabies transmission, impact on public health, livestock industry and alternatives to an indiscriminate reduction of bat population. *Journal of Veterinary Medicine* B 50:469–72.

Mazumder, A., D. R. S. Lean, and W. D. Taylor. 1992. Importance of microzooplankton in the energy transfer efficiency of Lake Ontario. *Journal of Great Lakes Research* 18:456–66.

Mazumder, A., D. J. McQueen, W. D. Taylor, and D. R. S. Lean. 1988. Effects of fertilization and planktivorous fish predation on size-distribution of particulate phosphorus and assimilated phosphate. *Limnology and Oceanography* 33:430–37.

Morens, D. M., G. K. Folkers, and A. S. Fauci. 2004. The challenge of emerging and re-emerging infectious diseases. *Nature* 430:242–49.

Paperna, I. 1969. Study of an outbreak of schistosomiasis in the newly formed Lake Volta in Ghana. In *Zeitschrift für Tropenmedizin und Parasitologie* 21 (4): 411–25.

Patz, J. A., P. Daszak, G. M. Tabor, A. A. Aguirre, M. Pearl, J. Epstein, N. D. Wolfe, et al. 2004. Unhealthy landscapes: Policy recommendations on land use change and infectious disease emergence. *Environmental Health Perspectives* 112:1092–98.

Peeters, M., V. Courgnaud, B. Abela, P. Auzel, X. Pourrut, F. Bibollet-Ruche, S. Loul, et al. 2002. Risk to human health from a plethora of simian immunodeficiency viruses in primate bushmeat. *Emerging Infectious Diseases* 8 (5): 451–57.

Petersen, L. R., and E. B. Hayes. 2004. Westward ho? The spread of West Nile virus. *New England Journal of Medicine* 351:2257–59.

Petersen, S. O., K. Henriksen, G. K. Mortensen, P. H. Krogh, K. K. Brandt, J. Sørensen, T. Madsen, J. Petersen, and C. Gron. 2003. Recycling of sewage sludge and household compost to arable land: Fate and effects of organic contaminants, and impact on soil fertility. *Soil and Tillage Research* 72:139–52.

Rapport, D. J. 2007a. Healthy ecosystems: An evolving paradigm. In *Sage handbook on environment and society,* ed. J. Pretty, A. Ball, T. Benton, J. Guivant, D. Lee, D. Orr, M. Pfeffer, and H. Ward. Thousand Oaks, CA: Sage. pp. 431–441.

———. 2007b. Sustainability science: An ecohealth approach. *Sustainability Science,* in press.

Rapport, D. J., R. Costanza, and A. McMichael. 1998. Assessing ecosystem health: Challenges at the interface of social, natural, and health sciences. *Trends in Ecology and Evolution* 13 (10): 397–402.

Rapport, D. J., W. Lasley, D. E. Rolston, N. O. Nielsen, C. O. Qualset, and A. B. Damania, eds. 2003. *Managing for healthy ecosystems.* Boca Raton, FL: Lewis Publishers.

Rapport, D. J., and V. Lee. 2003. Ecosystem approaches to human health: Some observations on North/South experiences. *Ecosystem Health* 3:26–39.

Rapport, D. J., and D. Mergler. 2004. Expanding the practice of ecosystem health. *Eco-Health* 1 (Suppl. no. 2): 4–7.

Rapport, D. J., and A. Singh. 2006. An EcoHealth-based framework for state of environment reporting. *Ecological Indicators* 6:409–28.

Rapport, D. J., and O. Ullsten. 2006. Managing for sustainability: Ecological footprints, ecosystem health and the forest capital index. In *Sustainable development indicators in ecological economics,* ed. Philip Lawn, 268–87. Current Issues in Ecological Economics. Northampton, MA: Edward Elgar Publishing.

Reynolds, K. A. 2004. Cyanobacteria: Natural organisms with toxic effects. *Water Conditioning and Purification* 46:78–80.

Schmidt, K. A., and R. S. Ostfeld. 2001. Biodiversity and the dilution effect in disease ecology. *Ecology* 82:609–19.

Smolinski, M. S., M. A. Hamburg, and J. Lederberg. 2003. *Microbial threats to health: Emergence, detection, and response.* Washington, DC: National Academies Press.

Southgate, V. R. 1997. Schistosomiasis in the Senegal River Basin: Before and after the construction of dams at Diama, Senegal and Manantali, Mali and future prospects. *Journal of the Helminthological Society of Washington* 71:125–32.

Stover, J., N. Walker, G. P. Garnett, J. A. Salomon, K. A. Stanecki, P. D. Ghys, N. C. Grassly, R. M. Anderson, and B. Schwartlander. 2002. Can we reverse the HIV/AIDS pandemic with an expanded response? *Lancet* 360 (9326): 73–77.

Switzer, W. M., M. Salemi, V. Shanmugam, F. Gao, M. E. Cong, C. Kuiken, V. Bhullar, et al. 2005. Ancient co-speciation of simian foamy viruses and primates. *Nature* 434:376–80.

Thomas, F., F. Renaud, and J. F. Guégan. 2005. *Parasitisms and ecosystems.* Oxford: Oxford University Press.

Torchin, M. E., and C. E. Mitchell. 2004. Parasites, pathogens and invasions by plants and animals. *Frontiers in Ecology and the Environment* 2:182–90.

Vitousek, P. M., H. A. Mooney, J. Lubchenco, and J. M. Melillo. 1997. Human domination of Earth's ecosystems. *Science* 277:494.

Vittor, A. Y., R. H. Gilman, J. Tielsch, G. Glass, T. Shields, W. Sanchez Lozano, V. Pinedo-Cancino, and J. A. Patz. 2006. The effect of deforestation on the human-biting rate of *Anopheles darlingi,* the primary vector of falciparum malaria in the Peruvian Amazon. *Journal of Tropical Medicine and Hygiene* 74 (1): 3–11.

Weiss, R. A., and A. J. McMichael. 2004. Social and environmental risk factors in the emergence of infectious diseases. *Nature Medicine* 10:S70–S76.

Weller, D. M., J. M. Raaijmakers, and B. B. M. Gardener. 2002. Microbial populations responsible for specific soil suppressiveness to plant pathogens. *Annual Review of Phytopathology* 40:309.

Wolfe, N. D., W. M. Switzer, J. K. Carr, V. B. Bhullar, V. Shanmugam, U. Tamoufe, A. T. Prosser, et al. 2004. Naturally acquired simian retrovirus infections among central African hunters. *Lancet* 363:932–37.

4

Biodiversity and Human Health: The Decision-Making Process

Camille Parmesan, Suzanne M. Skevington, Jean-François Guégan, Peter Jutro, Stephen R. Kellert, Asit Mazumder, Marie Roué, and Manju Sharma

Activities that alter biodiversity are performed for many reasons, and environmental assessment is frequently extensive. However, analysis of potential long-term health and welfare implications of these activities is often not formally incorporated into the decision-making process. Conversely, many projects are implemented specifically to improve health, but such projects typically consider only the short-term benefits to health and the short-term costs of implementation. There may, however, also be a long-term cost of human activities in terms of lost or degraded ecosystem services, which, in turn, impact human health. In this chapter we explore existing approaches from the social sciences that may provide useful methodologies for assessing strategies that maximize human health in the context of sustainable biodiversity (see this volume, chapter 5). We adopt a broad view of human health, as stated in the preamble to the Constitution of the World Health Organization: "Health is a state of complete physical, mental, and social well-being and not merely the absence of disease or infirmity" (see also this volume, chapters 1, 6, and 7).

Most existing institutional frameworks act primarily on short time frames. Developmental plans are often based on economic principles for specified periods of time (typically 5 to 10 years). This is due partly to the fact that uncertainty in projected outcomes increases with an increasing future time frame. Thus policies are more likely to be effective in the short term, for which there is a greater certainty in the outcome. When long-term objectives emerge, decision makers often need to construct and implement policies for which the outcomes are uncertain in magnitude, or even in direction.

Economic theories are rich in approaches to aid policymakers in assessing the possible short-term as well as long-term consequences of multiple planning strategies (Rosenhead 1990). With a long planning horizon, though, it is incumbent on decision makers to evaluate their possible interventions across a wide range of possible futures. A complete evaluation then would search for robust strategies that work well across a sizable subset of futures, designing responsive strategies that can be adjusted as the future unfolds, hedging against the possibility of intolerable (or perhaps unattractive futures), or designing risk-spreading mechanisms that reduce their consequences. Each of these approaches has been well developed in the economics literature, and each has its own comparative advantages across different characterizations of risk. By adapting these existing approaches and expanding their scope beyond the boundaries of purely economic systems, it may be possible to develop a decision-making model of human health and well-being that not only exploits the efficacy of economic techniques but also allows their application to take account of social, psychological, and biological criteria.

A prerequisite for developing such models is clear identification of "currencies" of ecosystem and human health. There is a large literature that develops, compares, and contrasts various measures of ecosystem condition. Historically, this literature has focused on standard measures of growth, such as net primary productivity, but more recently the literature has emphasized the abilities of ecosystems to provide services (Harwell et al. 1999; Heinz Center 2002; Meyerson et al. 2005; NRC 2000). Because the topic of ecosystem services has been well reviewed and critiqued elsewhere, it will not be discussed further here.

In parallel, the medical community has developed assessments of human health from physical and psychological angles (WHO 2002). These assessments of physical and mental health can be accomplished by relatively objective measurements that have been standardized across countries. The World Health Organization also recognizes quality of life (QOL) as an integral part of human health (see this volume, chapters 1 and 7). However, measurement of QOL is inherently a subjective process, and standardization across socioeconomic and cultural boundaries is a challenge (see this volume, chapters 6 and 7). There are many quality-of-life measures. Most deal with the complexity of assessment by targeting the study design to serve a specific purpose. Few have integrated generic health and environmental aspects in a rigorous way; even fewer enable international (cross-cultural) comparisons of results; and none have explicitly extended their scope to include aspects of natural biodiversity. By combining expertise from different disciplines, adaptation of existing metrics (for both physical health and quality of life) to incorporate impacts of biodiversity change on short- and long-term human health should be eminently doable with existing tools.

Figure 4.1 illustrates the conceptual framework guiding this chapter. We start by briefly discussing the difficulties of imposing long-term biodiversity and health planning on institutional systems designed for action on shorter time frames and for other objectives. Two case studies illustrate how conventional institutional responses ultimately harm the health of the population by ignoring the long-term effects of biodiversity deg-

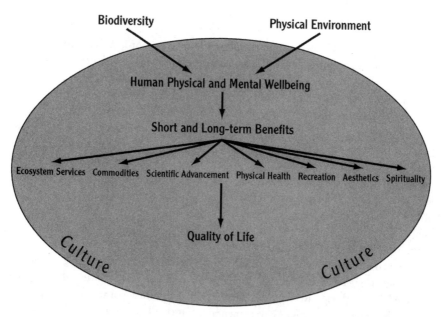

Figure 4.1. Relationships among biodiversity, the physical environment, and quality of life. Both biodiversity and the physical environment influence humans against a backdrop of local culture. When the physical environment is in good condition (e.g., there are good water storage and filtration; a stable climate regime; and healthy soil, water, and air chemistry) and biological diversity is in a healthy state (e.g., there are high native species diversity, few exotic species, and a properly functioning ecosystem), together they foster human physical and mental well-being by producing a range of short- and long-term economic and noneconomic benefits. These benefits include ecosystem services such as waste decomposition, crop pollination, and soil fertility; commodities such as fisheries and wood products; gains in scientific knowledge and understanding; physical health due to few environmental contaminants and ample opportunities for physical recreation; and aesthetic and spiritual satisfaction. When these benefits are harmonized with local culture, they lead to a high quality of life for humans as well as the natural world.

radation. In the next section, an assessment of health and quality of life that is holistic and integrated with biodiversity conservation is formulated. Resources available from economic theory are discussed for their applicability to a novel type of economic assessment that includes both QOL and conservation values. Specifically, we discuss how economic models might be adapted to include both biodiversity and human health in the decision-making process. A key point of this discussion is the articulation of an existing method from social psychology that could be easily adapted to measure the impacts of natural biodiversity on broad-sense health and perceived quality of life. The chapter ends with suggested future directions for research.

Institutional Issues

Since the 1972 UN Conference on the Human Environment in Stockholm, there has been growing concern for the protection of global biodiversity. The 1992 Rio Conference on Environment and Development and the 2002 Johannesburg Declaration on Sustainable Development have reinforced the global commitment to protect the natural resources of the planet Earth. Global initiatives have resulted in several highly successful treaties and programs. The 1992 Convention on Biological Diversity (CBD) has over 180 parties and has resulted in global programs to protect biodiversity (e.g., the IUCN's Species Survival Commission and Red List of Threatened and Endangered Species). Further, implementation of the CBD in many countries has led to national legislation providing legal protection for rare and endangered species (e.g., the U.S. Endangered Species Act of 1973 and Canada's Species at Risk Act of 2002). The Global Environmental Facility, established in 1991, helps developing nations to finance environmental protection measures, including preservation of native biodiversity.

These commitments have encompassed both developed and developing nations on every continent. Yet, on a global scale, natural ecosystems have continued to become degraded. Political scientists have identified a suite of institutional and political circumstances characteristic of various governmental forms that have the potential to frustrate the implementation of effective approaches to biodiversity management (Von Schirnding 2002; World Bank 1997). In the decision-making process, some actions have consequences that are immediately apparent. When consequences are both immediate and well defined, existing institutions have little difficulty in minimizing negative outcomes. Delayed consequences—which are apparent some time after action—may also be incorporated into the planning process when they are anticipated and known. However, delayed consequences are frequently anticipated but indeterminate, or unanticipated altogether. These latter two circumstances present the greatest challenges to sustainable biodiversity management. Some of the characteristics of current institutional frameworks that hinder the inclusion of long-term, imprecise, or unanticipated outcomes into the decision-making process are as follows:

- *Political frameworks are inherently short-term.* Political cycles tend to be short, creating governments that have difficulties in dealing with intergenerational issues. This can lead to decisions in favor of actions with immediate benefits, even when undesirable future effects are clear. Although economic mechanisms for dealing with anticipated futures exist (e.g., discounting), no general governmental mechanisms exist for weighing or incorporating the interests of those who will bear the consequences of decisions in the future.
- *Political boundaries don't match biodiversity boundaries.* Governmental and ecosystem boundaries rarely correspond. International treaties offer a solution, but they can be broadly worded, leaving room for multiple interpretations in national or local implementation of those treaties.

- *International agreements are at the national level, while implementation is often at the local level.* According to international law, under which environmental treaties are negotiated, only nations can be parties to treaties. Yet natural resource responsibility often lies with provincial governments, and land use decisions frequently lie in the hands of local governments.

- *Conflicts exist in the aims of different governmental entities.* Laws and programs that are the responsibility of one ministry are frequently in conflict with those that are the responsibility of other ministries. For example, a ministry for agricultural development may have ill-informed, and even conflicting, goals with respect to those of a ministry for conservation.

- *Conflicts exist within a governmental entity.* The same ministry may have responsibilities for both resource protection and resource exploitation—for example, the U.S. Department of the Interior is responsible both for protection of biodiversity via a system of national parks and wilderness areas, and for exploitation of timber, water, and mining resources on government lands in general (e.g., designated National Forest or Bureau of Land Management lands).

- *Health protection and biodiversity conservation have different historical roots.* Issues of concern may have arisen at different historical times and may have developed independent legal structures and bureaucracies. In the United States, for instance, land use conservation entered the national dialogue in the 1880s, whereas the public health aspects of environmental protection (clean air and clean water) have grown to prominence only since the 1970s. Since advocacy for the health of people and for the preservation of the environment grew out of distinct camps, and were often seen as competing goals, federal investment in one was frequently made at the expense of the other. Clean air, clean water, and ecosystem health are inextricably interrelated, yet planning is usually not synchronized (Davies and Mazumder 2003).

- *Political instability is an obstacle to health and biodiversity protection.* As the 21st century begins, we face a global situation in which large areas of the world are outside effective national control. Large areas within biodiverse countries are in the throes of political unrest, whereas others are conceded, formally or informally, as under "tribal" or indigenous control. Other areas are sites of active military conflict, within which the concept of national environmental management is little more than a dream. For these areas, political instability is intrinsically incompatible with biodiversity protection.

Short-Term Gains with Long-Term Negative Consequences: Examples from Real Life

The following two case studies illustrate how traditional institutional frameworks have succeeded in providing short-term foreseen solutions to health and service problems—

solutions, however, that have led to unforeseen, negative long-term consequences for human well-being.

Case Study 1—Unexpected Impacts: Dams, Biodiversity, and Health

THE CREE INDIANS OF CANADA

Environmental changes have often had consequences on human health that were not foreseen in the planning of development projects. Roué and Nakashima (1999) documented the impacts of the James Bay hydroelectric megaprojects on the health of the subarctic Cree Indians of Chisasibi (Quebec, Canada), traditionally a society of hunters, fishers, and trappers. This example shows that even apparently appropriate short-term solutions to societal needs can engender unforeseen long-term problems for both the physical health and the quality of life of local populations. When the government of Quebec announced the damming of the La Grande River in 1971, it was done without an environmental impact assessment of the effects on biodiversity upstream and downstream of the reservoir, contamination of the aquatic resources of aboriginal communities, and the health of the population (Desbiens 2004). A quote from a Cree elder in regard to the water says much about the health concerns generated by the construction of the reservoir: "I do not drink it. . . . There is so much construction there and there is so much stuff going into the water. . . . This is why people get sick so often." Compounding compromised water quality and loss of vegetation and wildlife was the contamination of fish and wildlife with methyl mercury generated by the flooded land and the decomposing soil and vegetation (Roebuck 1999).

In the 1970s, shortly after the creation of artificial reservoirs for the La Grande project, lake productivity increased significantly, with associated changes in fish community biomass and relative diversity. However, inundation of the land released mercury contained naturally in the soil. Decomposition of flooded vegetation also released methyl mercury, leading to a rapid increase in the concentration of mercury in all fish species, but even more so in the predatory (fish-eating) fish and, in turn, in their predators (e.g., ospreys, *Pandion haliaetus*, and bears, *Ursus* spp.) (DesGranges et al. 1998; Lucotte et al. 1999). Counter to expectation, higher mercury concentrations were also found in fish below the dams, as well as in invertebrates (Schetange, Doyon, and Fournier 2000). The likely cause was that fish that were normally "prey" were feeding on the minced remains of large predatory fish (e.g., the northern pike, *Esox lucius*, and various species of trout, *Oncorhynchus* spp. and char, *Salmo* spp.) that had been drawn into the turbines. This transformed nonpredatory fish into superpredators, biomagnifying mercury as would normally only occur in organisms at the top of the food chain.

Because mercury is recognized as a serious human health risk in terms of neurobehavioral deficits and motor performance (Lucotte et al. 1999; Mergler 2002), remedial measures to reduce mercury exposure were implemented by state authorities and the developer. Consumption of fish was severely restricted and, in certain stretches of the river, completely prohibited. The local diet of the Cree Nation went very quickly from containing 30%–50% local river fish toward unhealthy alternatives such as commercial frozen dinners. Locals also became more sedentary. As a consequence, obesity and diabetes have become the new diseases in the community. Perhaps a more serious consequence from the perspective of the Cree peoples has been the disruption of their symbolic relationship with nature (Desbiens 2004). The diminution of fishing has undermined networks of sharing and exchange, caused the erosion of social practices, and diminished the transmission of traditional knowledge to the youth. All this has led to a weakening of the social cohesion of the group, that in turn has provoked psychological problems and violence.

For the Cree, as for many other indigenous peoples from the north, the animals give themselves generously to the people who eat them. As long as the people respect the animals, consume only what they need, and do not waste any food, this spiritual relationship between humans and wildlife continues. The sudden transformation of local animals from generous partners into harbingers of death was especially detrimental for Cree youth. Scientific facts delivered by governmental health agencies contradicted traditional relationships with nature. The very measures taken by the central government to restrict risks on local health itself has led to a long-term eroding of local representation, which had historically allowed people to use their resources in a sustainable way (Desbiens 2004; Penn 1996; Roebuck 1999).

THE YANGTZE RIVER IN CHINA

The Cree Indians are not an isolated example of how short-term solutions can have long-term negative effects on health. The Three Gorges Dam, in the main channel of the upper Yangtze River in China, is the biggest hydroelectric dam in the world. Its impacts on biodiversity; ecological processes; contamination of fisheries and wildlife; disease; and community health are causing serious concerns worldwide (Wu et al. 2004). Even before the dam becomes fully operational in 2011, early data are already showing a significant increase in nutrients that promote eutrophication and an associated deterioration in water quality and health (see this volume, chapter 8). Besides the diffuse impacts of the Three Gorges Dam on health via deterioration of water quality and urbanization of the reservoir and its watershed, direct impacts on physical health are also expected. From the knowledge developed from other similar reservoir construction (e.g., the La Grande project), it is anticipated that the concentrations of mercury in traditional fisheries resources will increase, with significant health implications for regional communities.

The Yangtze River supports one of the greatest diversities of fish species in the Palearctic region, represented by 350 species of fish, of which 112 are endemic to the river ecosystem (Chang 2001). It is expected that the Three Gorges Dam will entirely modify the upper Yangtze River and that about 54% of the existing species will lose their natural habitats for feeding and reproduction (Park et al. 2003). The anadromous Chinese sturgeon (*Acipenser sinensis*), an endemic and endangered species in the river, has declined since the dam blocked its migratory route (Zhang et al. 2003). It is also anticipated that, as the reservoir fills, the new environment will allow proliferation of exotic species at the expense of existing endemic species, as has been documented in most post-dam systems. It has also been suggested that the dam environs will increase habitat for the freshwater snail *Oncomelania hupensis,* a known intermediate host for schistosomiasis. With more suitable habitat, the density of this snail is expected to increase, which would increase the spread of schistosomiasis in the region (Seto et al. 2002).

Risk to human health of water quality deterioration and mercury can be assessed through known impacts of similar projects on local peoples and traditional measures. Assessment of the future risk to health of increased incidence of schistosomiasis relies heavily on biological knowledge of the wild snail vectors and on ecological modeling to project its population growth, as well as that of the parasite. Such studies are rarely conducted a priori by governmental bodies, and when conducted post hoc are of explanatory use but do little for mitigating the current health crisis. Impacts of loss of native species, however, are more likely to be on quality of life of the local peoples, rather than on physical health. If the sturgeon is already endangered, it is not likely to be a commercial species (measurable in economic terms) but may still be important in local culture, perhaps as a food delicacy—caviar. Likewise, other native species may have cultural or spiritual significance that could be measured with QOL studies conducted prior to the dam's being built.

Case Study 2—Short-Term versus Long-Term Benefits: Wetlands and Malaria

Malaria (members of genus *Plasmodium*), carried by a mosquito vector (*Anopheles* spp.), is one of the most important diseases in tropical nations, affecting nearly 3 million people annually in India alone (Sharma 1999). Insecticides such as DDT (dichloro-diphenyl-trichloroethane) and malathion have been very effective in checking the spread of malaria and containing epidemics. However, both of these are broad-spectrum insecticides and have considerable nontarget effects. DDT remains in the environment for a long time, ultimately traveling up the food chain. Spraying of DDT in the swamplands of the United States caused the extinction of the brown pelican, *Pelecanus occidentalis,* in Louisiana—the official state bird (Blus 1982; Blus et al. 1979; Hickey and Anderson 1968). While malathion is considered relatively "safe" for application because of its short life and low toxicity for humans, during its short active time, it effectively kills all insect species that are exposed, causing massive local loss of biodiversity. Further, it requires spraying year after year, resulting in resistant *Anopheles* vector and *Plasmodium* parasite populations and diminishing returns in the long run (Sharma 1999). In sum, spraying operations are costly and unsustainable, and insecticides eventually produce adverse health consequences for humans as well as wildlife.

Drug treatment has proven no more sustainable than use of insecticides. Chloroquine, the primary drug used for prophylactic immunity as well as postinfection treatment of malaria, has greatly diminished in effectiveness. Many pathogen populations have evolved partial or complete resistance to this drug, and few alternatives exist. Thus vector control continues to remain one of the best options for reducing human health impacts.

Recognition that mainstream methods of vector control were failing to stem malaria deaths while causing severe harm to local biodiversity has led to global efforts to develop more environmentally friendly, sustainable approaches. In 1998, the World Health Organization instituted the Roll Back Malaria Partnership, which emphasized biocontrol and physical approaches (Saiprasad and Banergee 2003; WHO 1999). Limited application of malathion only to interior walls and netting over beds was encouraged. This directed, topical application has proved to be effective for malaria control while minimizing nontarget effects and evolution of resistance in the wild host/parasite populations. Biocontrol has primarily been through the introduction of fish predators that eat mosquito larvae.

A traditional method given new promotion by Roll Back Malaria was the purely physical approach of destroying mosquito habitat (Snellen 1987). Draining or creation of flowing water to eliminate stagnant water bodies effectively destroys mosquito breeding grounds. This strategy has eliminated malaria from many regions, including the southern United States, the North American lower Great Lakes, and Israel/Palestine (Kitron 1987). In India, methods of building drains and scooping out standing vegetation and built-up detritus are labor intensive and involve the active participation of the local community. These physical and targeted forms of vector control, integrated with early case detection and prompt treatment, have eliminated parasite reservoirs and decimated vector populations in many parts of India, resulting in a sharp decline in malaria cases (Sharma 1998). These strategies have been touted a success by health communities for stemming malaria deaths while simultaneously promoting local employment and community involvement (Nakajima 1998; Sharma 1987).

From an ecological standpoint, however, these "environmentally sensitive" approaches have been just as harmful to native biodiversity as old-school broad-scale DDT application. "Introduced" fish predators are often non-native *Gambusia* spp., whose diet is not specific to mosquito larvae. The undiscriminating palate and aggressive nature of most *Gambusia* renders it an invasive group of species that often decimates native aquatic fauna. More environmentally damaging has been the management of mosquito habitat through drainage and controlled water levels. This oldest of malaria control strategies has led to the reduction and elimination of wetland habitats worldwide, with associated losses to biodiversity. This ecological loss ultimately has been a major factor in global declines of two important human food resources—waterfowl (e.g., ducks) and fisheries.

The Need for Comprehensive Decision Making

These case studies exemplify the complexity of decision making and the failures of short-term responses to protect long-term human health and the environment. Actions in both cases were successful in accomplishing perceived goals in the short term: in case 1, the dams provided water and electricity for each region; in case 2, deaths from malaria were drastically reduced. However, biologically unanticipated outcomes occurred in each case. Negative impacts on natural biodiversity were immediate. Biodiversity preservation was not part of perceived goals, and mitigation of potential damage was not part of the decision-making process. It is only recently, as ocean fish stock and game birds have become noticeably depleted, that the vital role of wetlands as breeding grounds has been recognized. Draining wetlands succeeded in controlling malaria in the short term but in the long term has reduced local and global high-protein food supplies (see this volume, chapters 2 and 3).

Delayed consequences exist as well. One anticipated outcome—the leaching of mercury into dam waters—may be indeterminate in its impacts. It may be that when the decision to build the La Grande dam was made, the probability that this action would lead to the extreme outcome of inedible fish was perceived as being very low. Similarly, the invasive nature of *Gambusia* fish biocontrols for prevention of malaria has only become apparent with postintroduction monitoring. The case of the Cree highlights a particularly poignant delayed scenario: the collapse of their local social structure was both unanticipated and probably irreversible.

Valuation of Well-Being

A comprehensive decision-making process should incorporate valuation of long-term human well-being as well as quality of life. While conventional measures of human health are a vital component of well-being, they are not sufficient. Valuation of human well-being stemming from biodiversity, and the environment in general, requires assessments across a variety of disciplines, using both quantitative and qualitative measures, as discussed in detail in chapters 6 and 7. Briefly, the notion of human well-being reflects a state of human health, performance, and productivity indicated by diverse physical, emotional, intellectual, and moral/spiritual criteria. Measures of well-being must therefore go beyond physical health: they must incorporate human needs and material resources relating to standard of living and quality of life (QOL) (see this volume, chapters 1, 6, and 7).

Basic human needs for survival and physical health include clean water and nutritious food, protective shelter, a nonhazardous environment, and physical security (especially in childhood). Sustainable development frameworks are typically concerned with availability of material resources, which in turn are related to the diversity of plant and animal life. However these needs are conventionally measured in economic terms, such as hectares of land per household, number of cattle per farmer, household goods available for trade and barter, and so forth.

Likewise, quantitative assessments of the benefits of biodiversity and the environment generally rely on economic, social/psychological, and biophysical measurement strategies. Economic approaches are useful in measuring biodiversity and environmental benefits that occur in conventional markets such as resources and commodities, but economic criteria tend to be deficient in capturing emotional, intellectual, and moral/spiritual benefits of biodiversity and the environment. Such diffuse benefits often occur when the benefits are long-term (e.g., nonmarket species, or religious and cultural values), or when property rights are unclear or ill defined (e.g., common property resources and environments, migratory and transnational species, and environments). In these cases, social-psychological and biological quantitative measures are often preferable, especially when used in complementary relation to economic indicators.

Social-psychological measures tend to rely on people's preferences revealed in structured surveys, observational studies, or secondary data (e.g., productivity measures, test scores, health records). Biophysical indicators have been used to quantitatively measure benefits of biodiversity reflected in the abundance and accessibility of preferred and commercially valuable species; the availability of healthy environments (e.g., clean air and water, energy, waste decomposition); and positive aesthetic and recreational conditions (e.g., species diversity, fast-flowing and oxygenated streams). The goal, then, of comprehensive decision making would be to develop a quantitative valuation procedure that combines economic, social-psychological, and biophysical indicators of the benefits of biodiversity and the environment.

It should also be noted that some important benefits of biodiversity and the environment are not amenable to quantitative assessment, particularly moral/spiritual and cultural values (Horwitz, Lindsay, and O'Connor 2001). Even the study of happiness (now fashionable in economics) leans more toward econometric evaluation—using, for example, the Human Development Index (HDI), which contains indicators of gross domestic product (GDP), life expectancy, and educational provision worldwide. The results from the HDI assume that happiness has been measured, but the components of the index do not include any direct evaluation of the happiness of people in each country of the world; to obtain this evaluation would involve asking individual people about their own state of happiness. The economic evaluation of health states—for example, disability adjusted life years (DALYs) for the disabled—has been an important source of information for the allocation of health care resources but represents the views of the professional experts much more than the views of the affected, *i.e.* from our example, those who are actually disabled.

Environmentalists should be wary of falling into the same philosophical trap. As studies in QOL have shown, the views of experts diverge from those of the people who are most affected, being correlated at only around $r = 0.3$. In health as in the environment, the person-centered approach has been largely ignored until recently. Subjective reports are often mistakenly seen as unreliable. In part, this is because many of the measures being taken, such as "satisfaction with life," are invisible to, and hence immeasurable by, the observer. Because the subjective approach sometimes produces different answers from those of the econometric approach, the former types of results are frequently seen as inferior and even worthless, when they may in fact be more authentic.

Peoples' quality of life is largely determined by their personal interpretations of features of their environment, and hence they may have idiosyncratic interpretations of environmental events. Some dimensions of QOL are less prone than others to accurate assessment by other people or proxies. How easy would it be for you to assess another's self-esteem or inner strength, for instance? Such apparent discrepancies between the results of subjective and "objective" methods of investigation in multimethods research tend to be the norm, for reasons outlined in chapter 7. Furthermore, this disjunction needs to be anticipated and accepted as normal in the planning of research and interventions, and in the eventual analysis and interpretation of their results. Answers about which results should be given priority in any particular situation will very much depend on the purpose of the investigation and the nature of the research questions posed. Subjective assessments of quality of life should probably be awarded the highest priority in assessing the needs of communities where biodiversity is threatened.

Second, where social change has been successful due to the introduction of innovative practices—for example, in projects on poverty, where loss to biodiversity is a pressing issue—this change has more often occurred and been maintained as a result of using participatory methods more than top-down approaches. Such participation must involve some form of assessment before strategies to engineer change are created and implemented, so that outcomes (both short- and long-term) from that change can be reliably monitored from a baseline. Involving researchers in the life of the community during grounding and piloting also tends to improve the quality of any structured data that are collected, and their eventual interpretation. It also provides qualitative material for thematic analysis that will be used to adapt the project to an appropriate cultural shape before implementation. As behavior with respect to the environment depends to some extent on people's subjective beliefs and value systems (see this volume, chapter 6), subjective methods must form a significant group of measures in any contemporary, comprehensive, and holistic study.

When quantitative measures of the effects of biodiversity and the environment are inappropriate or not possible, qualitative assessments may be employed. Valid and reliable mixed-method approaches have been developed in the social sciences (primarily psychology, sociology, and anthropology) that not only effectively combine quantitative and qualitative measures (WHO 2002) but also enable them to operate in complementary ways. While ethnomethodology may be needed to carry out this development work initially, once completed, such ethnically sensitive methodologies would enable a more accurate and detailed interpretation of scalar scores, for spirituality as well as for other areas. These mixed-method approaches will be essential to any satisfactory person-centered study of biodiversity and health. Furthermore, there is considerable scope for designing new and innovative methodologies that do more than merely integrate relevant information, but also transcend the environment-health divide.

Robust Decision-Making Frameworks

Economic theory contains a suite of modeling approaches designed to aid in the evaluation and implementation of planning strategies in the face of multiple options and consequences. These approaches can be lumped under the heading of "robust decision-making models"; all strive to find strategies that perform well under a large set of possible futures. Although these models are frequently used to optimize economic measures across a range of uncertain outcomes (e.g., Yohe, Andronova, and M. Schlesinger 2004; Yohe and Wallace 1996), they have been successfully adapted to a range of social and business situations (Rosenhead 1990).

We suggest that robust decision methodologies could be adapted to jointly optimize human health and biodiversity in both short- and long-term time frames. Thus the term "robust decision making" is viewed in this chapter as making choices (1) that try to *maximize* expected long-term human health, well-being, and productivity derived from biodiversity and other valued qualities of the environment, based on all relevant economic, social, psychological, and biological criteria, and (2) that try to *minimize* the variance of these valuations across a set of possible futures.

Uncertainties arise from many sources: from an inadequate understanding of the resilience of different ecosystems to perturbation (see the present chapter and this volume, chapters 2 and 3); from poor knowledge of the impacts of biodiversity loss or alteration on ecosystem functioning and services (see chapters 8–11); from unforeseen advances in technologies and social attitudes (see chapters 5, 12, and 13); from unanticipated long-term health repercussions of short-term policies (see chapter 14 and case studies in the present chapter); and from unrecognized impacts of biodiversity loss on quality of life (see the present chapter and chapters 6 and 7). The recent advent of inexpensive, powerful personal computers allows analysis of multiple scenario simulations. Thus uncertainties can be directly incorporated into the decision-making process through modeling a range of potential future scenarios so that literally hundreds to thousands of outcomes can be compared.

Global attention focused on climate change has highlighted some important capabilities of a robust decision-making approach. Recent concern over the uncertainties in future climate change scenarios has led to the search for "no regrets" strategies (Yohe, Andronova, and Schlesinger 2004)—that is, models aimed at finding strategies that work acceptably under a wide range of uncertain futures to avoid scenarios of high negative impact, even if these have low probability. Development of such strategies is particularly relevant given that many alterations to natural biodiversity are irreversible. Once driven extinct, a species cannot be brought back. There is considerable debate as to the extent to which species are replaceable in terms of fulfilling roles in ecosystem functioning (see part IV of this volume), but biological diversity in itself has been shown to be an integral part of human health, as exemplified in chapter 2 of this volume with respect to food, and in chapter 15 with respect to drug provisioning. The science of ecological res-

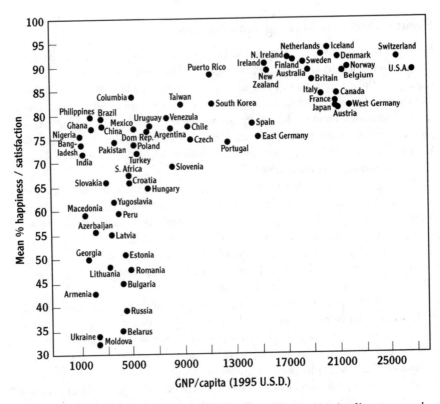

Figure 4.2. Relationship between national GNP and estimated levels of happiness and satisfaction of the human population. GNP data are from World Bank purchasing power parity estimates, given in terms of 1995 U.S. dollars. The *y* axis is the mean level of happiness or satisfaction with life as a whole for that population, estimated from QOL surveys. Adapted from Inglehart and Klingemann 2000.

toration has made tremendous gains in the past decade, but restoration is too costly to be considered a fallback strategy and also still relies on the existence of healthy reservoir populations of species and genotypes.

Adding Quality of Life to Decision Making

To make full use of emerging economic approaches requires finding quantitative measures of quality of life that can be used in all contexts. But how to measure quality of life? It is clear that use of standard economic measures fails as a surrogate for human well-being. For example, a 1995 study across 65 countries found only limited correlation between happiness/overall satisfaction with life and wealth of the country (Inglehart and Klingemann 2000) (figure 4.2). Countries with the lowest GNPs per capita (<US$5,000) varied widely; satisfaction was as low as 33% in Moldova, and as high as 80% for the Philippines. If you take out the former Soviet Union countries, there is only a weak relationship between happiness/satisfaction and wealth. The people of the Philippines and Brazil, with a GNP per capita of around US$5,000, were no more unhappy than people living in countries with a GNP per capita of greater than US$21,000, such as Japan, West Germany, and Australia. Clearly then, there is more to being satisfied with life than income and standard of living. The fact that national wealth is not a good predictor of individual happiness is a political conundrum for both developed and developing nations that want to improve quality of life, happiness, and well-being but find that the current economic model is inadequate. More direct measures are needed.

The types of estimates that have been made to evaluate biodiversity have generally not incorporated metrics that could be considered true measures of quality of life. The value of biodiversity has primarily been calculated in traditional economic terms (Costanza et al. 1997; Rapport, Costanza, and McMichael 1998), and these valuations typically focus on the market value of services and commodities stemming from intact ecosystems. Aesthetic value is sometimes tangentially measured—for example, through money spent on ecotourism. However, few health measures, even easily quantified health variables such as blood pressure, have been gathered. More esoteric measures of broad-sense health, such as human well-being, are exceedingly rare. Failure to incorporate broad measures of human health and well-being more likely stem from lack of collaboration between disciplines than from inherent lack of tools. Existing economic models contain considerable flexibility, opening the possibility of incorporation of alternative metrics. In addition, the Nobel Prize–winning work on behavioral economics by Kahneman and others (e.g., Kahneman, Diener, and Schwarz 1999) provides a relevant transdisciplinary analysis at the cutting edge of thinking about well-being.

A New Approach to the Measurement of Well-Being Related to Health: The WHO Quality of Life Assessment

The WHOQOL Group created a measure of QOL specifically to be used in applied health care (see this volume, chapter 7). It is broader than other QOL measures, embracing both the traditional concept of health and nontraditional measures of well-being. It also may be the first WHO project to empirically fulfill the aim of the last phrase of the WHO definition of health—that is to create a measure that is "not merely the absence of disease or infirmity" (WHO 1946). In other words, rather than framing measures of health around a lack of negative elements in a person's life, the WHO Quality of Life Assessment (WHOQOL) phrases conceptual elements (see domains of concern, this volume, chapter 7) and questions positively wherever possible. Such an affirmative approach contrasts with that taken by many policymakers, who frequently utilize a problem-centered orientation when working in environment, health, and other applied fields. A general emphasis on problem-solving (rather than striving for some ideal state) among end-users may begin to explain why so many QOL scales are framed in a negative way. The WHOQOL approach orientates users toward considering the upper end of their well-being continuum rather than merely assessing problems, with the ultimate aim of making a balanced and holistic assessment of the individual's life.

The WHOQOL provides a holistic and international concept for assessment, covering all areas of health-related well-being defined by WHO and extending it at the user's request to include spirituality, religion, and personal beliefs (Skevington et al. 2004a). A spiritual dimension can be influential where people live with infectious diseases—for example, malaria or HIV/AIDS. Spirituality often provides a distraction from an infectious condition and assists with coping. However, spirituality is more directly influential in the treatment of noncommunicable diseases through traditional healing where there is accompanying psychosocial distress. Including a spirituality domain for assessment adapts the measure particularly well to providing a better understanding of orthodox as well as indigenous approaches to healing. Because spirituality and the natural world are closely intertwined in many cultures, measures of the spiritual quality of life indirectly encompass people's relationships with biodiversity (see this volume, chapter 6).

Decision Making in Light of QOL

So how can the WHOQOL assessment help us to better study the relationships between biodiversity health and human health? Additional modules of selected items can be designed for use in special contexts and added to the international core WHOQOL items to provide a holistic concept of biodiversity. Similarly important national items can be added to round out the concept of QOL in a particular country (Skevington 2002; Skevington et al. 2004b). In adapting this process, modules could be constructed to assess important QOL issues pertaining to environmental situations in general and biodiversity issues in particular. Just as a spiritual module was added to fulfill an assessment need, through following the WHOQOL methodology it would be possible to develop and test a biodiversity module of items. Conceptually, we would expect this module to be closely aligned with the WHOQOL environment domain, and possibly scored with it. The WHOQOL instruments have rarely been used for environmental investigations before, and this would be an innovative departure if carried out cross-culturally.

A "person-environment fit" model for quality of life describes the complex interaction between how people feel internally (e.g., mood and physiology) and how they respond to their environment, including recent events (e.g., walking regularly in a park, having a range of nutritious foods to choose from). This psychological model has similarities to several environmental models (van Kamp et al. 2003). Although quality-of-life judgments are affected by mood, an assessment of mood cannot adequately replace a QOL measure. Furthermore, memory research shows that people have trouble recalling information accurately about how they feel. After about two weeks, this information becomes less reliable, so QOL assessments tend to inquire about a period in the recent past—for example, the last two weeks. Research that aims to compare estimates of QOL now with those of 40 years ago, when global biodiversity was generally in a much healthier state, could therefore not be supported by this method. However, it is possible for people to evaluate their ideal QOL, or to project what they would expect it to be under specified environmental conditions. In this way it is possible to obtain optimal subjective visions by examining the balance between good and poor aspects of QOL, using a profile like the WHOQOL. It is as important in studying biodiversity to be able to identify areas of QOL that are negatively affected as it is to identify those that are enhanced, particularly in situations where unanticipated outcomes from interventions turn out to be more harmful than beneficial.

Decision Making for Long-Term Health and Well-Being

The short-term focus of many decision frameworks is an artifact of the immediate goals of the involved sectors—for example, a business development, urban population expansion, or new resource exploitation. However, there is no technical impediment to incorporating long-term risk and gain into this process. Economic theory is rich in techniques for robust decision making that could be applied to long-term human and biodiversity health in the face of multiple scenarios and uncertainties. As we saw in the Cree and the Three Gorges Dam case studies earlier in this chapter, even known long-term consequences to physical health, such as the negative effects of mercury exposure, are not always incorporated a priori into the decision-planning process. Further, we argue not only for traditional measures of health to be included, but for the expansion of health measures to include quality-of-life measures.

By inspecting a profile of multiple QOL dimensions prior to initiating major environmental change, it would be possible to identify which aspects of local biodiversity are closely tied to good QOL for locals and would be affected by the project. Scenarios could then be developed that have greater or lesser impact on those facets of native biodiversity so that decision makers could see multiple possible outcomes. QOL profiles also provide a tool with which to monitor the ebb and flow of conflicts between the short-term goals of creating immediate environmental improvements, and longer-term strategies that would enhance the lives of future generations. The impacts of certain decisions on QOL could be documented by repeated sampling of the population with the same questionnaire over a period of time while transitions were in progress.

It is also possible to make a priori predictions about which combinations of QOL profile dimensions are most likely to change, given a particular set of circumstances. Even without impending environmental destruction, many regions have already suffered small to moderate biodiversity degradation. Use of a QOL profile in a study of such regions could indicate exactly which aspects of QOL are good, acceptable, and poor, and, moreover, could target causes of poor QOL for improvement. Results from an individual profiling technique used as input into a decision-making framework could assist researchers in figuring out the ultimate human impact of planning strategies that affect both short- and long-term ecosystem conditions. Such a proactive approach could be very useful in regions that are only moderately degraded, for which inexpensive restoration projects have high probability for success. A priori QOL studies could point to specific restoration projects that would have the highest impact on QOL of the local peoples while also improving conditions for local biodiversity.

Research Directions

Many suggestions for research directions are embedded in this chapter, but, to close, we would like to present some specific directions for future thinking:

- Development of reliable and valid economic, social-psychological, and biological valuation tools for quantitatively measuring the contributions of biodiversity and the environment to human health, productivity, and well-being
- Expansion of the use of scenarios in biodiversity management decision-making processes—for example, in considering the efficacy of different management options under various economic or political circumstances
- Expansion of true interdisciplinary research programs—for example, the joint Ecology of Infectious Diseases program of the National Institutes for Health and the National Science Foundation, which brings together ecologists and medical scientists on joint research projects
- Expansion of interdisciplinary education—for example, promotion of formal interactions between students at medical schools and traditional ecology and evolution programs via guest lectures, cross-referenced classes, and interdisciplinary degree plans
- Continued research into improved definitions and identification of ecosystem and biodiversity indicators
- Expanded use of decision-making frameworks that incorporate ecological, physical human health, and QOL measures so that both the short-term and long-term impacts of biodiversity change on human health and well-being can be considered
- Development of robust decision-making models that jointly maximize long-term human health (including QOL) and biodiversity while minimizing risk of harmful outcomes across a range of possible future scenarios

References

Blus, L. J. 1982. Further interpretation of the relation of organochlorine residues in brown pelican eggs to reproductive success. *Environmental Pollution* A 28:15–33.
Blus, L. J., E. Cromartie, L. McNease, and T. Joanen. 1979. Brown pelican: population sta-

tus, reproductive success, and organochlorine residues in Louisiana, 1971–1976. *Bulletin of Environmental Contamination and Toxicology* 22:128–35.

Chang, J. 2001. Conservation of endemic fish with catchment management of upper Yangtze. In *Flood risks and land use conflicts in the Yangtze Catchment, China and at the Rhine River, Germany: Strategies for a sustainable flood management*, ed. L. King, M. Metzler, and T. Jiang, 151–56. Frankfurt: Peter Lang Publishing.

Costanza, R., R. d'Arge, R. de Groot, S. Farber, M. Grasso, B. Hannon, K. Limburg, et al. 1997. The value of the world's ecosystem services and natural capital. *Nature* 387:253–60.

Davies, J-M., and A. Mazumder. 2003. Health and environmental policy issues in Canada: The role of watershed management in sustaining clean drinking water quality at surface sources. *Journal of Environmental Management* 68:276–86.

Desbiens, C. 2004. Producing north and south: A political geography of hydro development in Québec. *Canadian Geography* 48:101–18.

DesGranges, J-L., J. Rodriguez, B. Tardif, and M. Laperle. 1998. Mercury accumulation and bioaccumulation in ospreys (*Pandion haliaetus*) in the James Bay and Hudson Bay regions of Québec. *Archives of Environmental Contaminants and Toxicology* 35:330–41.

Harwell, M. A., V. Myers, T. Young, A. Bartuska, N. Gassman, J. H. Gentile, C. C. Harwell, et al. 1999. A framework for an ecosystem integrity report card. *BioScience* 49:543–56.

Heinz Center. 2002. *The state of the nation's ecosystems*. New York: Cambridge University Press. http://www.heinzctr.org/ecosystems.

Hickey, J. J., and D. W. Anderson. 1968. Chlorinated hydrocarbons and eggshell changes in raptorial and fish-eating birds. *Science* 162:271–73.

Horwitz, P., M. Lindsay, and M. O'Connor. 2001. Biodiversity, endemism, sense of place, and public health: Inter-relationships for Australian inland aquatic systems. *Ecosystem Health* 7 (4): 253.

Inglehart, R., and H. D. Klingemann. 2000. Genes, culture, democracy and happiness. In: *Culture and subjective well-being*, ed. E. Diener and E. M. Suh, 165–84. Cambridge, MA: MIT Press.

Kahneman, D., E. Diener, and N. Schwarz, eds. 1999. *Well-being: The foundations of hedonic psychology*. New York: Russell Sage Foundation.

Kitron, U. 1987. Malaria, agriculture and development: Lessons from past campaigns. *International Journal of Health Services* 17 (2): 295–326.

Lucotte, M., R. Schetange, N. Thérien, C. Langlois, and A. Tremblay, eds. 1999. *Mercury in the biogeochemical cycle: Natural environment and hydroelectric reservoirs of northern Quebec (Canada)*, 334. New York: Springer.

Mergler, D. 2002. Review of neurobehavioral deficits and river fish consumption from the Tapajós (Brazil) and St. Lawrence (Québec). *Environmental Toxicology and Pharmacology* 12:93–99.

Meyerson, L. A., J. Baron, J. M. Melillo, J. Naiman, R. I. O'Malley, G. Orians, M. A. Palmer, A. S. P. Pfaff, S. W. Running, and O. E. Sala. 2005. Aggregate measures of ecosystem services: Can we take the pulse of nature? *Frontiers in Ecology and the Environment* 3 (1): 56–59.

Nakajima, H. 1998. Editorial: United against malaria. *World Health* 51 (3): 3.

NRC (National Research Council). 2000. *Ecological indicators for the nation*. Washington, DC: National Research Council.

Park, Y.-S., J. Chang, S. Lek, W. Cao, and S. Brosse. 2003. Conservation strategies for endemic species threatened by the Three Gorges Dam. *Conservation Biology* 17:1748–58.

Penn, A. 1996. The North. Ch. 6 in: *Perspectives and Realities,* volume 4 of the Report of the Royal Commission on Aboriginal Peoples. Canada Communication Group. http://www.ainc-inac.gc.ca/ch/rcap/sg/sgmm_e.html.

Rapport, D. J., R. Costanza, and A. J. McMichael. 1998. Assessing ecosystem health. *Trends in Ecology and Evolution* 13 (10): 397–402.

Roebuck, B. D. 1999. Elevated mercury in fish as a result of the James Bay hydroelectric development: Perception and reality. In *Societal and environmental impacts of the James Bay hydroelectric project,* ed. J. F. Horing, 73–92. Montrèal: McGill-Queens University Press.

Rosenhead, J. 1990. Robustness analysis: Keeping your options open. In *Rational analysis for a problematic world: Problem structuring methods for complexity, uncertainty and conflict,* ed. J. Rosenhead, 193–218. New York: Wiley.

Roué, M., and D. Nakashima. 1999. The discourse of ecological correctness of dambuilders "rescuing" biodiversity for the Cree. In *Cultural and spiritual values of biodiversity: A complementary contribution to the global biodiversity assessment,* ed. D. A. Posey. Bourton on Dunsmore, UK: Intermediate Technology Development Group Publishing (now Practical Action Publishing).

Saiprasad, G. S., and A. Banerjee. 2003. Malaria control: Current concepts. *Medical Journal Armed Forces India* 59:5–6.

Schetange, R., J. F. Doyon, and J. J. Fournier. 2000. Export of mercury downstream from reservoirs. *Science of the Total Environment* 260:135–45.

Seto, E., B. Xu, S. Liang, P. Gong, W. P. Wu, G. Davis, D. C. Qiu, X. G. Gu, and R. Spear. 2002. The use of remote sensing for predictive modeling of schistosomiasis in China. *Photogrammetic Engineering and Remote Sensing* 68:167–74.

Sharma, V. P. 1987. Community-based malaria control in India. *Parasitology Today* 3 (7): 222–26.

———. 1998. Getting the community involved. *World Health* 51 (3): 14–15.

———. 1999. Current scenario of malaria in India. *Parassitologia* 41:349–53.

Skevington, S. M. 2002. Advancing cross-cultural research on QOL. *Quality of Life Research* 11:135–44.

Skevington, S. M., M. Lotfy, and K. O'Connell, for the WHOQOL Group. 2004a. The World Health Organisation's WHOQOL-BREF Quality of Life Assessment: Psychometric properties and results of the international field trial. *Quality of Life Research* 13:299–310.

Skevington, S. M., N. Sartorius, M. Amir, and the WHOQOL Group. 2004b. Developing methods for assessing quality of life in different cultural settings: The history of the WHOQOL instruments. *Social Psychiatry and Psychiatric Epidemiology* 39 (1): 1–8.

Snellen, W. B. 1987. Malaria control by engineering measures: Pre–World War II examples from Indonesia. In *ILRI annual report,* 8–21. Wageningen, Netherlands: International Institute for Land Reclamation and Improvement.

van Kamp, I., K. Leidelmeijer, G. Marsman, and A. de Hollander. 2003. Urban environmental quality and human well-being: Towards a conceptual framework and demarcation of concepts; a literature study. *Landscape and Urban Planning* 65:5–18.

Von Schirnding. 2002. Health and sustainable development. *Lancet* 360 (9333): 632–37.

WHO (World Health Organisation). 1946. *Preamble to the constitution of the World Health Organisation.* Geneva: WHO.

———. 1999. *Establishing a global partnership to roll back malaria.* Geneva: WHO.

———. 2002. *World health report 2002: Reducing risks, promoting healthy life.* Geneva: WHO.

World Bank. 1997. *World development report: The state in a changing world.* New York: Oxford University Press for the World Bank.

Wu, J. G, J. H. Huang, X. G. Han, X. M. Gao, F. L. He, M. X. Jiang, Z. G. Jiang, R. B. Primack, and Z. H. Shen. 2004. The Three Gorges Dam: An ecological perspective. *Frontiers in Ecology and the Environment* 2:241–48.

Yohe, G., N. Andronova, and M. Schlesinger. 2004. To hedge or not against an uncertain climate future? *Science* 306:416–17.

Yohe, G., and R. Wallace. 1996. Near term mitigation policy for global change under uncertainty: Minimizing the expected cost of meeting unknown concentration thresholds. *Environmental Modeling and Assessment* 1:47–57.

Zhang, S. M., D. Q. Wang, and Y. P. Zhang. 2003. Mitochondrial DNA variation, effective female population size and population history of the endangered Chinese sturgeon, *Acipenser sinensis. Conservation Genetics* 4:673–83.

5

Sustainable Allocation of Biodiversity to Improve Human Health and Well-Being

Laura A. Meyerson, Osvaldo E. Sala, Alain Froment, Carolyn S. Friedman, Kerstin Hund-Rinke, Pim Martens, Asit Mazumder, Aditya N. Purohit, Matthew B. Thomas, and Andrew Wilby

The concept of sustainable development attempts to reconcile the real conflicts between the economy and the environment, and between present needs and those of the future. Many explanations of sustainable development exist, but generally the term refers to sustaining intact ecosystems with their associated species while developing or improving ways to meet human needs for health, nutrition, and security (Kates, Parris, and Leiserowitz 2005). The Brundtland Commission defined sustainable development as the ability of humanity to ensure that it meets the needs of the present, without compromising the ability of future generations to meet their own needs (WCED 1987). Sustainable development attempts to raise awareness and to create links between human values, responsibilities, and environmental decisions that in many cases have been decoupled. Sustainability strives to create new solutions that are beneficial for all interests—both humans and the environment.

Strategies for sustainable development that conserve biological diversity while also meeting human needs must satisfy multiple, often conflicting demands. Ecosystems fulfill basic requirements associated with maintaining and improving human health, often in the form of ecosystem services harnessed by society. In some instances, there are win-win management opportunities that maximize several ecosystem services. In other cases there are trade-offs among services, some potentially difficult to evaluate because of the different nature of the services being compared or because the services occur at different temporal or spatial scales.

The conceptual framework that guided the March 2005 workshop in Paris on which this book is based maintains that biodiversity affects human well-being through four parallel paths: quality of life, provisioning of medicinal and genetic resources, spread of infectious disease, and provisioning of ecosystem services (see this volume, chapter 1). In this chapter, we explore trade-offs and synergisms in allocating biodiversity and resources to these four different determinants of human health and well-being. We first examine the issue of scale—both the interactions among human health determinants that can occur at different scales and the distinctions between local and global biodiversity. Explicit identification of scalar differences is critical because it may help to clarify potential allocation conflicts. We end the chapter with a discussion of trade-offs and synergisms in the allocation and use of land, water, and fertilizers.

Local and Global Scales at Which Biodiversity Affects Human Health

Human activity alters biodiversity as a result of habitat destruction, climate change, invasive species, and nitrogen deposition (Sala et al. 2000). Habitat destruction is currently the most important driver of biodiversity change, first operating by reducing habitat availability and extirpating local populations (Sala et al. 2000, 2005). When local extirpations and extinctions occur in several locations, they reduce the overall area of distribution of the species of interest. When the total species habitat becomes smaller than the minimum required for a viable population to survive, the species enters into a trajectory that eventually leads to global extinction (Tilman et al. 1994). However, global extinctions do not occur immediately after habitat availability falls below the specific threshold. Lags in the extinction response vary among functional groups and species. For example, a study of fragmentation in Kenya found that 50% of the bird species predicted to go extinct went extinct in the range between 23 and 80 years after habitat degradation (Brooks, Pimm, and Oyugi 1999). A similar study with plant species in the North American tallgrass prairie reported a range of 32 to 52 years and a loss of 8% to 60% of the original species (Leach and Givnish 1996). Both studies demonstrate the delay in the effects of land use change in terms of causing extinctions, as well as the variability in the time frames in which extinctions occur.

The four biodiversity drivers of human health that we have identified—quality of life, medicinal and genetic resources, constraint on infectious disease, and ecosystem services—respond differently to local and global extinctions. For example, the ability to discover new drugs or molecules that can be used to cure current or future human diseases is directly related to global biodiversity. However, the abundance of individuals of a desired species does not affect to a large extent the availability of drugs and genetic resources. A small viable population (even one under cultivation) may be sufficient to

Table 5.1. The scales at which the biodiversity
determinants of human health operate

Biodiversity Determinants of Human Health	Local Scale	Global Scale
Quality of Life	✓	
Medicinal Resources:		
Modern	✓	✓
Traditional	✓	
Constraint of Infectious Disease	✓	
Ecosystem Services:		
Provisioning	✓	✓
Regulating	✓	

maintain the genetic library and keep the possibility open to discover new drugs and molecules with interest for human health. Nonetheless, the extirpation of local resources of medicinal species may preclude their use by local people and cultures, and local benefits may be lost. Similarly, provisioning (e.g., food) and regulating (e.g., climate) services may be dependent on biodiversity at different scales. For example, globalization has made food production an ecosystem service that now operates on local, regional, and global scales. Global carbon and water dynamics may be dependent on large-scale biodiversity, whereas control of pests and disease may operate over local scales. Table 5.1 summarizes the spatial scales on which these biodiversity drivers operate.

Quality of Life

Standards for "quality of life" vary by country, culture, and economic expectation. However, there are common threads that run through all criteria for quality of life, including meeting human needs and improving human health and well-being, such as proximity to nature, maintenance of sociocultural links with biodiversity, and provision of sufficient food. Extinctions of species at the local scale impair quality of life in multiple ways. For example, chapters 6 and 7 of this volume describe the multiple positive effects of biodiversity on human health, healing, worker satisfaction, productivity, and intellectual performance. Humans require proximity to natural or seminatural environments to enhance their well-being. While global species extinctions can certainly diminish quality of life in a large sense, positive and negative effects on quality of life are determined for the most part by local extinctions and reductions in native biodiversity. Therefore, in many cases, self-interest and necessity may drive people to protect local biodiversity, which in turn may lead to global protection.

Political orientation may well affect whether a person views the environment from a perspective of self-interest or one of altruism. Most people express a combination of these views, but their particular environmental circumstance and access to information on local and global issues can predispose them to articulate one over the other. For some people, local environmental conditions affecting quality of life heighten awareness of biodiversity issues. This raising of awareness may then later extend to an interest in global biodiversity issues beyond their own society. In this way, self-interest may be converted into more altruistic beliefs and actions over a period of time—one psychological process that underpins progress on the sustainability front. When people care, and if they have sufficient resources to meet their basic needs, they are willing to allocate resources to preserve global biodiversity even though it may not affect their daily lives. Hence, reasons range from pragmatism to aesthetics and ethics but commonly drive the desire to preserve biodiversity and slow global extinctions.

Medicinal and Genetic Resources

A recent estimate approximates that more than 50,000 plant species are used medicinally worldwide (Schippman, Leaman, and Cunningham 2002). In China and India alone, more than 7,500 species are used in traditional therapies. Herbal remedies have become popular in developed nations, a market sector that has recently grown at 10%–20% annually in Europe and North America (ten Kate and Laird 1999). Modern medicine benefits from biodiversity by providing ingredients for pharmaceuticals. For example, plant species are estimated to be the resource for 121 drugs in current commercial use, such as Taxol (antitumor), podophyllotoxin (antimitotic and antitumor), chiratin (anti-inflammatory), allicin (antidiabetic), artemisinin (antimalarial), and ephedrine (central nervous system stimulant) (Conforto 2004). However, biodiversity operates at different scales for traditional and modern medicines.

Global-scale biodiversity primarily affects modern medical resources, whereas local-scale diversity affects traditional medical resources (table 5.1). Modern medicine uses biodiversity to find unique molecules that can cure current and future diseases of humans and domesticated plants and animals. At this scale, abundance and proximity do not matter because once the source organisms are identified, they can be artificially multiplied and the compounds synthesized. However, users of traditional medicine must harvest their medicinal resources from the region where they live, and the existence of this resource in a distant location may not satisfy local needs.

Native plants and animals often form the basis of traditional therapeutics that are effective in treating nonchronic and noninfectious diseases. They are also an important part of many cultures. Concern is growing about the effect of biodiversity loss on traditional medicinal resources because the raw materials are collected from the wild rather than cultivated. Out of the more than 50,000 medicinal plants currently in use, more than 4,000 (ca. 8%) are already threatened, many of which are used either in modern or in traditional medicines (Canter, Thomas, and Ernst 2005). Extraction from the wild specifically for pharmaceuticals or other therapies is unsustainable for many species, particularly those that come under the Convention on International Trade in Endangered Species of Wild Fauna and Flora (CITES) Appendices I and II, such as rhino, tiger claw, and bear bladders among animals and *Phodophyllum, Sausoria castus,* and *Taxus baccata* among plants. Where it is possible, sustainable cultivation of these species is desirable to provide societal health and economic benefits. Where cultivation is not possible, therapies that use imperiled or endangered species should be strongly discouraged and alternatives made available.

Modern medicinal and genetic resources depend on natural plants and some animals, but these still represent a minor portion of all medicinal resources, in part because of the high cost of developing and producing medicines. Screening every plant or animal for compounds to treat every disease is unrealistic since a plant may contain hundreds of compounds, and there are tens of thousands of plants to screen versus only hundreds of ailments. Using current technological capacities, a thorough cross-testing of all such compounds would undoubtedly prove economically and temporally impractical. However, it is likely that compounds exist in nature that hold the cure to some of the diseases that plague humans. Therefore, a sustainable approach to these resources requires that provisions for modern medicinal and genetic resources be made where they are needed *and* that native biodiversity is protected to ensure continued and sustainable access to potential medicinal resources.

Constraint of the Spread of Disease

Noninfectious diseases such as hypertension, cancer, and heart disease are strongly influenced by social and environmental factors but are likely to have limited (and only indirect) links to biodiversity. On the other hand, infectious diseases have a direct link to biodiversity since they are caused by living organisms and may additionally involve disease vectors and species that serve as reservoirs for pathogens. Consequently, changes in biodiversity at genetic, population, or ecosystem levels can have marked effects on the epidemiology of infectious diseases, although the effects and mechanisms can be complex.

Local biodiversity affects the spread of infectious diseases because disease dynamics are a local phenomenon (table 5.1). Disappearance of vertebrate predators, for example, may alter the abundance of rodent reservoirs and induce an increase of zoonoses. However, if the diversity of rodent reservoirs rises and there is an increase in the proportion of incompetent hosts, then disease impact may fall via dilution (see this volume, chapter 12, for further discussion). Such apparent conflict identifies the need for greater understanding of the role of biodiversity, ecosystem complexity, and local ecological context in the ecology and evolution of infectious diseases.

Conventional approaches to disease control aim principally to reduce diversity by targeting disease agents (e.g., with antibiotics) and, where appropriate, disease vectors (e.g., with pesticides). However, such approaches can overlook potential beneficial effects of biodiversity. For example, elimination of one vector species might lead to the emergence of a previously suppressed competitor with higher vector competence and could therefore lead to indirect nontarget effects elsewhere in the food web. Biodiversity is also a source of biological control agents that may play a role in the control of vector-borne disease in the future (Blanford et al. 2005).

Infectious diseases, like any other living organisms, have their own life history. There is evidence that some diseases that were historically severely pathogenic (e.g., the English sweating disease in Tudorian Britain) have completely disappeared, whereas other, entirely new diseases emerge regularly. Ecosystems rich in biodiversity, such as equatorial rainforests, can harbor hidden biohazards, especially among wild game. For example, consumption of bushmeat presents large risks, as illustrated by the infections with human T lymphocyte viruses (HTLVs), human immunodeficiency viruses (HIVs), foamy viruses, and Ebola virus (Walsh et al. 2003). Even though humans have inhabited forest ecosystems for millennia, many forests continue to generate seemingly new emergent diseases. These diseases may at first be confined within remote areas, but with the development of trade, travel, and transport, they can spread to larger scales.

Biodiversity and Ecosystem Services

The availability of provisioning ecosystem services, such as food production, depends on both local and global diversity (table 5.1). Human gatherers depend directly on local plant and animal diversity for food quality and quantity, and therefore local extinctions affect the nutrition and well-being of local human populations. Under these circumstances, there is little benefit from the maintenance of the plant and animal populations in distant locations, and the provisioning ecosystem service (human health and food availability) depends directly on local diversity.

Food production relates to global biodiversity because the ability to sustain and even increase food production depends in many cases on crop breeding programs that utilize wild genetic resources. Past experience has made it clear that reliance on agricultural monotypes is unwise and can, in fact, lead to disasters like the Irish potato famine of the mid-1800s. Crop breeding programs rely heavily on the availability of biotypes carrying traits and genes to be incorporated into improved varieties using traditional approaches or genetically modified organisms (GMOs). New molecular techniques allow for fast incorporation of traits into "improved" varieties, even inserting genes from unrelated species. However, our limited ability to prevent genetically engineered biotypes from escaping cultivation is a potent reminder that while technology can provide new solutions that boost production, it should be used judiciously to avoid creating long-term problems for the sake of short-term gains (Marvier and Van Acker 2005). Both traditional breeding and the development of GMOs rely on the existence of genetic combinations that may enable crops to adapt to a changing environment or acquire resistance to new pests and diseases. Improving and sustaining crops and food production in the future, particularly in the face of global change, will depend in large part on the maintenance of a rich global genetic library, both wild and engineered.

While agricultural production is central to human health and well-being, conversion of land for agriculture generally has a large negative impact on biodiversity. The extent of land conversion may be dramatic, such as the felling of forests or the drainage of wetlands, and the agroecosystem created may bear little resemblance to the ecosystem that it replaced (Hannah et al. 1994). Clearly, land conversion results in large losses of biodiversity and can conflict with biodiversity conservation, the maintenance of the genetic library, the determinants of quality of life, and constraint of the spread of diseases. However, the changes wrought by land conversion are not always dramatic—some conversions maintain the general characteristics of the ecosystem. Irrigated and rain-fed rice, for example, is commonly grown in floodplains so that the herbaceous wetland vegetation is replaced by similar vegetation (at least structurally), albeit at reduced levels of biodiversity. Paddy rice maintains an extremely high arthropod diversity (Way and Heong 1994). Therefore, sustainable agricultural production should be concentrated in areas and systems where impacts are minimized, or in regions where land conversion has already taken place.

Many crops are dependent on specific natural pollinators for fertilization and require specific insect morphologies for flower pollination, so a wide diversity of insects is a necessity to ensure food production. Organic matter, such as crop residues, is decomposed by soil biota such as mites, earthworms, and a large diversity of microorganisms. Complete decomposition thus requires a suite of detritovores and microbial species because it is a multiphase process. Pesticides are mainly biodegraded by specialized microorganisms—an ecosystem service that assists in reducing accumulation of potentially harmful compounds in plants, soils, and surface waters and that reduces leaching to groundwater. These and many other examples of biodiversity-dependent ecosystem services have been thoroughly reviewed by the Millennium Ecosystem Assessment program (MEA 2003).

Freshwater is essential for maintenance of life, and thus its allocation is a serious concern, particularly in association with climatic variations (e.g., drought). Freshwater is a provisioning ecosystem service affected by local biodiversity (table 5.1), as demonstrated by two examples from South Africa and New York City. The Working for Water program began in 1995 in South Africa to increase water security by removing high water-consuming invasive plants and replacing them with less thirsty native vegetation. In South Africa, many invasives directly threaten biological diversity via competition and also significantly decrease available freshwater supplies that support people, natural ecosystem functions, and agricultural production (WfW n.d.). New York City's drinking water supply has greatly benefited from local biological diversity in the Catskill Mountains watershed. Recognizing that water quality was declining due to development and that the cost of an artificial filtration plant for New York City would reach into the several-billion-dollar range, New York City instead chose to protect a natural ecosystem service. A suite of conservation acts, property purchases, and implementation of land use restrictions and septic system improvements on private property is protecting natural ecosystems and biodiversity and guaranteeing New Yorkers a continued supply of chemically untreated clean drinking water (NRC 2004).

Both estuarine and marine (especially nearshore) waters are important resources for recreation, maintenance of aquatic biodiversity, aquaculture, and fisheries. Sustainable development on a local and regional level therefore becomes challenging under the conflicting exploitation of limited water resources. In the United States, terrestrial agriculture and drinking water compete for the same resources (Pimentel et al. 2004a). World agriculture is estimated to use around 70% of all extracted freshwater annually, reducing availability of water for other uses by humans and other species. Aside from water utilization for food production and drinking water, the production of energy from hydroelectric dams is essential for the dispersal of waters for crop irrigation and drinking water, in addition to other energy production (e.g., electricity).

Water shortages have been linked with reductions in biodiversity in both terrestrial and aquatic ecosystems (MEA 2003; Pimentel et al. 2004b). A notable example is the construction of dams on large rivers, where the benefits of generating hydropower and increasing capacity for irrigation and drinking have been compromising the biodiversity and integrity of both downstream and upstream ecosystems. For example, the construction of the James Bay Dam in northern Quebec has produced economic benefits through the generation of hydropower but has resulted in mercury contamination of traditional fisheries of native communities and associated reductions in quality of life. In addition to biodiversity changes associated with alterations in water availability, the erosion and salinization of soils caused by crop and livestock (e.g., inland shrimp farming) irrigation are growing concerns (Bouwer 2002; FAO 1998).

One of the most important regulating ecosystem services is the control of pests and diseases that directly affect food production. However, these apparently negative elements of biodiversity have the potential to play a positive role as antagonists of pests and diseases themselves. This is most apparent in applied biocontrol approaches such as classical biocontrol, in which one or a few enemy species (i.e., predators, parasitoids, or pathogens) are introduced for the control of an exotic pest or weed. Thus potentially damaging species in one system could be beneficial in another, creating a possible conflict between the desire to reduce diversity of, for example, plant pathogens (or at least to accept genetic and species diversity loss) and a need to conserve diversity of such enemy species for future weed biocontrol programs.

A more subtle but possibly more important variation on this theme is the role that native natural enemies, particularly diseases, appear to play in the dynamics and impact of invasive species. Nearly all species are subject to attack from natural enemies, and one of the mechanisms identified in determining the postinvasion success of exotic species is the escape from indigenous natural enemies in the novel environment. Often parasites do not invade with their hosts, leading to a decrease in the number of parasite species and the proportion of hosts infected in the introduced range (Cornell and Hawkins 1993; Mitchell and Power 2003; Torchin 2004).

Reestablishing the link between invasive species and their key natural enemies is at the heart of classical biocontrol, as outlined above. However, it appears that some new associations between exotic animal and plant species and native parasites present in the exotic range may also reduce the spread and impact of invasive species. In the United States, for example, a negative relationship has been demonstrated between the noxiousness of an exotic weed and the number of native pathogens accumulated by an exotic weed in the introduced range (Mitchell and Power 2003). As such, parasites and pathogens, which on the one hand could represent damaging elements of biodiversity in their own right, could provide a valuable service by contributing to the biotic resistance of an ecosystem to invasion.

The role that diversity plays in biocontrol in general creates both possible synergies and conflicts between diversity conservation and pest control functioning. Natural biological control can benefit enormously from habitat diversification options (e.g., Landis, Wratten, and Gurr 2000 and references therein). Such options include the establishment of flower-rich field margins that provide essential nectar and pollen sources for many insect parasitoids and predators, such as hoverflies, and the installation of grass margins and "beetle banks" across large fields to act as reservoirs of carabid beetles and other ground-dwelling predators, and to aid their timely dispersal into crops in the spring.

Our understanding of the relationship between biodiversity (as affected by habitat manipulations) and pest control functioning remains poor, and the mechanisms through which natural enemies interact to determine the extent and stability of pest control are unclear. For example, in a recent study of the effect of landscape, habitat diversity, and management on species diversity in cereal systems, Weibull, Östman, and Granqvist (2003) revealed that there was no straightforward relationship between species richness of natural enemies at either the farm level or in individual cereal fields, and biological control.

Moreover, while there are examples of synergistic interactions between predators (e.g., foliar predators eliciting dropping responses in aphid prey, which increases their vulnerability to ground-foraging predators; Losey and Denno 1998) and examples of increased predator diversity increasing prey control because of functional complementarity (e.g., Riechert 1999; Wilby and Thomas 2002), processes such as intraguild predation can severely disrupt biological control (Rosenheim et al. 1995; Snyder and Ives 2001). These potential negative effects of increased diversity on pest control highlight a possible conflict between the goals of conservation and the goals of biological control (Finke and Denno 2004). Whether positive or negative effects of predator diversity on pest control predominate is a subject for further research.

Table 5.2. Trade-offs and synergisms in the allocation and use of land, water, and fertilizer

Allocation of Resources	Trade-offs	Synergisms
Land	Food vs. genetic resources and quality of life	Genetic resources and quality of life
	Food vs. water quantity and quality	Spread of disease, genetic resources, and quality of life
Water	Water for food production vs. genetic resources, quality of life, spread of disease	Genetic resources, quality of life, and spread of disease
Nitrogen / Phosphorous	Nitrogen for food production vs. genetic resources, quality of life, human health, spread of disease	Reduced genetic resources, quality of life, spread of disease

Trade-offs and Synergisms

Land conversion to agriculture and food production has both benefits and costs for human well-being (table 5.2). In addition to altering cover type, agriculture often requires a diversion of water resources and applications of fertilizers to boost productivity. While quality of life is enhanced through increased availability of food and improved nutrition, land conversion, water diversion, the application of fertilizers, and intensified agriculture can reduce local biodiversity (or lead to extinction in the case of endemic species), result in the loss of genetic and medicinal resources, and have serious health consequences if nitrogen is leached to drinking water. On the other hand, some genetic diversity that might otherwise be lost may be preserved through agriculture and animal husbandry if pressures on native wild species (e.g., bushmeat) are reduced and if contact with vectors of zoonotic diseases are minimized.

Although conflicts between agriculture and other land uses, such as biological reserves or recreation areas, depend jointly on the extent of conversion and historical patterns of land use, it is clear that the recent intensification of agriculture to meet production goals usually has a serious impact on both the aesthetic quality of landscapes and the biodiversity they house. This is partly because intensified management commonly follows a broad-brush approach having impacts reaching far beyond those intended. Thus biocides targeted at pestiferous species commonly affect nontarget species, and resource inputs (e.g., fertilizer) leach out of agricultural systems away from the species they were intended to support. A future imperative to reduce agricultural impact will be to better focus interventions so that nontarget effects and conflicts with biodiversity conservation are minimized.

Opportunities to more sustainably improve quality of life and human health and well-being exist. For example, technology transfers between nations can accomplish major improvements in energy efficiency while potentially reducing costs, resource use, and pollutant emissions. Other win-win strategies can be applied within and around protected areas. For example, an unanticipated result of the end of the decades-long civil war in Guatemala has been the increased colonization of the Maya Biosphere Reserve in the Petén. The region's occupation by guerrilla fighters during the war, ironically, served to protect the forests and biodiversity by forestalling development and internal migration.

Many migrant families who have subsequently illegally colonized the reserve subsist on slash-and-burn agriculture and hunting. They often have had little access to family planning, and therefore women have more children than they would otherwise choose and face increased maternal mortality and other reproductive health risks. The resultant rapid population growth also increases pressure on the reserve resources—including the taking of protected species. Greater availability of family planning and health resources would have the benefits of improving quality of life (e.g., reducing health risks to mothers, achieving desired family size, needing fewer resources to support a family) and of reducing pressures on protected areas. Cost-benefit analysis also suggests that family planning programs compare very favorably to other modes of reserve protection, particularly in the long run (F. Meyerson 2003).

While there is a trade-off with maintaining biodiversity, water that is used for irrigation results in increased food production and food security via a reduction in interannual variability in production. However, it has been clearly demonstrated that diversion of river water for irrigation reduces the river flow and can drastically reduce fish diversity (Xenopoulos and Lodge 2006). The reduction in fish species diversity in turn affects food availability for those dependent on fish resources and who may or may not benefit from the irrigation-related increased food production.

Fertilization with nitrogen has allowed one of the largest increases in food production in the history of humankind and has helped to fuel global human population growth. Loading of nitrogen and phosphorus to groundwater and aquatic ecosystems is a global concern because unregulated loading of these chemicals can cause excessive algal and bacterial growth, promote algal toxins, impair taste and odor, and require disinfection of affected waters (see this volume, chapter 8). As the concentrations of nutrients increase, algal communities increase biomass, especially cyanobacterial species that produce toxins detrimental to human health. These algae produce foul taste and odor, and, following decomposition, cause anoxia and associated fish kills. Nitrogen loading is also linked to "blue baby" syndrome and some cancers caused by high levels of nitrate in drinking water. On the other hand, increased agricultural productivity has reduced famine and therefore improved human health, particularly in developed nations. Developing countries are short of nitrogen and have serious human health problems related to malnutrition. Developed countries have the opposite problems with excess nitrogen fertilization and the human health negative impacts described above. Therefore, in many cases, the economic and societal benefits of nitrogen fertilization are countered by a trade-off of excess nutrients in soil and runoff that impairs water quality, productivity, and associated native biodiversity of aquatic ecosystems.

The Path to Sustainable Development

All human activity modifies biological diversity to one degree or another, and there are substantial and varied positive links between biodiversity and human health and well-being. Conservation of biological diversity will deliver benefits to quality of life, medicines, genetic resources, biological control, and constraints on infectious diseases. Inevitably, however, conflicts between maintaining biodiversity and benefits to human health and well-being will also arise since all biological resources are limited. Agriculture, human construction, and even recreation result in land conversion or the introduction of invasive species that are detrimental to native biodiversity.

Therefore, real and significant trade-offs exist between development—even when it is considered to be *sustainable development*—and conservation of biological diversity. Developing and developed nations alike have experienced increased trade, travel, and transport—thus increasing introductions of invasive species, which, while beneficial in some cases, can threaten natural ecosystems, agriculture, and human health, and pose threats to a nation's biosecurity (L. Meyerson and Reaser 2002).

Further modifications to biodiversity are inevitable as countries develop and strive to improve health, well-being, and the quality of life for their people. The path to undertaking this development in a sustainable way will be to minimize development impacts on biodiversity—both in the short term and the long term, and over multiple spatial scales. The impacts that will occur need to be carefully planned and considered, and gene banks and biotic specimen banks will need to be further expanded or established. To achieve conservation, protected areas and natural parks will have to be considered more holistically—in the matrix of the local, regional, and global area.

Conservation of biological diversity will require that people reevaluate what is truly necessary to fulfill human needs. The answers will be different across different societies and cultures and, to some extent, individuals. However, at the present time, the global human population continues to increase by more than 70 million people annually even though growth rates are declining. The United States and China have populations that are increasing by 3 million and 10 million people per year, respectively. Consumption of natural resources globally is sure to increase as populations grow and standards of living improve. Sustainable development offers a pathway for improving human health and well-being while minimizing impacts on biological diversity, but ultimately there are limits to growth, and difficult choices between conservation and development remain.

References

Blanford, S. B., H. K. Chan, N. Jenkins, D. Sim, R. J. Turner, A. F. Read, and M. B. Thomas. 2005. Fungal pathogen reduces potential for malaria transmission. *Science* 308:1638–41.

Bouwer, H. 2002. Integrated water management for the 21st century: Problems and solutions. *Journal of Irrigation and Drainage Engineering* 128:193–202.

Brooks, T. M., S. L. Pimm, and J. O. Oyugi. 1999. Time lag between deforestation and bird extinction in tropical forest fragments. *Conservation Biology* 13:1140–50.

Canter, P. H., H. Thomas, and E. Ernst. 2005. Bringing medicinal plants into cultivation: Opportunities and challenges for biotechnology. *Trends in Biotechnology* 23:180–85.

Conforto, D. 2004. Traditional and modern-day biopiracy: Redefining the biopiracy debate. *Journal of Environmental Law and Litigation* 19:357–96.

Cornell, H. V., and B. A. Hawkins. 1994. Patterns of parasitoid accumulation on introduced herbivores. In *Parasitoid community ecology*, ed. B. A. Hawkins and W. Sheehan, 77–89. Oxford: Oxford University Press.

FAO (Food and Agriculture Organization). 1998. Food balance sheets. 24 September 2000. http://armanncorn:98ivysub@ faostat.fao.org.

Finke, D. L., and R. F. Denno. 2004. Predator diversity dampens trophic cascades. *Nature* 429:407–10.

Hannah, L., D. Lohse, C. Hutchinson, J. L. Carr, and A. Lankerani. 1994. A preliminary inventory of human disturbance of world ecosystems. *Ambio* 23:246–50.

Kates, R. W., T. M. Parris, and A. A. Leiserowitz. 2005. What is sustainable development? Goals, indicators, values, and practice. *Environment* 47 (3): 9–21.

Landis, D. A., S. D. Wratten, and G. M. Gurr. 2000. Habitat management to conserve natural enemies of arthropod pests in agriculture. *Annual Review of Entomology* 45:175–201.

Leach, M. K., and T. J. Givnish. 1996. Ecological determinants of species loss in remnant prairies. *Science* 273:1555–58.

Losey, J. E., and R. F. Denno. 1998. Positive predator-predator interactions: Enhanced predation rates and synergistic suppression of aphid populations. *Ecology* 79:2143–52.

Marvier, M., and R. C. Van Acker. 2005. Can crop transgenes be kept on a leash? *Frontiers in Ecology and the Environment* 3 (2): 93–100.

MEA (Millennium Ecosystem Assessment). 2003. *Ecosystems and human well-being: A framework for assessment.* Washington, DC: Island Press.

Meyerson, F. A. B. 2003. Population, biodiversity and changing climate. In *Climate change and biodiversity: Synergistic impacts*, ed. L. Hannah and T. E. Lovejoy. *Advances in Applied Biodiversity Science* 4:83–90.

Meyerson, L. A., and J. K. Reaser. 2002. A unified definition of biosecurity. *Science* 295:44.

Mitchell, C. E., and A. G. Power. 2003. Release of invasive plants from fungal and viral pathogens. *Nature* 421:625–27.

NRC (National Research Council). 2004. *Valuing ecosystem services: Toward better environmental decision-making.* Washington, DC: National Academies Press.

Pimentel, D., B. Berger, D. Filiberto, M. Newton, B. Wolfe, E. Karabinakis, S. Clark, E. Poon, E. Abbett, and S. Nandagopal. 2004. Water resources: Agricultural and environmental issues. *Bioscience* 54:909–18.

Pimentel, D., and M. Pimentel. 2004. Global environmental resources versus world population growth. *Ecological Economics* 59:195–98.

Riechert, S. E. 1999. The hows and whys of successful pest suppression by spiders: Insights from case studies. *Journal of Arachnology* 27:387–96.

Rosenheim, J. A., H. K. Kaya, L. E. Ehler, J. J. Marois, and B. A. Jaffee. 1995. Intraguild predation among biological-control agents: Theory and evidence. *Biological Control* 5:303–35.

Sala, O. E., F. S. Chapin, J. J. Armesto, E. Berlow, J. Bloomfield, R. Dirzo, E. Huber-Sanwald, et al. 2000. Global biodiversity scenarios for the year 2100. *Science* 287:1770–74.

Sala, O. E., D. van Vuuren, H. Pereira, D. Lodge, J. Alder, G. S. Cumming, A. Dobson, V. Wolters, and M. Xenopoulos. 2005. Biodiversity across scenarios. In *Ecosystems and human well-being: Scenarios*, ed. S. R. Carpenter, P. L. Pingali, E. M. Bennett, and M. Zurek, 375–408. Washington, DC: Island Press.

Schippmann, U., D. J. Leaman, and A. B. Cunningham. 2002. *Impact of cultivation and gathering of medicinal plants on biodiversity: Global trends and issues*, 1–21. Rome: Inter-Department Working Group on Biology Diversity for Food and Agriculture, Food and Agricultural Organization of the United Nations.

Snyder, W. E., and A. R. Ives. 2001. Generalist predators disrupt biological control by a specialist parasitoid. *Ecology* 82:705–16.

ten Kate, K., and S. A. Laird. 1999. *The commercial uses of biodiversity*. London: Earthscan.

Tilman, D., R. M. May, C. L. Lehman, and M. A. Nowak. 1994. Habitat destruction and the extinction debt. *Nature* 371:361–65.

Torchin, M. E., and C. E. Mitchell. 2004. Parasites, pathogens and invasions by plants and animals. *Frontiers in Ecology and the Environment* 2:182–90.

Walsh, P. D., K. A. Abernathy, M. Bermejo, R. Beyers, P. De Wachter, M. E. Akou, B. Hujibregts, et al. 2003. Catastrophic ape decline in western equatorial Africa. *Nature* 422 (6932): 611–14.

Way, M. J., and K. L. Heong. 1994. The role of biodiversity in the dynamics and management of insect pests of tropical irrigated rice: A review. *Bulletin of Entomological Research* 84:567–87.

WCED (World Commission on Environment and Development). 1987. *Our common future*. New York: Oxford University Press.

Weibull, A.-C., Ö. Östman, and Å. Granqvist. 2003. Species richness in agroecosystems: The effect of landscape, habitat and farm management. *Biodiversity and Conservation* 12: 1335–55.

WfW (Working for Water Programme). n.d. http://www.dwaf.gov.za/wfw.

Wilby, A., and M. B. Thomas. 2002. Natural enemy diversity and natural pest control: Patterns of pest emergence with agricultural intensification. *Ecology Letters* 5:353–60.

Xenopoulos, M. A., and D. Lodge 2006. Going with the flow: Using species-discharge relationships to forecast losses in riverine fish biodiversity. *Ecology* 87:1907–14.

PART II
Biodiversity and Quality of Life
Beyond Physical Health

Human survival, social systems, and human culture have been inextricably linked to biological diversity over evolutionary time. Descriptions of the necessity and the nature of this connection have been controversial and have often been highly dependent on disciplinary perspectives. For example, social scientists have taken exception to the concepts of sociobiology, while natural scientists bemoan the lack of a quantitative approach to schools of thought like biophilia. While these disputes continue, they are overshadowed by a growing acceptance in the medical community that human physical health is inexorably linked to mental health, and that mental health is highly dependent on a person's physical, social, work, and family environments. Social psychologists have coined the term "quality of life" to embrace this broader perspective of human well-being. Researchers, governments, and organizations, including the World Health Organization, now commonly embrace the quality-of-life concept in measuring societal health and in evaluating the success of specific policies.

Part II of this book explores the relationships between biodiversity and quality of life that have developed over ecological and evolutionary time and suggests ways in which these relationships can be better defined through quantitative approaches. Stephen R. Kellert in chapter 6 reviews evidence and discusses the level of support for the notion that natural biodiversity is fundamentally necessary to human mental health. Further, he discusses some studies suggesting that exposure to the natural world may aid in the healing process, even in the context of modern medical treatment. Suzanne M. Skevington in chapter 7 explores the issues associated with attempts to quantify an inherently subjective measure like quality of life. She discusses the importance of cultural context in defining such measurements and in designing appropriate, unbiased studies. Skevington uses the quality-of-life index developed by the World Health Organization as a specific example of a multinational effort to develop a quantitative index of quality of life across different countries, cultures, and social systems.

6

Biodiversity, Quality of Life, and Evolutionary Psychology

Stephen R. Kellert

The fundamental contention of this chapter is that human physical and mental well-being, health, and productivity are partly a function of our species' evolved relational dependence on the natural environment. I describe this dependence as manifest in a range of genetically encoded environmental values reflecting the need for ongoing contact with natural diversity. This psychosocial dependency on experiencing biological diversity emerges from humans' having evolved in a biological, not artificial or human-constructed, world. Thus, even in modern, increasingly urban and technological society, contact with diverse natural systems continues to be critically important to the quality of human life, most particularly health, productivity, and physical and mental well-being.

Some clarification of critical terms is necessary before continuing. The terms *nature* and *biodiversity* can be vague and are used in various ways (Wilson 1992; Wohlwill 1983). In this chapter, I use the term *nature,* broadly, to refer to the nonhuman world, including self-sustaining nature, such as relatively undisturbed species and natural areas. However, the chapter also considers contact with highly managed and controlled aspects of natural diversity—for example, contact with nature in a park, zoo, or even a backyard or lawn—as well as the symbolic experience of the nonhuman world as revealed in art, language, story, video, and more. Thus the term *natural diversity* as used in this chapter encompasses any form of direct, indirect, or symbolic experience of the nonhuman world.

The chapter invokes the term *biophilia* to describe a weak innate human tendency to affiliate with nature, and explores biophilia in the context of nine presumably inherent values that people hold toward the natural environment. These genetically encoded values developed over evolutionary time because they conferred a range of adaptive benefits instrumental in human health, well-being, and productivity (Kellert 1997; Kellert and

Wilson 1993; Wilson 1984). Biophilic values can thus be seen as genetically programmed meanings people attach to, and benefits they derive from, biodiversity. However, evolutionary development does not assure that these inherent values are still functional. As Thornhill notes: "An adaptation was necessarily adaptive in the environments of its evolution. . . . The relationship between an adaptation and current reproduction depends on the similarity between the environment in which the adaptation is expressed and the environmental features that generated the selection that designed the adaptation. [Sometimes] this correlation no longer exists for contemporary organisms" (Thornhill 1999, 553). Still, this chapter contends that these biophilic values remain largely adaptive and functionally relevant in contemporary human physical, emotional, intellectual, and moral health and well-being, even in an increasingly modern urban society.

This chapter adopts the framework of evolutionary psychology—that is, the "understanding [of] the human mind . . . in evolutionary perspective" (Thornhill 1999, 549). An important qualification, however, is the author's view of these biophilic values as "weak" genetic tendencies highly contingent on experience, learning, and sociocultural support to become functionally manifest. In other words, these tendencies are not "hardwired" instincts, such as breathing and eating, but rather are genetically programmed inclinations toward the natural environment that depend on adequate learning and experience to functionally occur, although they can be easily triggered and acquired quickly as "genetically prepared learning rules." Still, as "weak" genetic tendencies dependent on adequate learning and experience, the values can be manifest dysfunctionally if they have not been sufficiently stimulated or developed.

The concept of biophilia is considered in more detail later in the chapter. To begin, I offer evidence to support the contention that contact with the natural environment enhances human health and well-being. This review is necessarily brief, and presents four types of evidence:

- Data pertaining to contact with relatively self-sustaining natural systems, such as parks and undeveloped natural areas.
- Contact with more domesticated aspects of nature, including gardens, "companion animals," and nature in the workplace.
- People's experience of natural diversity in ordinary communities and neighborhoods.
- The role of nature in childhood maturation and development.

Collectively, this evidence partially supports the following conclusions:

- Living in proximity or having contact with relatively undisturbed natural areas enhances human well-being when compared with little or no access.
- Contact with nature can positively affect human health and healing, including recovery from illness.
- Experiencing nature can foster beneficial social ties and relationships.
- Work settings with natural lighting, natural ventilation, and other biophilic attributes can foster worker satisfaction and productivity, and reduce stress.

- Contact with nature can stimulate intellectual performance and problem solving.
- People living in communities with relatively healthy natural systems more often possess positive environmental values and have a higher quality of life (defined later in this chapter) than those living in communities with degraded natural systems.
- Contact with nature fosters emotional and intellectual development during childhood.

Contact with Self-Sustaining Nature: Parks and Outdoor Recreation

Exposure to parks and gardens has long been assumed to exert a beneficial and even restorative effect on people (Glacken 1967; Schmidt 1990; Thomas 1983). Some of the claimed benefits include rest, relaxation, contemplation, restoration from illness, and spiritual renewal (Cooper Marcus and Barnes 1999; Hartig 1993; Hartig, Mang, and Evans 1991; S. Kaplan and R. Kaplan 1989; Relf 1992; Ulrich 1993). This assumption is reflected in the ancient world in the tradition of creating "sacred" groves and places (Chandrakanth and Romm 1991; Gadgil 1993; Ramakrishnan 1996; Sharma, Rikhari, and Palni 1999). These largely undisturbed areas were often perceived as areas where deities or "guardian spirits" reigned and where people could achieve physical, mental, and spiritual renewal. In more secular terms, in Europe and the United States, the establishment of national parks has often been tied to the presumed physical, psychological, and sometimes moral benefits of contact with these areas (Coates 1998; Glacken 1967; Oelschlaeger 1991). In urban areas, park designers such as Frederick Law Olmsted argued that people's physical and mental health depended on regular contact with attractive natural scenery in parklike settings (Fein 1972; Todd 1982).

The association of parks and protected areas with physical and mental well-being has been more systematically studied in recent decades, although the data remain sparse. Various investigations have documented diverse psychological and health-related benefits of contact with parks and open space. These benefits include stress relief, peace of mind, enhanced coping, improved physical fitness, and greater creativity and problem solving (Cooper Marcus and Barnes 1995; Hartig 1993; Kaplan and Kaplan 1995; Parsons 1991; Parsons et al. 1998; Relf 1992; Ulrich 1983, 1993; Ulrich and Parsons 1992). Even passive viewing of appealing natural features has been found to reduce stress, increase relaxation, foster "emotional balance," and improve cognitive functioning, including enhanced concentration, attentiveness, and problem solving (Hull and Harvey 1989; R. Kaplan 1985; S. Kaplan, R. Kaplan, and Wendt 1972; Ulrich 1984, 1993; Ulrich et al. 1991; Zube and Brush 1975). These data are often derived from social surveys, although some investigations have used physiological indicators such as reduced blood pressure and improved muscle tension. The results are largely consistent and impressive, with the positive effects often observed in parklike settings with especially preferred physical elements such as savanna-like meadows, large and mature trees with wide canopies, forest edges, brightly colored flowers and shrubs, and the presence of certain water features such as a clear pond or fast-flowing stream. All these attributes have been associated with adaptive evolutionary development in the human species.

An illustrative study was conducted by Hartig and colleagues, who examined the emotionally restorative and intellectually enhancing effects of parklike settings in urban areas (Hartig and Evans 1993; Hartig, Mang, and Evans 1991; Hartig et al. 1996). Their investigations focused on a sample of college students recruited to perform intellectually demanding and mentally fatiguing tasks. Students were randomly chosen and assigned to three groups: one group took a 40-minute walk in an urban park; another walked for an equivalent period in an attractive urban area dominated by buildings and human activity; and a third remained indoors in comfortable chairs listening to music and reading. After 40 minutes, each group completed a proofreading assignment. A survey administered following the assignment revealed that the greatest emotional restoration (higher levels of positive and lower levels of negative affect), as well as highest levels of attentiveness and concentration, occurred among students who walked in the park, with significantly lower levels of emotional and intellectual restoration found among the urban and indoor activity groups.

Other studies have also reported stress relief, emotional well-being, and intellectual functioning as frequent benefits of contact with parks, open space, and natural settings in nonurban and urban areas. Yet it would be an oversimplification to conclude that rest, relaxation, and restoration are the only positive benefits of parks and open space. For many people, parks and the outdoors are less opportunities for relaxation and more chances for stimulation, challenge, or the expression of physical prowess and competition (Driver, Nash, and Hass 1987; Ewert 1989; Knopf 1987). Whether the result is relaxation or stimulation, these findings have in common the capacity of parklike settings to engender a variety of physical, emotional, and intellectual rewards, depending on the circumstances and the kind of persons involved.

The benefits of parks and gardens are often related to their aesthetic qualities. The experience of natural beauty has been correlated with relief from stress and other physical and mental rewards (Appleton 1975; Hildebrand 1999; Heerwagen and Orians 1993; Tuan 1993; Ulrich 1993). These aesthetic responses to nature appear to be a universal biophilic tendency that will be discussed in more detail later in the chapter. This aesthetic inclination has been tied to various adaptive benefits, including rest, relaxation, curiosity, creativity, exploration, discovery, and problem solving.

Another body of literature has examined the effects of outdoor recreational activities on physical and mental well-being. There are many kinds of outdoor recreational activities, from hiking, camping, nature study, ecotourism, birding, and whale watching to fishing, hunting, and more. Sometimes these outdoor experiences occur in extraordinary natural settings, yet, not unusually, they also exert significant effects on people's emotional and intellectual well-being in more ordinary everyday circumstances. Both anecdotal evidence and a growing body of research has found that immersion in relatively undisturbed natural settings can exert significant, even life-changing, effects on participants, especially late adolescents in the company of peers (Driver and Brown 1976; Driver, Nash, and Hass 1987; Driver et al. 1999; Easley, Passineau, and Driver 1990; S. Kaplan and R. Kaplan 1989; S. Kaplan and Talbot 1983; Kellert and Derr 1998; Levitt 1988; Schreyer, Williams, and Haggard 1988; Ulrich et al. 1991).

For example, a review of research in this area by Ewert (1989) found that outdoor challenge programs often produce significant physical, psychological, social, and educational benefits that affect participants long after the activity has occurred. Major social-psychological benefits include enhanced self-esteem, self-confidence, and personal effectiveness, as well as greater independence and risk taking. Significant improvements in coping skills include enhanced initiative, sense of personal responsibility, organizational ability, and the capacity to see tasks through to completion. Substantial gains in cognitive performance include enhanced critical thinking and problem-solving skills, increased curiosity and creativity, and better performance at school and work. The physical benefits of these programs include enhanced strength, fitness, stamina, endurance, coordination, and acquired outdoor skills such as hiking and orienteering. Interpersonal and social gains include greater ability to cooperate, work in teams, avoid conflict, respect others, exercise leadership skills, and make new friends. Major changes in personal values are greater peace of mind, acquiring a "philosophy of life," and spiritual well-being.

A detailed national study of outdoor programs was conducted by the author and colleagues focusing on Outward Bound, the National Outdoor Leadership School, and the Student Conservation Association (Kellert 2002a, 2005; Kellert and Derr 1998). Over 800 persons from all socioeconomic strata were included in this investigation, with data collected through surveys, interviews, and direct observation. Two related investigations occurred, a retrospective study of participants of the organization's programs over a 20-year period, and a longitudinal investigation of participants immediately before, just after, and 6 months following the program.

Three-quarters of respondents reported the experience as one of the most important in their lives and as exerting a major impact on their personal, intellectual, and emotional development. These results occurred in the retrospective and longitudinal studies and across all three organizations. Dramatic improvements were observed in self-confidence, self-esteem, and "self-concept." The great majority of participants reported significant gains in independence, autonomy, optimism, peace of mind, and ability to cope with stress. Less pronounced but striking gains occurred in interpersonal skills, including the capacity to work with—and tolerance for—others, cooperation, and the ability to meet new people. Substantial improvements were noted in cognitive skills such as problem-solving, resourcefulness, seeing tasks to completion, and making difficult and complex decisions. Finally, significant majorities reported greater appreciation, concern, and respect for, as well as spiritual connection to, nature.

Contact with Domesticated Nature: Recovery from Stress and Illness, Companion Animals, Nature in the Workplace

Certain habitats such as gardens, seashores, hot springs, mountains, and deserts have long been connected with stress-relieving and even curative effects. For example, the presence of flowers and other aesthetically appealing vegetation is often tied to healing and calming effects among the sick and is practically normative even in modern hospital and health care settings. Several studies have reported that plants in hospitals and exposure to "therapeutic gardens" can result in symptom-relieving benefits (Baird and Bell 1995; Cimprich 1993; Frumkin 2001; Graham 1994; Hartig, Mang, and Evans 1991; Hartig et al. 1996; Hull and Michael 1995; S. Kaplan and Peterson 1993; Korpela and Hartig 1996; Mooney and Nicell 1992; Nakamura and Fuji 1992; O'Connor, Davidson, and Gifford 1991; Parsons 1991; Parsons and Hartig 1999; Parsons et al. 1998; Relf 1992; Taylor 1976; Taylor, Kuo, and Sullivan 2001; Ulrich 1984, 1986, 1992, 1993; Ulrich and Lunden 1990; Ulrich et al. 1991). Several studies also report that patients generally prefer vegetation in hospital rooms because of its presumed healing effects. One investigation reported that contact with nature in hospital rooms, views of the outside, and access to gardens were among the most widely stated health care preferences of patients (Cooper Marcus and Barnes 1999).

Several papers by Ulrich report that the sight and sound of nature can reduce stress, relieve tension, and enhance recovery among patients suffering from clinically diagnosed disorders (Ulrich 1979, 1983, 1986, 1991, 1993). One well-known investigation by Ulrich and colleagues focused on patients recovering from gall bladder surgery (Ulrich 1984, 1993). The subjects were matched demographically and following surgery were randomly assigned two types of hospital rooms—one having a window view of trees and vegetation, the other looking out at a brick wall. Patients with the nature views had significantly faster recovery rates, more positive responses to treatment, and substantially less need for medication than those with the brick wall view. The researchers reported that "patients with the nature window view had shorter post-surgical hospital stays, . . . fewer minor post-surgical complications, . . . and received far fewer negative evaluative comments in nurses' notes. . . . The wall view patients required far more injections of potent narcotic pain killers, . . . while the nature group took more oral doses of weak pain drugs" (Ulrich 1993, 107).

Ulrich and colleagues conducted another study in Sweden of 160 patients recovering from heart surgery (Ulrich 1986). The patients were randomly assigned to three types of recovery rooms—one with wall pictures of water and trees, the second with wall pictures of abstract art, and the third with blank walls. Less anxiety and fewer demands for strong pain medication occurred among patients with pictures of water and trees, while the highest stress levels were found among patients with walls decorated with abstract art. In a related study of Swedish psychiatric patients, Ulrich and colleagues (Ulrich and Lunden 1990) exposed patients to representational art, one group viewing pictures of flowers and an appealing landscape, the other abstract art. The patients who looked at the nature art demonstrated more positive affect; in contrast, patients who viewed the abstract art showed more negative responses, and some normally nonaggressive patients attacked the pictures.

Another well-controlled study of persons facing surgery examined three groups of patients—one group was shown "serene" depictions of nature, a second highly active outdoor scenes such as ocean surfing, and a third no pictures at all (Ulrich 1986). Significantly lower blood pressure levels were observed among patients who had been shown the serene nature scenes. Another study conducted by Katcher and colleagues examined patients facing dental surgery (Katcher, Segal, and A. Beck 1984). Of the three randomly selected groups, one viewed fish in a tank, a second observed pleasant pictures of nature, and a third looked at a featureless wall. Nearly all patients revealed some degree of stress, although far lower levels of physical discomfort and "treatment aversive behaviors" occurred among patients exposed to the fish tank, with similar levels of stress reduction found among patients who received more conventional stress-relieving therapies, such as hypnosis.

Another form of relationship to the nonhuman world associated with enhanced human health and recovery from illness is contact with pets or "companion animals." A wide body of research has examined the physical and emotional effects of companion animals under normal and stressful circumstances (for example, Anderson, Hart, and Hart 1984; Beck and Katcher 1996; Fogle 1981; Katcher and Beck 1983; Katcher and Wilkins 1993; Rowan 1989; Serpell 1986). These studies generally find that companion animals enhance calm, peace of mind, physical and mental restoration, and healing. Companion animals have also been found to increase social interaction, even among strangers (Messent 1983).

These benefits are especially evident in circumstances of stress and disorder. Various studies have documented the positive physical and psychological effects of companion animals on persons suffering from mental impairment, isolation, and diverse maladies. Various studies report major improvements in physical and emotional symptoms, more rapid healing, and faster recovery among persons who experience companion animals (Allen 1997; Corson and Corson 1977; Draper et al. 1990; Fine 2000; Friedmann et al. 1980; L. Hart et al. 1987; Katcher 2002; Katcher and Beck 1991; Levinson 1969, 1972; D. Moore 1984; Serpell 1986; Siegel 1990). Occasionally negative findings occur, especially when the companion animals are chosen inappropriately, yet the overwhelming majority of findings report that contact with companion animals can produce significant therapeutic gains, including stress relief, enhanced recovery, and greater well-being.

A carefully conducted study by Katcher, Friedmann, and others examined the effect of companion animals on persons recovering from heart attacks (Friedmann and Thomas 1995; Katcher et al. 1983). The patients included persons who received conventional treatment and an experimental group exposed to companion animals; both groups were matched demographically and symptomatically. The researchers reported that the presence of companion animals significantly increased survival and recovery rates. Specifically, they found that "mortality rates among people with pets [was] one-third that of patients without pets" (Friedmann and Thomas 1995, 310).

A recent study by Katcher and Wilkins (1993) examined boys suffering from attention deficit hyperactive disorder (ADHD), a malady that includes difficulty controlling emotional impulses, inadequate conflict management, and highly strained social relationships. Boys suffering from ADHD were randomly assigned to two activities involving contact with nature—an outdoor challenge program including canoeing, water safety, and rock climbing, and an animal interaction activity focusing on care for companion animals. The two groups switched midway through the study. Significant therapeutic gains were observed for both activities, although more positive and lasting gains were reported for the animal care group. Contact with companion animals resulted in greater symptomatic improvement, greater learning, and superior school performance. Caring for animals also resulted in greater speech gains, better nonverbal behavior, improved attentiveness, and the control of impulsive behavior. Moreover, these differences were evident six months following the program. The researchers concluded that a caring relationship with companion animals can relieve stress, improve social interaction, increase empathy, and contribute to task performance.

Benefits from contact with nature have also been demonstrated in the workplace. This contact typically involves exposure to plants, outside views, natural lighting and ventilation, and displays of nature-related pictures, decorative art, and ornamentation. The data are consistent and impressive, adding support to the contention that contact with nature even in the workplace fosters physical and mental well-being, health, and even productivity.

Most early research focused on negative problems associated with climatically controlled buildings marked by extensive artificial lighting, poor ventilation, and extensive use of artificial chemicals in furnishings, wall coverings, and other products and materials (Baechler and Hadley 1992; Godish 1994; Marcoin, Seifert, and Lindball 1995; Spiegel and Meadows 1999). Poor lighting and ventilation, high levels of toxicants, and the presence of molds were often linked to symptoms such as respiratory and skin disorders, excessive fatigue, and various physical and psychological ailments, all resulting in greater absenteeism, poorer morale, and lower productivity.

More recent studies have focused on the positive effects of exposure to nature in the workplace. This literature has found that increases in natural lighting, natural ventilation, natural materials, direct contact with nature, and the representational expression of nature in the workplace can enhance worker comfort, satisfaction, physical and mental well-being, and even at times productivity (Boubreki, Hull, and Boyer 1991; Browning and Romm 1998; Heerwagen 1990a and b, 2000a and b; Heerwagen, Durbin, and Macaulay 1997; Heerwagen and Hase 2001; Heerwagen and Orians 1986; Heerwagen et al. 1995a and b; Heschong and Mahone 1999; R. Kaplan 1993, 1995; Katts 2003;

Kuller and Lindsten 1992; Larsen et al. 1998; Leather et al. 1998; Lohr, Pearson-Mims, and Goodwin 1996; E. Moore 1982; Rocky Mountain Institute et al. 1998; Sensharma, Woods, and Goodwin 1998; Stewart-Pollack 1996; Tennessen and Cimprich 1995; Veitch and Newsham 1998; Wise 1997).

A study of European office and factory workers found that simply viewing nature reduced job-related stress and improved emotional well-being. Another European investigation of workers in a windowless office found fewer allergies among persons randomly assigned plants. An American study reported that office workers with a window view had less work-related frustration and better physical and mental health than those lacking outside views. Another investigation reported that employees with windows had fewer physical symptoms than those working within the building's windowless core. Another study found that office workers with no windows were far more likely to decorate their workplaces with pictures of nature and potted plants. One productivity study reported fewer errors and more efficient performance on a computer task among workers assigned offices with plants. Another investigation found that workers with plants had lower blood pressure and better attention levels than those lacking plants. Several studies have reported that workers with improved natural ventilation and natural lighting have more focused attention and superior cognitive performance compared with those in artificially lit and poorly ventilated work settings. Finally, a prison investigation reported that inmates confined to cells with outside views had fewer sick calls, headaches, and digestive problems than those lacking views.

Few of these studies have been rigorously conducted. A major exception is a study of office and manufacturing workers at a major furniture company in central Michigan conducted by Judith Heerwagen and her colleagues (Heerwagen 2000a and b; Heerwagen et al. 1996). This study examined workers prior to, immediately following, and nine months after their moving from facilities with minimal environmental amenities to new facilities that included greater natural lighting, natural ventilation, natural materials, improved energy efficiency, restored wetland and prairie habitats, outdoor sitting areas, and walking trails. The researchers reported that most employees rated the new facilities as superior in lighting, air quality, "healthiness," and aesthetic appeal. Workers also responded positively to enhanced opportunities for contact with the outside environment, improved exterior views, and restored wetlands and prairie habitats. These changes resulted in significant improvements in work performance, with a reported 20% gain in productivity nine months following the move to the new facilities, as well as significant improvements in job satisfaction, physical health, and relaxation. A minority of employees reported little change, and improvements were generally greater and more lasting among office workers than among manufacturing workers. Still, the results of this carefully conducted investigation are impressive, indicating substantial improvements in physical and mental well-being and productivity among workers with greater contact with nature.

Natural Diversity in Neighborhoods and Communities

Most of the studies reviewed in this chapter have not considered the effects of experiential contact with nature in people's homes, neighborhoods, and communities, and surprisingly few data exist on the subject. Some relevant findings emerge from studies of residential developments designed to lessen the adverse environmental effects of home construction, as well as to provide homeowners with more positive contact with the natural environment. For example, valuable data have emerged from studies of the Village Homes housing project in Davis, California, constructed during the mid-1980s and consisting of 220 modest-sized homes of 600 to 3,000 square feet on a 68-acre site (Corbett and Corbett 2000). This development was deliberately designed to include a large proportion of open space despite high housing density. Approximately one-quarter of the development was left in open space, including an "edible" (agricultural) landscape, a greenbelt with houses separated by narrow, winding pedestrian and bike paths, and extensive plantings of flowers, shrubs, and trees that occupy more space than the vehicular roads. The streets were also designed to be narrower than usual and positioned behind the homes along the periphery in highly vegetated areas. And, instead of conventional sewers and pipes, aboveground vegetated channels or swales were designed to control storm water and mimic the landscape's natural contours. Sustainable energy design features of the development included tight building construction, extensive insulation, solar orientation of the homes, solar hot water collectors, and plant shading.

Enhanced contact with the natural environment resulted from extensive open space, pedestrian and bike paths, widespread natural creeks and drainage channels, and agricultural areas that fostered outdoor activity and increased resident interaction. Village Homes consequently reported a strong sense of community, encouraged by an active homeowners association, community center, and common playing fields. More specifically, a study of Village Homes residents (Corbett and Corbett 2000, 94) found that they knew, on average, "forty neighbors, compared with seventeen in the standard development, and had three or four close friends in the neighborhood, compared with one in the control group." There has also been relatively little housing turnover and significantly higher resale values at Village Homes. The developers report that "in the beginning, housing prices in Village Homes were comparable to prices elsewhere in Davis. However, at this writing, calculated by the square footage, Villages Homes is the most expensive place in Davis. . . . The homes come on the market less frequently . . . and they sell twice as quickly" (Corbett and Corbett 2000, 78).

Another related study of two residential developments in Massachusetts also found a strong relationship between property values and open space (Lacy 1990; Rocky Mountain Institute et al. 1998). The two housing developments were constructed at roughly the same time with similar on-average 1,600-square-foot homes and comparable housing densities. One major difference was the amount of open space in each. One development had homes sited on small lots, with the remaining area devoted to open space comprising roughly one-half of the development, while the other had significantly larger individual housing lots but little open space. Twenty years following construction, property values were significantly higher in the development with the much larger proportion of open space. The researchers reported (Lacy 1990, 12) that "in the mid-1970s, homes in both developments sold for about $26,600. After 20 years, both had appreciated significantly, but the average price of the [subdivision with little open space was] ... $134,200, while [that of the homes in the development with half in open space] was ... $151,300."

The positive effects of open space and increased contact with nature are assumed to be relevant mainly for people who can afford economically to appreciate such amenities. Logically, it might be thought to follow that natural diversity is marginally relevant to impoverished and socially oppressed people such as the urban poor, minorities, and those who must cope continuously with basic needs of security and survival. This assumption is countered, however, by the results of studies conducted by Frances Kuo and her colleagues in Chicago public housing projects occupied by largely poor African Americans (Coley and Kuo 1997; Kuo 2001; Kuo et al. 1998a and b; Sullivan and Kuo 2002; Taylor, Kuo, and Sullivan 1998). Residents of two housing projects were compared, with the only physical design difference between the projects being that one had vegetation, mainly grass and a few trees, while the other was largely devoid of vegetation. Residents had been randomly assigned to the housing projects and had no control over the landscaping, and both buildings were architecturally identical. In one housing project, the vegetation had been totally removed to ease maintenance; in the other, sparsely situated trees, shrubs, and a modest covering of grass remained. Residents in the housing project with vegetation were found to have significantly higher levels of physical and emotional well-being, greater coping capacity, better conflict management, and superior cognitive functioning. The researchers reported that "attentional performance [was] systematically higher in individuals living in greener surroundings; [and] management of major issues [were] systematically more effective for individuals ... living in greener surroundings. ... It is striking that the presence of a few trees and some grass outside a 16-story apartment building could have [such a] measurable effect on its inhabitants' functioning" (Kuo 2001, 25).

In another study of a related public housing project, Kuo and her colleagues (2001) reported major differences among residents of two buildings, one surrounded by concrete and asphalt, the other by trees and grass. Residents of the vegetated building had significantly better social ties, superior interpersonal relationships with neighbors and strangers, a greater feeling of safety and security, and a stronger sense of community. Additionally, substantially less violence and crime occurred in the vegetated housing project compared with the one surrounded by concrete and asphalt.

Most of these studies are limited by small sample sizes, few numbers and types of communities studied, and a lack of environmental quality and quality-of-life indicators. These limitations were addressed in an ambitious study (Kellert 2005) of 18 rural, suburban, and urban communities within a single watershed that examined the relationship of environmental functioning to human physical and mental well-being. The 18 studied communities or "subwatersheds" were located in south-central Connecticut in a watershed that encompassed three rivers that eventually converge on Long Island Sound at the city of New Haven. The watershed is roughly 400 square kilometers, including 275 streams, part or all of 17 towns, and a population of 500,000 people that is 13% urban, 24% suburban, 11% agricultural, and 41% forested.

Initial results revealed a connection between ecosystem functioning, human environmental values and quality of life. The results of two powerful multivariate statistics were especially revealing. Factor analysis found 25 variables particularly related to one another and falling into two distinct clusters of 11 biophysical variables and 8 socioeconomic variables. Together, these two factors explained two-thirds of the variability in the data collected. The 18 subwatershed communities were compared on these two factors. This comparison found that residents of communities with better environmental quality (measured by such factors as species diversity, biomass, nutrient flux, and others) were more likely to express a greater appreciation and awareness of the natural environment, reveal significantly more outdoor activity, superior household characteristics, and a substantially higher quality of life. Quality of life was assessed by using criteria identified by Suzanne Skevington (see this volume, chapter 7), and included measures of neighborhood and household quality; civic and cultural institutions; availability of schools, libraries, and museums; roads; health care; recreational facilities; levels of crime; open space and park; a semantic differential of contrasting community characteristics (e.g., clean versus dirty, safe versus dangerous, prosperous versus poor, secure versus stable, etc.); and others. Conversely, residents of environmentally perturbed and degraded subwatersheds were far more likely to reveal less environmental interest, poorer household conditions, and a lower quality of life. These correlations occurred in both rural and urban communities.

Redundancy analysis is a powerful statistical tool that can identify variables in one data set that are highly related to and predictive of variables in another data set. In this case, redundancy analysis was conducted to determine if particular biophysical factors were especially associated with and predictive of human social variables, as well as the reverse. My colleagues and I (Kellert 2005) found this to be the case. Specifically, "environmental affinity" (a 25-factor scale that measured participation in various outdoor recreational activities, nature-related reading and television viewing, environmental organization membership, and assorted nature-related educational activities) was strongly related to and a powerful predictor of environmental quality in the 18 subwatershed communities studied. Within the biophysical data, diversity of native tree species (a measure of biological diversity and indigenous versus denaturalized environments) was found to be a strong predictor of human socioeconomic conditions and environmental values. The relationship between environmental affinity and native tree species diversity was also observed to be independent of income and education. In other words, the connection between environmental quality and human experience occurred in both poor and affluent communities, and among persons of limited and advanced education. Redundancy analysis further revealed that few measures of environmental quality were directly related to human values and social conditions, but rather that these conditions were highly mediated by the intervening influence of salient landscapes features and land use practices. Specifically, ecosystem health indicators such as species diversity, chemical pollution, and nutrient cycling were rarely directly related to human socioeconomic factors such as neighborhood quality, environmental affinity, environmental knowledge, or quality of life. Instead, these biophysical and socioeconomic factors were strongly related to intervening landscape features such as road quality, presence or absence of large trees, amount of open space, appearance and quality of streams, and other human land use variables. This finding made sense on reflection. Most people lack awareness of indicators of ecosystem function such as biological diversity or hydrological flow, but most are highly appreciative of prominent landscape features and land use practices that are the consequence of better ecosystem functioning, such as the presence of large trees, fast-flowing streams, and abundant and attractive open space.

To summarize, the findings of this study revealed a strong relationship between environmental quality, human values toward nature, and physical and mental well-being in diverse urban and nonurban communities. The major findings included:

- Communities with higher environmental quality had more positive environmental values and a higher quality of life, while lower-environmental-quality communities revealed less environmental interest and a lower quality of life.
- The relationship between environmental quality, values toward nature, and socioeconomic conditions occurred in both nonurban and urban communities.
- Particular environmental factors, mainly diversity of native tree species, were strongly related to and an excellent predictor of human socioeconomic conditions, while certain social factors, mainly affinity for the natural environment, were powerfully related to and predictive of environmental health and functioning.
- The relationship between environmental and social variables occurred independent of income and education.
- The connection between natural and human factors was highly mediated by the intervening influence of salient landscape and land use features from a human perspective.

The Role of Nature in Childhood Maturation and Development

The basic contention of this chapter is that the human affinity for nature is biologically encoded, revealed in a range of basic values toward nature, and related to human physical and mental well-being. It stands to reason that even for a human species capable of lifelong learning, the most important period for the development of inherent tendencies is childhood. Some research supports this contention of contact with natural diversity as an important aspect of childhood maturation and development, although the data are sparse. The psychiatrist Harold Searles suggested in this regard that "the non-human environment, far from being of little or no account to human personality development, constitutes one of the most basically important ingredients of human psychological existence" (Searles 1960, 3).

The maturational importance of experiential contact with nature during childhood is suggested by Rachel Sebba (1991), who found that 96.5% of a randomly sampled adult population reported contact with the outdoors as their most critical environment during childhood. Such settings were typically modest and simple, such as a backyard or nearby park. Most of the adults cited these areas as highly treasured and a motivational source even during adulthood. In a similar fashion, Edith Cobb (1977) reported, based on her study "The Ecology of Imagination in Childhood," that most gifted adults use childhood memories of largely natural settings as an emotional source for creative production as adults. She described childhood experiences of nature as enabling the gifted and perhaps others "to renew the power and impulse to create at its very source" (Cobb 1977, 32). Rachel Carson (1998) similarly claimed that a child's capacities for wonder, exploration, and discovery rest on early experiences in nature.

How might the natural environment function as strong source of emotional and intellectual development for children? Rachel Sebba (1991) cited several features of the natural environment that typically exercise a powerful maturational influence on children. First, she suggested that the natural environment is extraordinarily diverse and variable, exercising a highly stimulating effect on the child's senses of sight, sound, smell, touch, and feeling. Children are rarely if ever far from the sight of plants, the feel of wind, the smell of soil, the sounds of animals, or even the disturbing presence of spiders. These experiences are typically unavoidable and present in most settings. Second, the child's experience of nature tends to be highly dynamic as well as continuous. Nature is constantly shifting across both space and time. Third, most children typically encounter the natural world in random and unpredictable ways. This uncertainty and change demands alertness, adaptation, and attention, as well as the ability to cope with new and frequently challenging situations. Finally, nature for the child is replete with animate and lifelike features. This quality of aliveness fundamentally distinguishes the natural from the built environment. No degree of finely fabricated or artificial product can ever fully replicate the vital qualities of living nature. For the child, the natural world is the singular place where life is born, grows, feeds, feels, seemingly thinks, dies, and gives birth to new life. Even the nonliving elements of nature, including soils, waters, ecosystems, and landscapes, often strike children as lifelike, if not exactly alive.

Various studies offer partial evidence to support the importance of experiential contact with nature in childhood maturation, particularly during the period of middle childhood, roughly 6 to 10 years of age (Bettelheim 1977; Chawla 1992, 2001; Cobb 1977; R. Hart 1979; Kahn 1999a and b; Kahn and Kellert 2002; Kellert 1996; R. Moore 1986; Nabhan and Timble 1994; Pyle 1993, 2002; Searles 1960; Shepard 1978; Sobel

1993). David Sobel, addressing the importance of contact with nature during middle childhood, remarked that "middle childhood is a critical period in the development of the self and in the individual's relationship to the natural world. . . . Middle childhood appears to be the time when the natural world is experienced in highly evocative ways. It appears to be the time when children strike out, alone or with peers, to explore an ever-expanding repertoire of reachable places, in search of new experiences and adventure" (Sobel 1993, 159). Other studies suggest that middle childhood is instrumental in facilitating several stages of both cognitive development (knowledge, comprehension, application, analysis; Bloom 1956; Maker 1982) and affective development (responding, receiving, valuing; Krathwohl, Bloom, and Masia 1964).

Children encounter natural diversity in direct, indirect, and symbolic ways. Several studies suggest that direct and spontaneous contact is especially critical in children's maturation and development, although there appears to have been a marked decline in modern society in direct encounters with nature in favor of more indirect (e.g., visits to nature centers, museums) and vicarious (e.g., television, film, video) experiences. This change reflects a marked deterioration in access to and availability of open space, as well as time children are allowed to play outdoors. White observed that "between 1981 and 1997, the amount of time children ages 6 to 8 in the U.S. played decreased 25%. . . . 70% of mothers in the U.S. played outdoors everyday when they were children, compared with only 31% of their children, and that when the mothers played outdoors, 56% remained outside for three or more hours compared to only 22% of their children" (White 2004, 12). As David Orr contends (2002, 291), contemporary urban society increasingly relies on indirect and vicarious forms of contact with nature in place of direct and spontaneous experience with problematic consequences: "The reigning political economy has shifted the lives and prospects of children from direct contact with nature to an increasingly abstract and symbolic nature; routine and daily contact with animals to contact with things; . . . direct exposure to reality to abstraction and virtual reality."

Diminished direct contact with nature among children today prompted Robert Pyle (1993) to label this phenomenon an "extinction of experience." Pyle asserted that the decline in direct contact with nature today constitutes a profound loss of psychological and social bearings. He argued that "the loss of neighborhood species endangers our experience of nature. . . . Direct, personal contact with living things affects us in vital ways that vicarious experience can never replace. . . . One of the greatest causes of the ecological crisis is the state of personal alienation from nature in which many [children] live. . . . The extinction of experience . . . implies a cycle of disaffection. . . . As cities and metastasizing suburbs forsake their natural diversity, and [children] grow more removed from personal contact with nature, awareness and appreciation retreat" (Pyle 1993, 145). The journalist Richard Louv has also described a similar profound diminution of children's contact with nature that he has referred to as 'nature-deficit disorder' (Louv 2008).

Biophilic Values: Explaining the Relationship between Biodiversity and Quality of Life

The several kinds of data presented in this chapter partially support the contention that human physical and mental well-being, health, and productivity remain highly contingent on the quantity and quality of people's contact with natural process and diversity. How can we explain this seemingly continuing relational dependency? Any answer is necessarily based on incomplete knowledge, requiring a book in itself. Still, part of the explanation offered here, drawing from the logic of evolutionary psychology, focuses on the concept of *biophilia* (Wilson 1984; Kahn 1999a and b; Kellert 1997; Kellert and Wilson 1993). As previously described, biophilia assumes that human beings possess innate tendencies to affiliate with nature, as manifest in nine inherent values that people hold in relation to the natural environment, each conferring adaptive advantages that foster human physical and mental well-being (Kellert 1997, 2002a, 2005).

These biophilic values, as also previously noted, are viewed as "weak" genetic tendencies highly dependent on learning, experience, and social and cultural support to become functionally manifest (Kellert 2005; Lumsden and Wilson 1981; Wilson 1993). In people lacking adequate experience, these values will atrophy or remain undeveloped, resulting in emotional and intellectual deficits. The biophilic values are thus "biocultural" phenomena whose functional development depends on learning and group support. As a consequence, the biophilic values can be functionally or dysfunctionally expressed. Because each value is the product of both genetics and learning, considerable diversity can occur in their content and intensity among individuals and groups because of variations in experience and culture. Yet, this variability is not infinite, but rather is bounded by the limiting requirements of biological fitness. Thus each value hypothetically occurs along a continuum in which considerable "normal" variation is found but in which dysfunction is encountered at either extreme—atrophied or insufficient development at one end, and inordinate or unduly exaggerated expression at the other, both tendencies typically leading to maladaptive behavior.

I will briefly describe nine biophilic values, each reflecting a range of physical, emotional, and intellectual benefits or advantages that each potentially confers (Kellert 1997, 2002a). These nine values, or perspectives in relation to nature, collectively reflect the richness of the human reliance on natural diversity for fitness and security, a web of relational dependency so pronounced that an ethic of environmental concern can emerge from a greatly expanded understanding of human self-interest in sustaining a healthy and abundant environment.

The Utilitarian Perspective

The utilitarian perspective values natural diversity primarily for the physical, material, and commodity advantages it confers. The term *utilitarian* is something of a misnomer, however, as all the biophilic values advance human welfare. *Utilitarian* is used here in the narrow sense of material benefits derived from exploiting the natural environment. The utilitarian perspective reflects the recognition of natural diversity as a source of agricultural, medicinal, industrial, and other commodity advantages. Both traditional and modern peoples rely on natural diversity for food, medicine, building materials, and other services. Moreover, this material dependence is likely to expand in the future because of advances in molecular genetics and bioengineering that allow for the greater exploitation of the natural environment. People further possess a utilitarian inclination to value nature in the absence of necessity because it fosters physical, emotional, and intellectual rewards and benefits. Cultivating proficiencies like gardening or harvesting wildlife can yield a host of adaptive skills, including physical fitness, self-sufficiency, and self-confidence.

The Dominionistic Perspective

The dominionistic perspective values natural diversity for expressing and developing mastery abilities and skills. Benefits include the ability to achieve safety and security, to demonstrate independence and autonomy, to take risks and cope with adversity, and to show resourcefulness. People can hone their physical and mental fitness through subduing and mastering nature. By competing, outwitting, and contesting species and habitats, they can emerge surer of themselves and their ability to deal with challenge and the unknown. Modern society may no longer rely on besting prey, eluding menacing predators, or surviving in the wild, but the strengths obtained from confronting the natural world continue to function as a source of adaptive fitness.

The Naturalistic Perspective

The naturalistic perspective values natural diversity for its engendering of curiosity, imagination, and discovery. These rewards often derive from immersion in nature—above all, its most conspicuous plants, animals, and landscapes. Through deep participatory involvement, various creatures and habitats become a source of knowledge and discovery. Adaptive benefits include heightened awareness, increased attentiveness, exploratory urges, a sense of wonder, a desire to discover, and greater inventiveness. Deep participatory involvement in nature can also foster a sense of timelessness and boundlessness. The dull and colorless rocks can become far more varied, the amorphous vegetation loaded with meaning, and the stillness of the landscape transformed into a mosaic of sensations and experiences. This heightened awareness frequently fosters an enhanced sense of clarity, interest, and curiosity.

The Scientific Perspective

The scientific perspective values natural diversity as a source of empirical knowledge and understanding. People possess a need to know and understand their world with authority. This tendency is often fostered through observing, identifying, labeling, and categorizing the natural environment. Among the cognitive abilities cultivated through the study of nature are enhanced critical-thinking, problem-solving, observational, and analytical skills. Empirically observing and systematically studying the natural world is sometimes regarded as the characteristic of modern life and dependent on rigorous training. Scientific valuing of nature is, however, a likely characteristic of all peoples throughout human history. Scientific tendencies can be observed among all preliterate peoples, who devote considerable time to recognizing, classifying, and identifying species and habitats within their home ranges, even when these species lack immediate material advantage (Berkes and Folke 1998; Diamond 1993; Maffi 2001; Nelson 1993). It appears that all people have an urge to understand and classify their world because it facilitates critical-thinking, problem-solving, and intellectual capacities. Carefully observing even a fraction of natural diversity offers countless opportunities for developing cognitive competence. The natural world remains an unrivaled source of intellectual stimulation, probably the most information-rich context people ever encounter. Moreover, the knowledge gained by chance and over time often yields tangible advantage.

The Symbolic Perspective

The symbolic perspective values natural diversity as a source of language, imagination, and communication. People employ the sights, sounds, and symbols of nature to expedite the exchange of information through various means, including language, story, imagery, and other representational forms (Abrams 1996; Bettelheim 1977; Lawrence 1993; Shepard 1978, 1996). This symbolizing of nature occurs in obvious ways, but often in an abstract, subtle, and highly disguised fashion. Adaptive advantages include enhanced capacities for language, communication, and psychosocial development. Symbolizing nature is especially critical in language development, among the most prized characteristics of the human species. Language depends on ever more refined distinctions and taxonomic classifications. This linguistic capacity is facilitated through children distinguishing emotionally relevant subjects, which often focuses on living creatures and lifelike natural objects. Natural diversity offers emotionally powerful and intellectually salient sources of imagery for imposing classifications and distinctions. Symbolizing nature occurs in many children's stories and narratives that engage critical maturational issues of identity and social development in often disguised and more palpable ways. The attribution of human motives to the natural world, or anthropomorphism, is frequently employed to assist in confronting complex, often threatening, and painful subjects in a tolerable fashion. Finally, symbolizing nature is frequently used to facilitate communication in everyday life. People employ natural imagery in ordinary language, in the marketplace, and in oratory and debate. These images are often simple and crude but are occasionally the source of eloquence and an ability to render language more vivid, colorful, and imaginative.

The Aesthetic Perspective

The aesthetic perspective values natural diversity as a source of physical attraction and beauty. This valuation of nature can enhance people's capacities for curiosity, imagination, and creativity, as well as their ability to recognize order, harmony, symmetry, and balance. Few experiences in human life exert as consistent and powerful an impact on people as the aesthetic attraction of nature. People are incapable of resisting an aesthetic response to particular features of nature, such as a beautiful rose, a striking sunset, a large pyramidal mountain, and more. The aesthetic inclination toward nature has become genetically encoded because it contributed in many ways to human fitness and security. An aesthetic response is fundamentally one of attraction that fosters interest and curiosity. With cultivation and experience, this attraction and curiosity can become more complex and lead to exploration and discovery. The aesthetic attraction to nature can facilitate the capacities for observation, imagination, and eventually creativity. Moreover, nature's aesthetic appeal can be associated with achieving safety, sustenance, and security (Appleton 1975; Heerwagen and Orians 1993; Hildebrand 1999). People are often aesthetically drawn to environmental features instrumental in human welfare, such as clean, flowing water; promontories fostering sight and mobility; areas offering refuge and shelter; and bright flowering colors, frequently signifying food. Finally, beauty in nature can engender an enhanced sense of balance, symmetry, and harmony. People are inspired and instructed by recognizing order and unity. The natural world often serves as a model or prototype that enables people, through mimicry and invention, to capture analogous advantages in their own lives.

The Humanistic Perspective

The humanistic perspective values natural diversity as a source of emotional relationship and attachment. Developing this tendency fosters the capacities for giving and receiving affection, forming intimate and companionable bonds, and expressing trust and cooperation. These capacities are all adaptive benefits for a largely social species such as the human. Emotionally bonding with nature typically focuses on familiar and domesticated animals, often viewed as friends and companions. These animals can become the subjects of intense loyalty, commitment, caring, and affection. By contrast, aloneness constitutes a heavy burden for humans who usually crave companionship and relationship. By identifying with animals, people achieve a sense of belonging and interpersonal connection. The affection of another creature rarely substitutes for other people, but it sometimes provides a valued complement for experiencing closeness and intimacy. Bonding with nature can nurture the capacities for cooperation and sociability. Caring and seemingly being cared for by another creature can also enhance self-confidence and self-esteem. These benefits accrue under normal circumstances but often become acute during moments of crisis and disorder. When distressed, people often seek the therapeutic powers of nature, including companion animals, gardens, and other "healing" landscapes.

The Negativistic Perspective

The negativistic perspective views natural diversity with avoidance, anxiety, aversion, and sometimes fear. Many elements in nature provoke this response with little provocation, including snakes, spiders, large predators, swamps, steep precipices, lightning, and others (Ulrich 1993). This inclination may seem the antithesis of biophilia, but biophilia is the inherent tendency to affiliate with natural diversity, including apprehension and avoidance as much as interest and affection. Aversive reactions to natural diversity can provoke abusive behavior, but more typically this inclination is moderately expressed and adaptive. People benefit from isolating and at times eliminating aspects of natural diversity. Human well-being depends on acquiring skills associated with distancing from potentially injurious features of the natural world. Lacking this awareness, people often take unnecessary risks and behave naively, ignoring their inevitable vulnerability before powerful natural forces. Furthermore, fear of nature can provoke feelings of awe and respect. People defer to and respect powers greater than themselves. Nature stripped of its power usually becomes a subject of superficial amusement and condescension. A lion, wolf, or bear confined to a small, barren cage rarely evokes much awe and respect, and habitats utterly subdued do not provoke much admiration and ethical regard.

The Moralistic Perspective

Finally, the moralistic perspective appreciates natural diversity as a source of moral and spiritual inspiration. The philosopher Holmes Rolston remarked: "Nature is a philosophical resource, as well as a scientific, recreational, aesthetic, or economic one. We are programmed to ask why and the natural dialectic is the cradle of our spirituality" (Rolston 1986, 88). People obtain a feeling of purpose and meaning by cultivating their sense of connection and relation to creation. This sense of being connected to creation can engender a feeling of unity and serve as a cornerstone of spiritual and religious belief. Despite the extraordinary diversity of nature, religion and science have also identified an astonishing commonality uniting much of life on Earth, including shared molecular, chemical, and genetic structures, analogous circulatory and reproductive parts, and parallel bodily features. This web of relationship unites a beetle on the forest floor, a fish in the ocean, an ungulate on the savanna, and a human in a modern metropolis. Faith and confidence are nurtured by perceiving this connection, recognizing symmetries and commonalities that transcend a single individual isolated in time. This perspective can further encourage the inclination to conserve nature, since people temper their tendencies to destroy when they perceive a connection that binds them to other elements of the natural world. The willingness to protect natural diversity can emerge as much from moral and spiritual inspiration as from economic calculus or regulatory mandate.

Each of the nine biophilic values can thus be tied to various physical and mental advantages instrumental in human welfare over time. The nine values collectively affirm the human reliance on natural diversity for achieving physical, material, emotional, intellectual, and moral well-being. Because the biophilic values are rooted in human biology, they represent an ethical argument for conserving natural diversity based on a greatly expanded understanding of human self-interest viewed over the long term.

References

Abrams, D. 1996. *The spell of the sensuous: Perception and language in a more-than-human world.* New York: Pantheon.

Allen, D. 1997. Effects of dogs on human health. *Journal of the American Veterinary Medicine Association* 210:1136–39.

Anderson, R. K., B. L. Hart, and L. A. Hart, eds. 1984. *The pet connection.* Minneapolis: University of Minnesota Press.

Appleton, J. 1975. *The experience of landscape.* London: Wiley.

Baechler, M., and D. Hadley. 1992. *Sick building syndrome: Sources, health effects, mitigation.* New York: Noyes.

Baird, C., and P. Bell. 1995. Place attachment, isolation, and the power of a window in a hospital environment. *Psychological Reports* 76:847–50.

Beck, A., and A. Katcher. 1996. *Between pets and people: The importance of animal companionship.* West Lafayette, IN: Purdue University Press.

Berkes, F., and C. Folke, eds. 1998. *Linking social and ecological systems.* Cambridge: Cambridge University Press.

Bettelheim, B. 1977. *The uses of enchantment: The meaning and importance of fairy tales.* New York: Vintage Books.

Bloom, M. 1956. *Taxonomy of educational objectives: The classification of educational goals.* Handbook 1, *Cognitive domain.* New York: Longman.

Boubekri, M., R. B. Hull, and L. L. Boyer. 1991. Impact of window size and sunlight penetration on office workers' mood and satisfaction: A novel way of assessing sunlight. *Environment and Behavior* 23:474–93.

Browning, W., and J. Romm. 1998. Greening and the bottom line: Increasing productivity through energy efficient design. In *Proceedings of the Second International. Green Buildings Conference and Exposition,* ed. K. Whitter and T. Cohn. National Institute of Standards and Technology Special Publication 888. Gaithersburg, MD: National Institute of Standards and Technology.

Carson, R. 1998. *The sense of wonder.* New York: HarperCollins.

Chandrakanth, M., and J. Romm. 1991. Sacred forest, secular forest policies, and people's actions. *Natural Resources Journal* 41:741–56.

Chawla, L. 1992. Childhood place attachments. In *Place Attachments,* ed. I. Altman and S. Low. New York: Plenum Press.

———, ed. 2001. *Growing up in an urbanizing world.* Paris: UNESCO; London: Earthscan.

Cimprich, B. 1993. Development of an intervention to restore attention in cancer patients. *Cancer Nursing* 16:83–92.

Coates, P. 1998. *Nature: Western attitudes since ancient times.* Berkeley and Los Angeles: University of California Press.

Cobb, E. 1977. *The ecology of imagination in childhood.* New York: Columbia University Press.

Coley, R., and F. Kuo. 1997. Where does community grow? The social context created by nature in urban public housing. *Environment and Behavior* 29:468–94.

Cooper Marcus, C., and M. Barnes. 1995. *Gardens in healthcare facilities: Uses, therapeutic benefits, and design recommendations.* Martinez, CA: Center for Health Design.

Cooper Marcus, C., and M. Barnes, eds. 1999. *Healing gardens: Therapeutic landscapes in healthcare facilities.* New York: Wiley.

Corbett, J., and M. Corbett. 2000. *Designing sustainable communities: Learning from village homes.* Washington, DC: Island Press.

Corson, R., and E. Corson. 1977. The socializing role of pet animals in nursing homes. In *Society, stress, and disease*, ed. L. Levi. London: Oxford University Press.

Diamond, J. 1993. New Guineans and their natural world. In *The biophilia hypothesis*, ed. S. Kellert and E. O. Wilson, 251–74. Washington, DC: Island Press.

Draper, R. J., G. J. Gerber, and E. M. Layng. 1990. Defining the role of pet animals in psychotherapy. *Journal of Developmental Psychology.* 15:169–72.

Driver, B., and P. Brown. 1976. Probable personal benefits of outdoor recreation. In *President's Commission on American Outdoors: A literature review*, 63–67. Washington, DC: U.S. Government Printing Office.

Driver, B. L., D. Dustin, T. Baltic, G. Elsner, and G. Peterson, eds. 1996. *Nature and the human spirit: Toward an expanded land management ethic.* State College, PA: Venture Publishing.

Driver, B., R. Nash, and G. E. Hass. 1987. Wilderness benefits: A state-of-the-knowledge review. In *Proceedings of the National Wilderness Research Conference*, ed. R. C. Lucas. General Technical Report INT-220. Ft. Collins, CO: USDA Forest Service.

Easley, A. T., J. F. Passineau, and B. L. Driver. 1990. *The use of wilderness for personal growth, therapy, and education.* General Technical Report RM-193. Ft. Collins, CO: USDA Forest Service.

Ewert, A. 1989. *Outdoor adventure pursuits: Foundations, models, and theories.* Scottsdale, AZ: Publishing Horizons.

Fein, A. 1972. *Frederick Law Olmsted and the American environmental tradition.* New York: Braziller.

Fine, A., ed. 2000. *The handbook of animal assisted therapy: Theoretical foundations and guidelines for practice.* New York: Academic Press.

Fisk, W., and A. Rosenfeld. 1997. Estimates of improved productivity and health from better indoor conditions. *Indoor Air* 7:158–72.

Fogle, B. 1981. *Interrelations between people and pets.* Springfield, IL: C. C. Thomas.

Friedmann, E., A. H. Katcher, J. J. Lynch, and S. A. Thomas. 1980. Animal companions and one-year survival of patients discharged from a coronary care unit. *Public Health Reports* 95:307–12.

Friedmann, E., and S. A. Thomas. 1995. Pet ownership, social support, and one-year survival after acute myocardial infarction in the cardia arrhythmia suppression trial. *American Journal of Cardiology* 76:1213–17.

Frumkin, H. 2001. Beyond toxicity: Human health and the natural environment. *American Journal of Preventive Medicine* 20:234–39.

Gadgil, M. 1993. Of life and artifacts. In *The biophilia hypothesis*, ed. S. Kellert and E.O. Wilson, 365–80. Washington, DC: Island Press.

Glacken, C. J. 1967. *Traces on the Rhodian shore: Nature and culture in Western thought from ancient times to the end of the eighteenth century*. Berkeley and Los Angeles: University of California Press.

Godish, T. 1994. *Sick buildings: Definitions, diagnosis, and mitigation*. Boca Raton, FL: Lewis Publishers.

Graham, P. 1994. Green structures: The importance for health of nature areas and parks. *European Regional Planning* 56:89–112.

Hart, L. A., B. L. Hart, and B. Bergin. 1987. Socializing effects of service dogs for people with disabilities. *Anthrozoos* 1:41–44.

Hart, R. 1979. *Children's experience of place*. New York: Irvington.

Hartig, T. 1993. Nature experience in transactional perspective. *Landscape and Urban Planning* 25:17–36.

Hartig, T., A. Book, J. Garvill, T. Olsson, and T. Garling. 1996. Environmental influences on psychological restoration. *Scandinavian Journal of Psychiatry* 37:378–93.

Hartig, T., and G. Evans. 1993. Psychological foundations of nature experience. In *Behavior and environment: Psychological and geographical approaches*, ed. T. Garlking and R. Golledge, 427–57. Amsterdam, Netherlands: Elsevier Science.

Hartig, T., M. Mang, and G. W. Evans. 1991. Restorative effects of natural environment experiences. *Environment and Behavior* 23:3–26.

Heerwagen, J. 1990a. Affective functioning, light hunger, and room brightness preferences. *Environment and Behavior* 22:608–35.

———. 1990b. The psychological aspects of windows and window design. *Environmental Design Research Association* 21:269–80.

———. 2000a. Do green buildings enhance the well being of workers? Yes. *Environmental Design and Construction* (July): 24–34.

———. 2000b. Green buildings, organizational success, and occupant productivity. *Building Research and Information* 28:353–67.

Heerwagen, J., N. Durbin, and J. Macaulay. 1997. Do energy efficient, green buildings spell profits? *Energy and Environmental Management* 3:29–34.

Heerwagen, J., and B. Hase. 2001. Building biophilia: Connecting people to nature. *Environmental Design and Construction* (March).

Heerwagen, J., J. G. Heubach, J. Montgomery, and W. C. Weimer. 1995. Environmental design, work and well being. *American Association of Occupational Health Nurses Journal* 43:458–68.

Heerwagen, J., and G. Orians. 1986. Adaptations to windowlessness: A study of the use of visual décor in windowed and windowless offices. *Environment and Behavior* 18:623–30.

———. 1993. Humans, habitats, and aesthetics. In *The biophilia hypothesis*, ed. S. Kellert and E. O. Wilson, 138–72. Washington, DC: Island Press.

Heerwagen, J., J. Wise, D. Lantrip, and M. Ivanovich. 1996. A tale of two buildings: Biophilia and the benefits of green design. Paper presented at the U.S. Green Buildings Council Conference, San Diego, CA.

Heschong, L., and D. Mahone. 1999. Daylighting in schools: An investigation into the relationship between daylighting and human performance. Report submitted to the Pacific Gas and Electric Company on behalf of the California Board for Energy Efficiency Third Party Program.

Hildebrand, G. 1999. *The origins of architectural pleasure*. Berkeley and Los Angeles: University of California Press.

de Hollander, D. 2002. Health, environment and quality of life: An epidemiological perspective on urban development. *Landscape and Urban Planning* 89:1–10.

Hull, R., and A. Harvey, 1989. Explaining the emotion people experience in suburban parks. *Environment and Behavior* 21:323–45.

Hull, R., and S. Michael. 1995. Nature-based recreation, mood change, and stress reduction. *Leisure Science* 17:1–14.

Kahn, P. 1999a. Developmental psychology and the biophilia hypothesis. *Developmental Review* 17:1–61.

———. 1999b. *The human relationship with nature*. Cambridge, MA: MIT Press.

Kahn, P., and S. Kellert, eds. 2002. *Children and nature: Psychological, sociocultural, and evolutionary investigations*. Cambridge, MA: MIT Press.

Kaplan, R. 1973. Some psychological benefits of gardening. *Environment and Behavior* 5:142–52.

———. 1985. Nature at the doorstep: Residential satisfaction and the nearby environment. *Journal of Architectural Planning and Research* 2:115–27.

———. 1993. The role of nature in the context of the workplace. *Landscape and Urban Planning* 26:193–201.

Kaplan, S., and R. Kaplan. 1995. Urban forestry and the workplace. In *Managing urban and high use recreation settings,* ed. P. Gobster. General Technical Report NC-163. Chicago: USDA Forest Service.

Kaplan, R., and S. Kaplan. 2002. Adolescents and the natural environment: A time out? In *Children and nature: Psychological, sociocultural, and evolutionary investigations,* ed. P. Kahn and S. Kellert, 227–58. Washington, DC: Island Press.

Kaplan, S., and R. Kaplan 1995. The restorative benefits of nature: Toward an integrative framework. *Journal of Environmental Psychology* 12:169–82.

Kaplan, S., and R. Kaplan, 1989. *The experience of nature*. New York: Cambridge University Press.

Kaplan, S., R. Kaplan, and J. S. Wendt. 1972. Rated preference and complexity for natural and urban visual material. *Perception and Psychophysics* 12:354–56.

Kaplan, S., and C. Peterson. 1993. Health and environment: A psychological analysis. *Landscape and Urban Planning* 26:17–23.

Kaplan, S., and J. Talbot. 1983. Psychological benefits of a wilderness experience. In *Behavior and the natural environment,* ed. I. Altman and J. Wohlwill. New York: Plenum.

Katcher, A. 2002. Animals in therapeutic education: Guides into the limnal state. In *Children and nature,* ed. P. Kahn and S. Kellert. Cambridge, MA: MIT Press.

Katcher, A., and A. Beck, eds. 1983. *New perspectives on our lives with companion animals*. Philadelphia: University of Pennsylvania Press.

Katcher, A., and A. Beck. 1991. Animal companions: More companion than animals. In *Man and beast revisited,* ed. R. Robinson and L. Tiger. Washington, DC: Smithsonian Press.

Katcher, A., E. Friedmann, A. M. Beck, and J. J. Lynch. 1983. Looking, talking and blood pressure: The physiological consequences of interaction with the living environment. In *New perspectives on our lives with companion animals*, ed. A. Katcher and A. Beck. Philadelphia: University of Pennsylvania Press.

Katcher, A., H. Segal, and A. Beck. 1984. Comparison of contemplation and hypnosis for the reduction of anxiety and discomfort during dental surgery. *American Journal of Clinical Hypnosis* 27:14–21.

Katcher, A., and G. Wilkins. 1993. Dialogue with animals: Its nature and culture. In *The biophilia hypothesis*, ed. S. Kellert and E. O. Wilson, 173–200. Washington, DC: Island Press.

Katts, G. 2003. *The costs and financial benefits of green buildings*. Sacramento, CA: California Sustainable Building Task Force.

Kellert, S. 1996. *The value of life: biological diversity and human society*. Washington, DC: Island Press.

———. 1997. *Kinship to mastery: Biophilia in human evolution and development*. Washington, DC: Island Press.

———. 2002a. Experiencing nature: Affective, cognitive, and evaluative development in children. In *Children and nature: Psychological, sociocultural, and evolutionary investigations*, ed. P. Kahn and S. Kellert, 117–52. Cambridge, MA: MIT Press.

———. 2002b. Values, ethics, and spiritual and scientific relations to nature. In *The good in nature and humanity*, ed. S. Kellert and T. Farnham, 49–64. Washington, DC: Island Press.

———. 2005. *Building for life: Designing and understanding the human-nature connection*. Washington, DC: Island Press.

Kellert, S., and V. Derr. 1998. *National study of outdoor wilderness experience*. New Haven: Yale University School of Forestry and Environmental Studies.

Kellert, S., and E .O. Wilson. 1993. *The biophilia hypothesis*. Washington, DC: Island Press.

Knopf, R. 1987. Human behavior, cognition, and affect in the natural environment. In *Handbook of environmental psychology*, ed. D. Stokols and I. Altman. New York: John Wiley.

Korpela, K., and T. Hartig. 1996. Restorative qualities of favorite places. *Journal of Environmental Psychology* 16:221–33.

Krathwohl, D., B. S. Bloom, and B. B. Masia. 1964. *Taxonomy of educational objective: The classification of educational goals. Handbook 2, Affective domain*. New York: Longman.

Kuller, R., and C. Lindsten. 1992. Health and behavior of children in classrooms with and without windows. *Journal of Environmental Psychology* 12:305–17.

Kuo, F. 2001. Coping with poverty: Impacts of environment and attention in the inner city. *Environment and Behavior* 33:5–34.

Kuo, F., W. C. Sullivan, R. L. Coley, and L. Brunson. 1998a. Fertile ground for community: Inner-city neighborhood common spaces. *American Journal of Community Psychology* 26:823–51.

———. 1998b. Transforming inner-city landscapes: Trees, sense of safety, and preference. *Environment and Behavior* 30 (1): 28–59.

Lacy, J. 1990. *An examination of market appreciation for clustered housing with permanently protected open space*. Amherst, MA: Center for Rural Massachusetts, University of Massachusetts.

Larsen, L., J. Adams, B. Deal, B. S. Kweon, and E. Tyler. 1998. Plants in the workplace: The effects of plant density on productivity, attitudes, and perceptions. *Environment and Behavior* 31:261–81.

Lawrence, E. 1993. The sacred pig, the filthy pig, and the bat out of hell: Animal symbolism as cognitive biophilia. In *The biophilia hypothesis*, ed. S. Kellert and E. O. Wilson, 301–44. Washington, DC: Island Press.

Leather, P., M. Pyrgas, D. Beale, and C. Lawrence. 1998. Windows in the workplace: Sunlight, view, and occupational stress. *Environment and Behavior* 30:739–62.

Levinson, B. 1969. *Pet-oriented child psychotherapy*. Springfield: C. C. Thomas.

———. 1972. *Pets and human development*. Springfield, IL: C. C. Thomas.

Levitt, L. 1988. Therapeutic value of wilderness. In *Wilderness benchmark 1988: Proceedings of the National Wilderness Colloquium*. General Technical Report SE-51, 156–68. Asheville NC: USDA Forest Service.

Lohr, V., C. H. Pearson-Mims, and G. K. Goodwin. 1996. Interior plants may improve worker productivity and reduce stress in a windowless environment. *Journal of Environmental Horticulture* 14:97–100.

Louv, R. 2008. Last child in the woods: Saving our children from nature-deficit disorder. New York: Algonquin Books.

Lumsden, C., and E. O. Wilson. 1981. *Genes, mind, and culture: The coevolutionary process*. Cambridge, MA: Harvard University Press.

Maffi, L. ed. 2001. *On biocultural diversity: Linking language, knowledge, and the environment*. Washington, DC: Smithsonian Press.

Maker, C. 1982. *Teaching models of the gifted*. Austin: Pro-Ed.

Marcoin, M., B. Seifert, and T. Lindball. 1995. *Indoor air quality*. New York: Elsevier.

Messent, P. 1983. Social facilitation of contact with other people by pet dogs. In *New perspectives on our lives with companion animals*, ed. A. Katcher and A. Beck. Philadelphia: University of Pennsylvania Press.

Mooney, P., and P. Nicell. 1992. The importance of exterior environment for Alzheimer residents. *Healthcare Management Forum* 5:23–29.

Moore, D. 1984. Animal-facilitated therapy: A review. *Children's Environment Quarterly* 1 (3): 37–40.

Moore, E. 1982. A prison environment's effect on health care service demands. *Journal of Environmental Systems* 11:17–34.

Moore, R. 1986. *Childhood's domain: Play and space in child development*. London: Croom Helm.

Nabhan, G., and S. Trimble. 1994. *The geography of childhood*. Boston: Beacon.

Nakamura, R., and E. Fuji. 1992. A comparative study of the characteristics of the electroencephalogram when observing a hedge and a concrete block fence. *Journal of the Japanese Institute of Landscape Architects* 55:139–44.

Nelson, R. 1993. Searching for the lost arrow: Physical and spiritual ecology in the hunter's world. In *The biophilia hypothesis*, ed. S. Kellert and E. O. Wilson, 201–28. Washington, DC: Island Press.

O'Connor, B., H. Davidson, and R. Gifford. 1991. Window view, social exposure and nursing home adaptation. *Canadian Journal of Aging* 10:216–23.

Oelschlaeger, M. 1991. *The idea of wilderness: From prehistory to the age of ecology*. New Haven: Yale University Press.

Orr, D. 2002. Political economy and the ecology of childhood. In *Children and nature:*

Psychological, sociocultural, and evolutionary investigations, ed. P. Kahn and S. Kellert, 415–40. Cambridge, MA: MIT Press.

Parsons, R. 1991. The potential influences of environmental perception on human health. *Journal of Environmental Psychology* 11:1–23.

Parsons, R., and T. Hartig. 1999. Environmental physiology. In *Handbook of psychophysiology*, ed. J. Caccioppo. New York: Cambridge University Press.

Parsons, R., L. G. Tassinary, R. S. Ulrich, M. R. Hebl, and M. Grossman-Alexander. 1998. The view from the road: Implications for stress recovery and immunization. *Journal of Environmental Psychology* 18:113–40.

Pyle, R. 1993. *The thunder tree: Lessons from an urban wildland*. Boston: Houghton Mifflin.

———. 2002. Eden in a vacant lot: Special places, species, and kids in the neighborhood of life. In *Children and nature: Psychological, sociocultural, and evolutionary investigations*, ed. P. Kahn and S. Kellert, 305–28. Washington, DC: Island Press.

Ramakrishnan, P. 1996. Conserving the sacred: From species to landscapes. *Nature and Resources* 32:11–19.

Relf, D., ed. 1992. *The role of horticulture in human well-being and social development*. Portland, OR: Timber Press.

Rocky Mountain Institute, A. Wilson, J. L. Uncapher, L. McManigal, L. H. Lovins, M. Cureton, and W. D. Browning. 1998. *Green development: Integrating ecology and real estate*. New York: Wiley.

Rolston, H. 1986. *Philosophy gone wild: Environmental ethics*. Buffalo, NY: Prometheus Books.

Rowan, A., ed. 1989. *Animals and people sharing the world*. Hanover, NH: University Press of New England.

Schmidt, P. J. 1990. *Back to nature: The Arcadian myth in urban America*. Baltimore: Johns Hopkins University Press.

Schreyer, R., D. R. Williams, and L. Haggard. 1988. *The role of wilderness in human development*. General Technical Report SE-51. Ft. Collins, CO: USDA Forest Service.

Searles, H. 1960. *The nonhuman environment: In normal development and in schizophrenia*. New York: International Universities Press.

Sebba, R. 1991. The landscapes of childhood: The reflections of childhood's environment in adult memories and in children's attitudes. *Environment and Behavior* 23:395–422.

Sensharma, N. P., J. E. Woods, and A. K. Goodwin. 1998. Relationship between the indoor environment and productivity: A literature review. *ASHRAE Transactions* 104.

Serpell, J. 1986. *In the company of animals: A study of human-animal relationships*. Oxford, UK: Basil Blackwell.

Sharma, S., H. C. Rikhari, and L. M. S. Palni. 1999. Conservation of natural resources through religion: A case study from central Himalaya. *Society and Natural Resources* 12:599–622.

Shepard, P. 1978. *Thinking animals: Animals and the development of human intelligence*. New York: Viking.

———. 1996. *The others: How animals made us human*. Washington, DC: Island Press.

Siegel, J. 1990. Stressful life events and use of physician services among the elderly: The moderating role of pet ownership. *Journal of Personality and Social Psychology* 58:1081–86.

Sobel, D. 1993. *Children's special places: Exploring the role of forts, dens, and bush houses in middle childhood*. Tucson: Zephyr Press.

Spiegel, R., and D. Meadows. 1999. *Green building materials*. New York: Wiley.

Stewart-Pollack, J. 1996. The need for nature. Part 1: How nature determines our needs for and responses to environments. *Isdesign,* September–October.

Sullivan, W., and F. Kuo. 2002. Do trees strengthen urban communities, reduce domestic violence? Forestry Report R8-FR 55, Technical Bulletin No. 4. *USDA Forest Service.* Athens, GA: USDA Forest Service Southern Regions.

Taylor, A. 1976. The therapeutic value of nature. *Journal of Operative Psychiatry* 12:64–74.

Taylor, A., F. E. Kuo, and W. C. Sullivan. 2001. Coping with ADD: The surprising connection to green places. *Environment and Behavior* 33:54–77.

Tennessen, C., and B. Cimprich. 1995. Views to nature: Effects on attention. *Journal of Environmental Psychology* 15:77–85.

Thomas, K. 1983. *Man and the natural world: A history of the modern sensibility*. New York: Pantheon.

Thornhill, A. 1999. Darwinian aesthetics. In *Evolutionary psychology: The new science of the mind*, ed. D. Buss. London: Allyn and Bacon.

Todd, J. 1982. *Frederick Law Olmsted*. Boston: Twayne Publishers.

Tuan, Y. 1993. *Passing strange and wonderful: Aesthetics, nature, and culture*. Washington, DC: Island Press.

Ulrich, R. 1979. Visual landscapes and psychological well-being. *Landscape Research* 4:17–23.

———. 1983. Aesthetic and affective response to natural environment. In *Human behavior and environment*, ed. I. Altmann and J. F. Wohlwill. New York: Plenum.

———. 1984. View through a window may influence recovery from surgery. *Science* 224:420–21.

———. 1986. *Effects of hospital environments on patient well-being*. Research Report Series 9, No. 55. Trondheim, Norway: University of Trondheim, Department of Psychiatry and Behavioral Medicine.

———. 1991. Stress recovery during exposure to natural and urban environments. *Journal of Environmental Psychology* 11:201–30.

———. 1992. Effects of health facility interior design on wellness: Theory and recent scientific research. *Journal of Health Care Design* 3:97–109.

———. 1993. Biophilia, biophobia, and natural landscapes. In *The biophilia hypothesis*, ed. S. Kellert and E. O. Wilson, 73–137. Washington, DC: Island Press.

Ulrich, R., and O. Lunden. 1990. Effects of nature and abstract pictures on patients recovering from open heart surgery. Paper presented at the First International Congress of Behavioral Medicine, Uppsala, Sweden.

Ulrich, R., and R. Parsons. 1992. Influences of passive experience with plants on individual well-being and health. In *The role of horticulture in human well-being and social development*, ed. D. Relf, 93–105. Portland, OR: Timber Press.

Ulrich, R., R. F. Simons, B. D. Losito, E. Fiorito, M. A. Miles, and M. Zelson. 1991. Stress recovery during exposure to natural and urban environments. *Journal of Environmental Psychology* 11:201–30.

Veitch, J., and G. Newsham. 1998. Lighting quality and energy efficient effects on task performance. *Journal of Illuminating Engineering Society* 27:107–29.

White, R. 2004. *Young children's relationship with nature: Its importance to children's development and the earth's future*. Kansas City, MO: White Hutchinson Leisure and Learning Group.

Wilson, E. O. 1984. *Biophilia*. Cambridge: Harvard University Press.

———. 1992. *The diversity of life*. Cambridge: Harvard University Press, Belknap Press.

———. 1993. Biophilia and the conservation ethic. In *The biophilia hypothesis*, ed. S. Kellert and E. O. Wilson, 31–41. Washington, DC: Island Press.

Wise, J. 1997. How nature nurtures: Buildings as habitats and their benefits to people. *Heating/Piping/Air Conditioning* 78:48–51.

Wohlwill, J. F. 1983. The concept of nature: A psychologist's view. In *Behavior and the natural environment*, ed. I. Altman and J. Wohlwill. New York: Plenum.

Zube, E., and R. O. Brush, eds. 1975. *Landscape assessment: Values, perceptions, and resources*. Stroudsburg, PA: Dowden, Hutchinson, and Ross.

7

Quality of Life, Biodiversity, and Health: Observations and Applications

Suzanne M. Skevington

Biodiversity and health are high on the international agenda to promote sustainable development, together with water, energy, and agriculture (Annan 2002). In building a sustainable society, one of the objectives of the United Nations Environment Programme (UNEP) has been to improve the quality of life (UNEP 1992). While economic growth is part of development, it cannot go on in itself, and go on indefinitely. As stated by UNEP: "Development should enable people to realize their potential, and lead dignified and fulfilled lives" (UNEP 1992, 21).

Adding to this debate, von Schirnding (2002) makes the case for a shared health, environment, and development agenda to address the burden of disease that is environmentally related (and vice versa), and for new tools to be developed to ensure intersectoral action. Health-related quality-of-life assessment could plausibly be one of these new tools.

Biodiversity, quality of life, and health can be interlinked through several identifiable models drawn from the existing literature. Perhaps the simplest model is that in which biodiversity is viewed as having a direct effect on quality of life (QOL), although the empirical evidence is generally poor and sparse beyond anecdotal and political levels. However, some model predictions can be derived from existing empirical sources. Kellert (1993) cited nine values that people hold in respect to wildlife and nature (see also this volume, chapter 6), some of which could be mapped onto known QOL outcomes. For example, biodiversity can affect, and be affected by, aesthetic values, and these have cognitive as well as spiritual dimensions, such as feelings of awe and wonder. Positive feelings of happiness and contentment are integral to the humanistic and moralistic values that are expressed as an affinity with nature. Spiritual connection with other people in the past or future, or with other worlds—perhaps natural ones—is linked to moral values in some communities. The negativistic values identified by Kellert would more likely be linked with the presence of negative feelings, such as anxiety and depression. Most of these values should be reflected in people's perceptions of their physical environment, irrespective of whether they live in urban or rural settings or in straw huts, under plastic sheeting, or in a terraced house. These values would be expected to affect specific appraisals of different dimensions of environmental quality of life, such as pollution of all types, home environment, and so forth.

Biodiversity may also affect quality of life indirectly where the model indicates that QOL is mediated through changes to health. In this model, food diversity and sufficient clean water, for instance, are seen as vital to good human health, which in the absence of disease usually provides a better QOL. However, this is a reflexive process, since having the strength and motivation to obtain a rich and varied diet is in turn dependent on biodiversity (Wahlquist 2003; see also this volume, chapter 2). Loss to biodiversity, along with climate change, degradation of the land, and reductions to ocean fishery stocks and terrestrial aquifers, is known to weaken life support systems and pose serious risks to health (McMichael and Beaglehole 2000; see also this volume, chapter 3). Climate change increases the likelihood of some illnesses and death, especially among vulnerable groups such as older adults and children. Extreme weather conditions—for example, flooding and heat waves—can result in the consequences of more waterborne and foodborne diseases, more broadly distributed malaria, ozone depletion accelerating skin cancer, and weakened immune systems (see this volume, chapter 14). Where physical and mental health deteriorates due to biodiversity loss, we can expect QOL to be reduced as well. However. it is also worth noting that mental health can deteriorate and QOL be damaged simply by *perceiving* a threat to a loss of biodiversity, even if that threat itself is not carried out. An example is the proposal by a multinational corporation to fell forests in the Olympic Peninsula in Washington State, thereby threatening the QOL and health of local communities.

However, we should not necessarily assume that QOL is directly and positively associated with health; it shows only modest to poor correlation with standard indices of health and illness, such as number and/or intensity of symptoms, severity and length of illness, mortality rate, and life expectancy. This is because the *meaning* of a person's condition and situation varies enormously, depending on the person's history, experience, knowledge, personality, mood, and so forth. Two people walking through the same rainforest together will have very different interpretations of what is objectively the same event, and this is as true for environmental events as it is for health conditions. Even though they watch the same macaws nesting in the Peruvian rainforest canopy, this will have different meanings for each, and will be reflected in the different accounts they tell over supper. In health care, one patient may be happy, relieved, and back to work quickly following hip replacement surgery, while another may be immobilized, depressed, and chronically unable to function following the same technical event at the hands of the same surgeon. These subjective interpretations of events are as important as "objective" conditions. Unfortunately, such interpretations have often not been collected systematically in environmental settings from those people who are most affected.

Furthermore, many areas of quality of life are not open to visible inspection. While it is possible to see whether people are sufficiently mobile by looking at their ability to walk, it is much more difficult to observe whether their sleep is refreshing, whether they are afraid for their physical safety, or satisfied with their home environment without directly asking them. Quality of life is like an iceberg, with only a small tip of its concept open to inspection by others. For this reason, we argue that the best way of finding out about a person's QOL is simply to ask them, because assumptions made by others will necessarily be erroneous. While there is a moral argument to support the view that we should also be considering the QOL of "countless other life forms on whom we unwittingly depend" (Tickell 1992), the means of obtaining that information is beyond the boundaries of the human methodology proposed here. There is a rapidly growing body of evidence to demonstrate links between biodiversity changes and health, and a substantial literature on the relation of health to QOL. However, sound empirical evidence that reliably tracks the projected causal and presumed linear relationship from biodiversity through health to subjective QOL is so far largely absent. In the continued absence of robust empirical data on this model, it is difficult to convince policymakers of the need to act.

A third model linking biodiversity, health, and quality of life is through a consideration of those species that could contribute to the maintenance and treatment of human well-being and health (see this volume, chapter 15). A rich array of natural resources potentially provides treatment opportunities for the future, and during the 1990s international pharmaceutical companies investigated the potential of bioexploitation. For example, Taxol, which can be extracted from the Pacific yew tree, is efficacious in the treatment of ovarian and advanced breast cancers (Walsh and Goodman 1999). Without effective treatments, QOL would be generally poorer among those living with untreatable and chronic illnesses, so this eye-catching application provides a useful mechanism for raising awareness about biodiversity among the public. "A desire for good health has been the motivator for humans trying thousands of plants as remedies for various ailments" (Perlman and Adelson 1997, 32).

However, many of these naturally occurring substances were remedies only known to traditional healers (e.g., Tabuti, Dhillion, and Lye 2003), and many more are still undiscovered (see this volume, chapter 15). In a study of nine rural communities in Nepal, 113 local remedies derived from 58 species in 40 families were identified, giving some idea of the scope of this endeavor. These remedies were used to treat common conditions like colds, coughs, digestive and skin problems, fever, and headache (Shrestha and Dhillion 2003). Such remedies required careful screening to identify the most promising plants, and to investigate the provision of a continuous supply (Farnsworth 1988). Related to this, the use of good harvesting and manufacturing practice is essential to the safe, controlled production of herbal drugs (Mukherjee 2002). Access to robust phytotherapeutic evidence on efficacy for many species is also urgently needed (Shrestha and Dhillion 2003). In this sense, biodiversity provides a vital natural resource in managing health and illness, and, ostensibly, it also promotes good quality of life.

Environmental Quality of Life

Quality of life has been a hot political issue since Lyndon B. Johnson first used the phrase while electioneering in the early 1960s, and a large body of empirical research is now available to support the concept and its measurement. In a historical review, Sirgy (2001) notes that for "quite some time, many [researchers] thought that when people talked about QOL they meant the quality of the environment." This implies that, in the public's mind, at least, general QOL was seen to be synonymous with environmental QOL. To illustrate this point, Fisk (1988) defines QOL as "the perception of well-being, resulting from each person's interaction with their environment." Sirgy (2001) observes that while this conception of QOL is "quite limited," it [still] "remains valid." He identified more than 15 types of theories, conceptual frameworks, and philosophical approaches to QOL (Sirgy 2001), which can be broadly separated into those taking an ecological perspective and those taking a health perspective. Within the ecological group, models take either an anthropocentric approach, in which the nonhuman world is viewed as being in the service of humans, helping them to achieve and sustain their goals, or an ecocentric approach, in which nonhumans are seen as important in their own right, being symbiotic with, and interrelated to, humans (Iyer 1999). These two philosophies underpin and direct thinking about different solutions to environmental problems. Thus the anthropocentric goal of business ethics is to provide technological solutions and innovative products that would not reduce human QOL. Ecocentrics, on the other hand, seek to reorient business policies, operations, and products to ensure that the ecosystem is healthy.

In taking an overview, it is clear that researchers working in the fields of health-related and environmental QOL have largely ignored each other's theories, concepts, and methods. Consequently, they continue to work relatively independently in quite separate traditions, and to use the term *QOL* somewhat differently (Lercher 2003). For this reason, the World Health Organization Quality of Life Assessment (WHOQOL) provides an unusual measure in the health-related QOL field because, unlike other popularly used generic measures such as the Short Form 36 Health Survey (SF-36) (Ware, Kosinski, and Keller 1994), the WHOQOL instrument provides a substantial domain of environmental QOL for assessment. In this way, it uniquely fuses health and environmental concerns. Furthermore, in a major review of generic cross-cultural instruments for the measurement of health-related QOL, Bowden and Fox-Rushby (2003) concluded that when criteria relating to psychometrics, translation methods, and other cross-cultural requirements were taken into account, there were only two measures that attained acceptable standards and were available in multiple language versions covering all the major regions of the populated world. One of these is the SF-36 (Ware, Kosinski, and Keller 1994), which does not include any assessment of QOL relating to the environment, and the other is the WHOQOL. The WHOQOL instrument has been critically and independently reviewed from an environmentalist perspective by Lercher (2003) and has been judged to have considerable potential in the study of environmental matters. For this reason, a detailed account of the WHOQOL is provided below. Many reviews containing hundreds of measures of QOL are now published in the journal *Quality of Life Research,* and are available through existing compendia—for example, McDowell and Newell (1996); Bowling (1995); and five volumes edited by Salek (1998). It is not the purpose of this chapter to review again all these measures, but rather to illustrate some pertinent issues by providing an accessible account of one of the more recent measures available for cross-cultural research.

The World Health Organization Quality of Life Assessment (WHOQOL)

In the early 1990s, an international collaboration of scientists, researchers, and clinicians (the WHOQOL Group) came together at the World Health Organization in Geneva to discuss whether QOL was a salient concept in their culture, what the dimensions of it might be, and how this might relate to health care in their society. The WHOQOL Group defined quality of life as "an individual's perception of their position in life, in the context of the culture and value systems in which they live and in relation to their goals, expectations, standards and concerns" (WHOQOL Group 1994).

The collaboration identified many aspects of QOL that might be universal and tested these following the construction of an internationally agreed-on protocol. The latter eventually formed a manual from which other new language versions of the WHOQOL could be created. Using a unique "spoke-wheel" methodology (Skevington 1999), 15 centers worldwide simultaneously investigated internationally agreed-on dimensions in focus groups of users: patients, health professionals, and community members. This information was later applied to produce a generic instrument for use by all types of sick people and also those who were healthy. The value of this work was that people's QOL could be compared in different cultures, as greater semantic and conceptual equivalence between different language versions had been achieved. It would also be possible to track people's QOL from one situation to another, and across time, such as the life span. These focus group participants were typical of the people who would eventually use the scale, and they assisted in formulating questions that might be included in the measure. Transcribed discussions about QOL made it possible to adopt the language and concepts that they used in the wording of these items. The WHOQOL was therefore devised through a "bottom-up" approach so that it would be meaningful to users living in that culture.

Qualitative and quantitative research confirmed a high level of consensus about a concept called QOL in 15 countries worldwide, and about the importance of the dimensions or facets that were important constituents of the concept (WHOQOL Group 1994). Facets were first clustered in five domains—physical health, psychological state, social relations, independence, and environment. But the international focus groups made a compelling case for the inclusion of a sixth domain of QOL on spiritual, religious, and personal beliefs. This structure of six domains is reliable and valid, as confirmed by survey data obtained in many countries worldwide, and through the use of psychometric analyses including structural equation modeling (e.g., WHOQOL Group 1998a, 1998b). Respondents answered these items using a five-point (Likert) interval scale (WHOQOL Group 1994, 1998a, 1998b). From a survey of 4,800 sick and well people in 15 countries, 100 items were selected and organized into 25 facets of QOL, with 4 items for each facet (table 7.1). More recently, a short-form questionnaire of 26 items—the WHOQOL-Bref—has been extracted from the WHOQOL-100 and is standardized for international use in 23 countries (n = 11,800) (Skevington, Lotfy, et al. 2004). Field testing showed that the WHOQOL-Bref has good properties of reliability and validity, although sensitivity to change is still being assessed. In Britain, well people can complete the WHOQOL-Bref in five minutes (see also Skevington, Sartorius, et al. 2004).

The WHOQOL provides a useful instrument for the comprehensive and subjective measurement of QOL. Questions and concepts are phrased in ways that put QOL in a positive, rather than negative, light wherever possible. This has enabled the upper limits of well-being to be addressed, in addition to the more usual lower limits demanded by the problem-centered clinical consultation. Items are also phrased in ways that are designed to accommodate the widest range of cross-cultural situations possible. For example, in the home environment facet, people are asked about "the place where you live," not a "house" or "home," as many people around the world live under canvas or in cardboard boxes, or are homeless.

The unusual "environment" domain contains eight facets that assess subjective perceptions of physical safety and security; home environment; financial resources; access to health and social care; opportunities to acquire new information and skills; opportunities to participate in recreation and leisure; physical environment, including noise and air pollution, climate, traffic, and so forth; and transport. Large-scale multinational surveys using the WHOQOL (e.g., Skevington, Lotfy, et al. 2004) show that the environment domain forms an important and integral part of the QOL concept. In setting the minimum content requirements for any instrument to be acceptable to the investigation of environmental health, Lercher (2003) states that contents should cover all the main domains and facets of health. The instrument should also include positive health as well as being problem centered. It must be sensitive enough to test the research question, and should include subjective and objective indicators. The WHOQOL well satisfies Lercher's first two requirements and the last two partially; the SF-36 partly fulfills two. When research questions on biodiversity, health, and QOL are being posed, it will be necessary to test the sensitivity of the WHOQOL within different environmental contexts, paying particular attention to the performance of the environment domain.

The subjective nature of the WHOQOL suits its use alongside suitable objective indicators where needed. However, using health indicators in relation to biodiversity has not always been entirely successful. Huynen, Martens, and De Groot (2004) found little

Table 7.1. The six domains and twenty-five facets of the WHOQOL

25. Overall QOL & General Health

Domain 1. Physical Health	Domain 2. Psychological	Domain 3. Independence	Domain 4. Social Relationships	Domain 5. Environment	Domain 6. Spirituality Religion, & Personal Beliefs
1. Pain & discomfort	4. Positive feelings	9. Mobility	13. Personal relationships	16. Physical safety & security	24. Meaning of Life+
2. Sleep & rest	5. Thinking, learning, memory, & concentration	10. Activities of daily living	14. Social support	17. Home environment	
3. Energy & fatigue	6. Self-esteem	11. Dependence on medication & treatment	15. Sex	18. Financial resources	
	7. Body image & appearance	12. Working capacity		19. Health & social care: accessibility & quality	
	8. Negative feelings			20. Opportunity to acquire new information & skills	
				21. Participation in & opportunity for recreation & leisure	
				22. Physical environment (pollution, noise, traffic, climate)	
				23. Transport	

Source: WHOQOL Group 1995.

convincing empirical evidence to support the view that global loss of biodiversity was associated with poorer human health. They included several measures of biodiversity loss—in terms of percentage of threatened species, percentage of remaining original forests, and percentage of land highly disturbed by human use—as well as several measures of health, including life expectancy rate, disability-adjusted life years (DALYs), infant mortality rate, and percentage of low-birth-weight babies. Although these negative results are explained in terms of statistical anomalies and a shortage of suitable indicators, it is possible that indicators do not provide a sufficiently sensitive method for tackling this problem, and that, for metric reasons, a different, more person-centered approach is required.

The WHOQOL approach implicitly addresses another form of diversity through its definition—human cultural diversity—as it places culture at the very center of QOL interpretation and understanding. For this reason, it has important conceptual resonances with the biodiversity approach. It acknowledges the need for a version of QOL assessment adapted to each language, as language is embedded in the culture of the respondent. Although a sophisticated iterative translation methodology of translation and back-translation has been developed by the WHO within this project (Sartorius and Kuyken 1994), the WHOQOL approach allows investigators creating a new translation to adapt the item wording to their culture while still retaining meaning to create equivalence with the source (English). Furthermore, as every new language version is developed, users are invited to identify important aspects of QOL that were not included in the original international core but which need to be included as vital national issues to "round out" the concept of QOL in their culture—for example, skin color in India, feeling "fed up" in Britain, physical safety and security in Israel. Cultural diversity therefore serves the purpose of helping people to adapt to changing environmental conditions. In addition to language, cultural diversity is also represented by art, music, social structure, land management practices, crop selection, diet, and religious and spiritual beliefs (UNEP 1992) are important issues that influence QOL which are likely to affect the answers people give to WHOQOL questions.

People are concerned about their environment. Boehnke et al. (1998) found that the environment was one of seven major life domains that people were most worried about. In other studies, the environment is featured as one of the two most intense and important worries reported by younger people in 12 countries (Schwartz and Melech 2000). Specifically, younger people were concerned about the destruction of the environment, damage to nature (animals, forests), population explosion in developing countries, and air, water, noise and trash pollution in the immediate neighborhood. The presence of these environmental features can affect people's perceptions of their QOL. But even when these worries are not confirmed by objective reality, worry in itself threatens a good QOL because it engenders negative feelings. The intensity of negative feelings caused by worry will depend on whether these environmental changes are temporary or chronic. Depression with anxiety can pervade every aspect of QOL (e.g., Skevington and Wright 2001); thus if people are worried about their environment, we can expect their

QOL to be affected on at least some dimensions. Because the WHOQOL provides a profile of QOL facets, it is possible to make a priori predictions about which of the 25 facets would be most likely to be significantly changed by a particular intervention. Targeting enables focused remedial action to be taken should QOL be found to be poor in these areas. In this way, a profile like WHOQOL has more flexible uses than other QOL measures that produce a single score or index.

Potential Uses of Quality-of-Life Information in Environmental Health

The WHOQOL has so far not been widely applied in studies of the environment, although it appears to have some uses in large-scale international projects—for example, in urban planning through the WHO's Healthy Cities Initiative. One of the few studies that has explicitly focused on the environment domain looked at neighborhood income inequality and socioeconomic deprivation relating to psychological QOL in Dutch families. Where considerable socioeconomic deprivation was perceived, poorer environmental QOL was also reported (Drukker, Feron, and van Os 2004). Other QOL assessments have been successfully used for different environmental purposes. For example, one study investigated whether good health-related QOL was determined by lower population density in Norwegian cities (Cramer, Torgersen, and Kringlen 2004). Some of these topics would bear replication using the common metric provided for all dimensions by the WHOQOL, having been carried out with batteries of scales or aggregates of single items that use very different metrics.

Having been developed with the evaluation of clinical trials in mind, QOL assessment is well adapted to assess the effectiveness of single interventions—for example, horticulture in India to replace shifting cultivation (Gadgil 1993). In another example, holistic resource management (HRM) was introduced in the United States to enhance goal setting, decision making, and monitoring by integrating social, economic, and ecological factors. Participants considered holistic goals such as improving QOL, how different forms of production support QOL, and landscape planning to protect and enhance biodiversity. The results showed that fewer than 9% of farmers and ranchers had thought about biodiversity before knowing about HRM. Afterward, 95% perceived increases to biodiversity, 80% saw more profits from their land, and 91% reported improvements to their QOL due to changes in their time budget (Stinner, Stinner, and Martsolf 1997). Thus biodiversity enhancement may be fundamental to sustainable profit under HRM.

QOL could also be useful comparatively, to assess the relative merits of several different types of intervention. Quality of life assessment may also be useful in evaluating the provision of a new service—for example, to determine whether a given sustainable management program is complete and of the highest quality. It can be used to monitor the effect of policy changes on individuals—for example, the effect of implementing particular conservation policies by local or national governments or nongovernmental organizations. More radically, it seems plausible that changes to QOL detected by human communities may turn out to provide early warning signs of biodiversity loss. Human intelligence is well tuned to detect and report such changes where they occur consistently, or impact significantly upon people's lives. As such, QOL could be tested for its potential as a psychomonitoring agent, in much the same way that lichens serve as successful biomonitoring agents (Swaminathan 2003), and the results used to activate prevention.

Lastly, where resources are available, one-to-one feedback about a person's QOL based on assessment results might plausibly be used to promote awareness among individuals about their own QOL. An aim would be to motivate and promote self-management so that each person would improve his or her own environment. Alternatively, in indigenous communal societies, it might be more culturally appropriate to first raise awareness among "opinion leaders"—for example, tribal chiefs or traditional healers—about any local areas of poor QOL. Subsequent widespread action to improve QOL might then be promoted through community meetings, whose discussions would include local knowledge of traditional healing methods and their relationship to biological resources (Timmermans 2003). Success would depend on the community's designing an appropriate and feasible plan of action for implementation within that culture. This QOL approach is very useful where there is a need for surveys of individuals in a population or communities, or for use at a one-to-one individual level—that is, where individual perceptions, values, beliefs, and habits need to be established. However, the QOL approach has some limitations and is no panacea. Caution is needed in its application, as it is less well suited to the sensitive assessment of broader political and economic factors—for example, globalization or consumerism.

References

Annan, K. 2002. Towards a sustainable future. American Museum of Natural History annual environmental lecture, New York.

Boehnke, K., S. H. Schwartz, C. Stromberg, and L. Sagiv. 1998. The structure and dynamics of worry: Theory, measurement and cross-cultural replications. *Journal of Personality* 66: 745–782.

Bowden, A., and J. A. Fox-Rushby. 2003. A systematic and critical review of the process of translation and adaptation of generic health-related QOL measures in Africa, Asia, Eastern Europe, the Middle East and South America. *Social Science and Medicine* 57 (7): 1289–1306.

Bowling, A. 1995. *Measuring disease: A review of disease-specific quality of life measurement scales*. Philadelphia: Open University Press.

Cramer, V., S. Torgersen, and E. Kringlen. 2004. QOL in a city: The effect of population density. *Social Indicators Research* 69 (1): 103–16.

Drukker, M., F. J. M. Feron, and J. van Os. 2004. Income inequality at neighborhood level and QOL: A contextual analysis. *Social Psychiatry and Psychiatric Epidemiology* 39 (6): 457–63.

Farnsworth, N. R. 1988. Screening plants for new medicines. In *Biodiversity*, ed. E. O. Wilson. New York: National Academy of Sciences.

Fisk, G. 1988. Marketing systems analysis of economic development and the QOL. In *Marketing: A return to the broader dimensions*, ed. S. Shapiro and A. H. Wale, 283–85. American Marketing Association.

Gadgil, M. 1993. Biodiversity and India's degraded lands. *Ambio* 22 (2–3): 167–72.

Huynen, M. M., P. Martens, and R. S. De Groot. 2004. Linkages between biodiversity loss and human health: A global indicator analysis. *International Journal of Environmental Health Research* 14 (10): 13–30.

Iyer, G. R. 1999. Business, consumers and sustainable living in an interconnected world: a multilateral eco-centric approach. *Journal of Business Ethics* 20 (July 11): 273–88.

Kellert, S. R. 1993. The biological basis for human values of nature. In *The Biophilia Hypothesis*, ed. S. R. Kellert and E. O. Wilson, 42–72. Washington, DC: Island Press.

Lercher, P. 2003. Which health outcomes should be measured in health-related environmental quality of life studies? *Landscape and Urban Planning* 65 (1–20): 65–74.

McDowell, I., and C. Newell. 1996. *Measuring health: A guide to rating scales and questionnaires*. 2nd ed. New York: Oxford University Press.

McMichael, A. J., and R. Beaglehole. 2000. The changing global context of public health. *Lancet* 356 (9228): 495–99.

Mukherjee, P. K. 2002. Problems and prospects for good manufacturing practice for herbal drugs in Indian systems of medicine. *Drug Information Journal* 36 (3): 635–44.

Neal, J. D., J. M. Sirgy, and M. Uysal. 2004. Measuring the effect of tourism services on travelers' quality of life: Further validation. *Social Indicators Research* 69 (3): 243–77.

Perlman, D. L., and G. Adelson, eds. 1997. *Biodiversity: Exploring values and priorities in conservation*. Oxford, UK: Blackwell Science.

Salek, S., ed. 1998. *Compendium of quality of life instruments*. Chichester, UK: Wiley.

Sartorius, N., and W. Kuyken. 1994. Translation of health status measures. In *Quality of life assessment: International perspectives*, ed. J. Orley and W. Kuyken, 1–18. Heidelberg: Springer-Verlag.

Schwartz, S. H., and G. Melech. 2000. National difference in micro and macro worry: Social, economic and cultural explanations. In *Culture and subjective well-being*, ed. E. Diener and E. M. Suh, 219–56. Cambridge, MA: MIT Press.

Shrestha, P. M., and S. S. Dhillion. 2003. Medicinal plant diversity and use in the highlands of Dolakha district, Nepal. *Journal of Ethnopharmacology* 86 (1): 81–96.

Sirgy, J. 2001. *Handbook of quality-of-life research: An ethical marketing perspective*. Social Indicators Research Series. Dordrecht, The Netherlands: Kluwer Academic.

Skevington, S. M. 1999. Measuring quality of life in Britain: An introduction to the WHOQOL-100. *Journal of Psychosomatic Research* 47 (5): 449–59.

———. 2002. Advancing cross-cultural research on quality of life: Observations drawn from the WHOQOL development—World Health Organisation Quality of Life assessment. *Quality of Life Research* 11:135–44.

Skevington, S. M., M. Lotfy, and K. O'Connell, and the WHOQOL Group. 2004. The World Health Organisation's WHOQOL-BREF Quality of Life assessment: Psychometric properties and results of the international field trial. *Quality of Life Research* 13:299–310.

Skevington, S. M., N. Sartorius, M. Amir, and the WHOQOL Group. 2004. Developing methods for assessing QOL in different cultural settings: The history of the WHOQOL instruments. *Social Psychiatry and Psychiatric Epidemiology* 39 (1): 1–8.

Skevington, S. M., and A. Wright. 2001. Changes in the QOL of patients receiving antidepressant medication in primary care: Validating the WHOQOL-100. *British Journal of Psychiatry* 178:261–67.

Stinner, D. H., B. R. Stinner, and E. Martsolf. 1997. Biodiversity as an organizing principle in agro-ecosystem management: Case studies of holistic resource management practitioners in the USA. *Agriculture Ecosystems and Environment* 62 (2–3): 199–213.

Swaminathan, M .S. 2003. Biodiversity: An effective safety net against environmental pollution. *Environmental Pollution* 126 (3): 287–91.

Tabuti, J. R. S., S. S. Dhillion, and K. A. Lye. 2003. Traditional medicine in Bulamogi country, Uganda: Its practitioners, users and viability. *Journal of Ethnopharmacology* 85 (1): 119–29.

Tickell, C. 1992. The quality of life: What quality? Whose life? *Interdisciplinary Science Reviews* 17 (1): 19–25.

Timmermans, K. 2003. Intellectual property rights and traditional medicine: Policy dilemmas at the interface. *Social Science and Medicine* 57 (4): 745–56.

UNEP (United Nations Environment Programme). 1992. *Global biodiversity strategy: Guidelines for action to save, study and use the earths biotic wealth sustainability and equity.* Geneva: World Resources Institute, World Conservation Union.

Von Schirnding, Y. 2002. Health and sustainable development: Can we rise to the challenge? *Lancet* 360 (9333): 632–37.

Wahlquist, M. L. 2003. Regional food diversity and human health. *Asia Pacific Journal of Clinical Nutrition* 12 (3): 304–8.

Walsh, V., and J. Goodman. 1999. Cancer chemotherapy, biodiversity, public and private property: The case of the anti-cancer drug Taxol. *Social Science and Medicine* 49 (9): 1215–25.

Ware, J. E., M. Kosinski, and S. D. Keller. 1994. *SF-36 physical and mental health summary scales: A user's manual.* Boston: The Health Institute, New England Medical Center.

WHOQOL (World Health Organisation Quality of Life) Assessment Group. 1994. Development of the WHOQOL: Rationale and current status. *International Journal of Mental Health* 23 (3): 24–56.

———. 1995. The World Health Organisation's Quality of Life assessment: position paper from the World Health Organisation. *Social Science and Medicine* 41 (10): 1403–1409.

———. 1998a. Development of the WHOQOL-BREF QOL assessment. *Psychological Medicine* 28:551–58.

———. 1998b. The World Health Organisation QOL assessment (WHOQOL): Development and general psychometric properties. *Social Science and Medicine* 46 (12): 1569–85.

PART III
Decay of Ecosystem Services Following Biodiversity Change, and Consequent Impacts on Human Health

In 1997, a report titled *Nature's Services* was among the first to publish a description of the relationship of ecosystem functioning to ecosystem services. While not explicitly aimed at human health, this report developed a list of those services that would decline if ecosystem functions were degraded, including pest control, pollination, fisheries, soil formation, and recycling of waste. More recently, the Millennium Ecosystem Assessment (MEA), an international program, published a several-volume series concerning the consequences of ecosystem changes for human well-being and outlined options for responding to those changes. Looked at from a purely anthropocentric view, ecosystem changes have contributed to substantial net gains in human well-being and economic development. Since 1960, global population has doubled, food production has more than doubled, food prices in much of the world have declined, wood harvest for pulp has tripled, hydropower has doubled, and economic activity has increased sixfold.

While on the surface these achievements appear to be societal gains, they have been achieved at growing environmental costs that, unless addressed, will substantially diminish benefits for future generations. For example, increased population growth and increased agricultural production since 1960 has also led to a doubling of water use globally, which in turn has led to drastic regional water shortages and subsequent loss of agricultural productivity and hydroelectric power on a local scale.

Part III of this book focuses on these effects, synthesizing the evidence from scientific studies that report the effects of changing biodiversity on provisioning of ecosystem services, and how these biodiversity changes, in turn, have ultimately affected human health. Chapters 8, 9, and 10 focus on these relationships in terms of aquatic biodiversity, alterations in the global nitrogen cycle, and microbial biodiversity, respectively. Asit

Mazumder in chapter 8 argues that while extensive research has been done on major areas of water quality, and independently on aquatic biodiversity, very little rigorous quantitative research has been conducted that causally relates the two—that is, data that demonstrates how alterations in aquatic biodiversity affect water quality and in turn, human health.

One exception in which linkages are better understood is the promotion of toxic algal blooms following nitrogen loading in aquatic systems. This is no surprise, as the global nitrogen cycle has been altered more rapidly, and to a greater extent, than any other major biogeochemical cycle. Rising nitrogen inputs have been shown to alter the structure and functioning of not only freshwater ecosystems, but marine and terrestrial ecosystems as well, sometimes causing a cascade of ecological responses that include changes in species abundance and/or existence. Alan Townsend and his coauthors in chapter 9 take a hard look at the evidence that such cascade effects have extended to human diseases and their vectors. Townsend and his colleagues found frequent suggestions in the literature that eutrophication and terrestrial nitrogen loading could lead to increases in many vectors that act as hosts for major diseases, such as malaria, liver fluke, and schistosomiasis, but concluded that the evidence was often based on incomplete correlational data. Specific studies based on rigorous experimental studies are highlighted as exceptions.

Kerstin Hund-Rinke and Anne Winding bring the underlying theme of knowledge gaps to the forefront in chapter 10, where they review what is known about microbial diversity, its alteration by anthropogenic stressors, and the consequent effects on ecosystem services such as contaminant detoxification. They review the challenges in defining "species" in the microbial world, that make it practically impossible to determine the total number of species and the number of individuals of each species in an environmental sample. They argue that it would be a great achievement simply to quantify the biodiversity of microorganisms, much less correlate that diversity with related functions, ecosystem stability, resilience, and respective impacts on human health and well-being. In spite of this gap, many hypotheses abound on the relationships between microbial functional and genetic diversity and nutrient cycling, anthropogenic stress in general, and pollution detoxification in particular. Hund-Rinke and Winding examine these theories in turn, as well as reviewing supporting or conflicting studies, and present suggestions for filling in much-needed knowledge gaps.

8

Consequences of Aquatic Biodiversity for Water Quality and Health

Asit Mazumder

Water is the most precious natural resource needed for sustainable human and ecosystem health, for society, and for the economy. Yet the dependence of the quality of human life on the services provided by aquatic ecosystems in terms of clean water and aquatic resources is still not fully evaluated or appreciated (Chichilnisky and Heal 1998; Davies and Mazumder 2003; Firth 1998; Naiman, Magnison, and Firth, 1998; Policansky 1998). Freshwater ecosystems and their watersheds are being subjected to variable degrees of human-generated pressures, often associated with population growth and the related demand for natural resources. Thus satisfying the demand for clean water by managing and protecting the integrity of ecosystems has become a global challenge for the 21st century. Understanding the abilities and limits of ecosystems to sustain clean and healthy water under changing human and natural interventions is a central issue for society and a challenge for scientists.

Another critical issue for the 21st century is the loss of biodiversity and its consequences for environmental and human health. Biodiversity of organisms in an ecosystem can be described in various ways, such as the genetic variability within a taxon or species, the number of taxa or species, the relative abundance of various species within a community assemblage, and the number of trophic levels or niches. Changes in any one or more of these biodiversity attributes or indicators may have significant consequences for water quality and the health of ecosystems and humans. In some instances, the reduction or disappearance of a single keystone species can cause dramatic changes in the structure, function, and quality of ecosystems (Power et al. 1996). However, a major challenge is to establish and quantify the linkages between biodiversity and water quality, and their consequences for ecosystem and human health. While extensive research has been done on major areas of water quality, and independently on aquatic biodiversity, very little research has been done to relate the two—that is, to document the impact of changes in aquatic biodiversity on water quality. Our knowledge of the extended linkages and feedbacks among aquatic biodiversity, water quality, and human health is even more limited. Some of the potential consequences and feedbacks among aquatic biodiversity, water quality, and health are through excessive concentrations of algae, many of which produce toxins; carcinogenic treatment (chlorination) by-products and taste and odor in potable water; and anoxia and fish kills. Another significant health consequence of changes in aquatic biodiversity and food web structure is through the accumulation of metals and persistent organic pollutants in fish and other traditional aquatic food resources.

This chapter provides a synthesis on how the changes in aquatic biodiversity may affect various criteria of water quality with consequences for ecosystem and human health. Specifically it addresses the following:

- Factors influencing aquatic biodiversity
- Consequences and feedback among aquatic biodiversity, water quality, and human health
- Health consequences of changes in biodiversity and water quality

Factors Influencing Aquatic Biodiversity

Land use for various human needs and related disturbances to plant and animal species, invasive and introduced species, and nutrient enrichments are considered as some of the most important causes of changes in species diversity (Dodson, Arnott, and Cottingham, 2000; Huston 1994; Kennedy et al. 2002; Power et al. 1996; Ritchie, Tilman, and Knoops 1998; Tilman 1999). Many of these disturbances are also associated with natural and anthropogenically modified productivity or enrichments of ecosystems, which have major implications for changes in diversity. A common finding is the productivity-biodiversity relationship across aquatic and terrestrial ecosystems, where the richness or diversity of species increases with increasing productivity (e.g., primary production, algal biomass, and total concentration of nutrients) from low productivity systems to moderate productivity systems, but seems to decline with further increase in productivity (Leibold 1999; Proulx et al. 1996). However, existing literature shows large variability in species richness as a function of ecosystem productivity, suggesting that aquatic biodiversity is regulated by complex interactions among various processes and factors. For example, about 50% of among-lake variability in zooplankton species richness is explained by lake size and lake productivity (Dodson, Arnott, and Cottingham 2000). It is not entirely clear if this pattern exists for or relates to other trophic levels, such as algae and fish (Chase and Leibold 2002; A. Downing and Leibold 2002). Post, Pace, and Hairston (2000) found that the number of trophic levels increases with lake size, which might suggest that species diversity of fish might also increase with ecosystem size (Barbour and Brown 1974). An emerging issue in aquatic diversity, one that will require extensive amounts of research, is how changes in the diversity of species at one trophic level or functional group translates into the diversity of other trophic levels and functional groups, and consequently the diversity, function, and quality of ecosystems.

Anthropogenic enrichment of aquatic ecosystems from increased loading of nutrients (phosphorus and nitrogen) is a global issue (Carpenter et al. 1998; Smith 1998). Once the nutrients from terrestrial or other extrinsic sources enter the aquatic environment, the processing of these materials by various bacterial and algal species results in variable productivity with frequent eutrophication of freshwater and coastal marine ecosystems. While increased phosphorus concentration in water does not have direct health consequence, by enhancing algal concentrations it can indirectly cause significant health consequences for ecosystems and humans (Cox et al. 2005; Lawton and Codd 1991; Martin and Cooke 1994; Smith 1998). On the other hand, increased concentrations of nitrogen in freshwater can have direct health consequences (see this volume, chapter 9).

While the loading of excessive nutrients (phosphorus and nitrogen) can cause major changes to aquatic biodiversity, the specific responses of aquatic ecosystems seem to depend on the physical, chemical, and biological characteristics of the receiving environment (J. Downing and Plante 1993; Horner-Devine et al. 2003; Mazumder 1994; Mazumder et al. 1988; Sarnelle 1993; Trimbee and Prepas 1987). Literature suggests that it is not only the total amount of nutrient loading, but also the composition of nutrients (e.g., the ratio of nitrogen to phosphorus), and the structure of consumer communities, that determine algal species responses. One well-recognized pattern in aquatic systems is the increasing algal biomass with increasing concentrations of N and P (figure 8.1B), but the composition of algae varies greatly with increasing concentrations of nutrients (Proulx and Mazumder 1998; Watson, McCauley, and J. Downing 1997). As the concentration of nutrients or productivity increases, the ratio of total nitrogen to total phosphorus (TN:TP) declines (figure 8.1A; J. Downing and McCauley 1992; Smith 1983). It has been suggested that a decline in TN:TP ratio could be due to differing sources of nutrients in nutrient-poor versus nutrient-rich ecosystems (J. Downing and McCauley 1992). Nutrient-poor ecosystems with high TN:TP ratios tend to occur in relatively undisturbed watersheds, where the diversity of plants and the nutrient retention capacity of soil may be high (reviewed in Tilman 1999). In these ecosystems, chlorophytes, diatoms, and chrysophytes seem to be the dominant groups of algae. In highly disturbed ecosystems with high loading of nutrients, the bloom-forming blue-green algae, especially the ones capable of fixing nitrogen from the atmosphere, become the dominant algal group, and algal diversity declines sharply (J. Downing, Watson, and McCauley 2001; Graham et al. 2004; Smith 1983; Watson, McCauley, and Downing 1997).

Growth- and loss-related processes are important in regulating the productivity, biodiversity, and quality of aquatic ecosystems. Organisms in aquatic ecosystems are composed of myriads of sizes and shapes, which are linked to their success in terms of competing for often limited nutrients or available resources and to their vulnerability to consumers (Carpenter, Kitchell, and Hodgson 1985; Mazumder et al. 1988). While aquatic food webs are extremely complex (figure 8.2), they have been categorized into broad trophic levels or functional groups of primary producers (algae) and bacteria; grazers of algae and bacteria (zooplankton such as cladocerans, copepods, and rotifers);

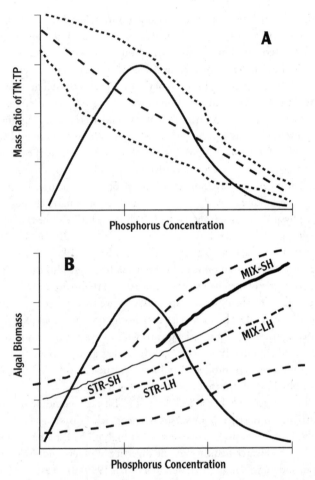

Figure 8.1. Schematic illustration of the unimodal productivity-biodiversity relationships observed in terrestrial and aquatic ecosystems. *A,* The thick dashed line shows the best-fitted line for the relationship between phosphorus (TP) and the ratio of total nitrogen to total phosphorus (TN:TP), with the thin dashed lines showing variability among ecosystems (adapted from J. Downing and McCauley 1992). As the TN:TP ratio declines among ecosystems with high nutrient loading, the potential for blue-green algal blooms increases. *B,* The dashed lines show the variability of algal biomass along the gradient of total phosphorus, and four fitted lines indicate four functionally different types of lake ecosystems: STR-SH = deep or stratified lakes with small herbivore dominance; STR-LH = deep or stratified lakes with large herbivore dominance; MIX-SH = mixed and shallow lakes with small herbivore dominance; and MIX-LH = mixed and shallow lakes with large herbivore dominance (adapted from Mazumder 1994). In both *A* and *B,* the hypothetical productivity-biodiversity curve (the solid line) is superimposed to illustrate that maximum biodiversity may occur (1) in aquatic ecosystems with moderate productivity; (2) perhaps among ecosystems where the functional types of ecosystems shift from shallow to deep water-column depth; and (3) perhaps also among those with highly variable food web structure.

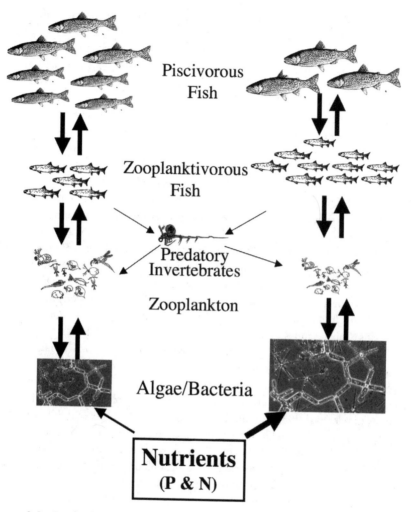

Figure 8.2. Food web structures and productivity in aquatic ecosystems receiving high versus low loading of nutrients. For the simplicity of presentation, the food web structures as shown here are much simpler than are observed among natural ecosystems. The thicker arrow from nutrients to algae/bacteria indicates higher nutrient loading; the thinner arrow, lower nutrient loading. The downward-pointing arrows between trophic levels illustrate consumer or predation effects (from fish and predatory invertebrates); the upward-pointing arrows, the transfer of nutrients, energy, and contaminants along the aquatic food web. This figure provides a conceptual framework of how the changes in the structure or diversity of each of the major trophic levels and changes in the length of the food chain due to exploitation of higher trophic levels, or to introduction/invasion by alien species, may modify the structure, function, and quality of aquatic ecosystems.

zooplanktivores (invertebrates and small fish) preying on zooplankton; and piscivores (large fish). Changes in the composition, diversity, and biomass at each of the trophic levels have been shown to cause major changes in the overall function and quality of aquatic ecosystems. For example, increased loading of nutrients (N and P), through enhanced growth and biomass of algae, translates into increased biomass and productivity of higher trophic levels. On the other hand, changes in the composition and biomass of higher trophic levels have also been shown to have cascading effects on the lower trophic levels. For example, a reduction of large predatory fish through overfishing and habitat loss allows their prey, the large invertebrates and small fish, to increase in density and biomass. Another cause of changes in the predation on lower trophic levels has been the invasion by large invertebrate predators like *Bythotrephes* and *Mysis* (Yan, Girard, and Boudreau 2002). With increased abundance of large invertebrates and small fish and the associated increase in the intensity of predation on large zooplankton grazers of algae and bacteria, the concentrations of algae and bacteria increase, causing water quality deterioration. Besides causing water quality deterioration through production of excessive algae and bacteria, changes in the structure of the aquatic food web in terms of the biomass at each trophic level and the length of the food chains have been shown to alter the accumulation of carcinogenic and neurogenic contaminants, such as persistent organic pollutants and heavy metals in higher trophic levels that are important sources of traditional food for humans (Cabana and Rasmussen 1994, 1996; Rasmussen et al. 1990).

While an extensive set of experimental and empirical research showed nutrients and their composition as important regulators of biodiversity in aquatic ecosystems, the last couple of decades of research have shown that diversity of aquatic ecosystems is driven by strong interactions between nutrients and consumers (Leibold 1989, 1999; Mazumder 1994; Mazumder et al. 1988; Proulx and Mazumder 1998; Proulx et al. 1996; Worm et al. 2002). A general agreement is that there are strong interactions between nutrient loading and food web structure in determining ecosystem responses and associated patterns of algal biomass and composition, toxic algal blooms, and accumulation of contaminants in higher trophic levels (Burkholder et al. 1997; Buskey et al. 1997; Kainz and Mazumder 2005; Pickhardt et al. 2005; Proulx et al. 1996; Rasmussen et al. 1990; Ridal, Mazumder and Lean 2001; Watson, McCauley, and J. Downing 1997).

Although extensive research has been done on the impacts of nutrients and food web structure on trophic-level biomass and composition and water quality, only a few attempts have been made to link nutrient–food web interactions with the biodiversity of aquatic ecosystems. By starting with a similar diversity of algal community, Proulx et al. (1996) demonstrated that individual and interactive effects of nutrient loading and grazing produced the hump-shaped productivity-diversity relationship shown in figure 8.3. A further synthesis of results from 30 experimental and empirical studies covering freshwater lakes and streams, coastal marine environments, and terrestrial ecosystems (Proulx and Mazumder 1998) found that in nutrient-poor systems, plant diversity declines with intense grazing, while in nutrient-rich systems, diversity increases with intense grazing pressure. It was proposed that in nutrient-poor systems, grazing suppresses diversity

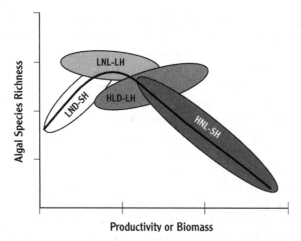

Figure 8.3. How nutrients and food web structure modify algal species richness under four contrasting nutrient–food web scenarios among north temperate aquatic ecosystems (adapted from Proulx and Mazumder 1998; Proulx et al. 1996).

because a limitation of resources prevents regrowth of species in systems with intense grazing pressure. On the other hand, in nutrient-rich systems, the inedible species probably take advantage of available resources and increase in dominance (Buskey et al. 1997; Leibold 1999; Proulx et al. 1996; Sarnelle 1993). It can be suggested that aquatic ecosystems, with a high loading of nutrients and a dominance of small and inefficient grazers, are most vulnerable to loss of diversity, formation of algal blooms, and production of toxins and taste and odor. As some of the pathogenic organisms, such as fecal coliform bacteria and protozoa (e.g., *Giardia* and *Cryptosporidium*), fall within the range of particles grazed by invertebrates like cladocerans, copepods, and insects (Mazumder et al. 1990b), it may be possible that aquatic ecosystems with a low density of efficient grazers may also produce more frequent outbreaks of pathogens of significant human and animal health consequences. As aquatic ecosystems face increased anthropogenic inputs of nutrients, and as the changes in food web structure due to fisheries exploitation, species invasion, and habitat degradation become more common, the changes in one or more of these disturbances or stressors will lead to significant changes in aquatic biodiversity, with considerable implications for water quality and health (Jeppesen et al. 2000; Leibold 1999; Proulx and Mazumder 1998).

Consequences and Feedbacks among Aquatic Biodiversity, Water Quality, and Human Health

Before we discuss linkages between biodiversity and water quality, and their consequences for ecosystem and human health, it is important to define the various criteria for water quality used for different purposes. There are various indicators of water quality, depending on the services provided by the aquatic ecosystems of reservoirs, lakes, rivers, streams, the marine environment, and aquifers. The aesthetic qualities of water, such as algal biomass, water transparency, and color of water, determine the desire of the public to use a water body for recreation and pleasure. From a human and ecosystem health perspective, the indicators of water quality are the concentrations of bacterial and protozoan pathogens, algal toxins, taste and odor, organic carbon, contaminants in aquatic organisms consumed by humans, and the quality of habitats required by organisms.

Algal biomass (concentration of chlorophyll) and total organic carbon (TOC) are the two important bulk indicators of water quality because they closely relate to ecosystem and human health. High algal biomass and TOC are often associated with oxygen depletion and associated loss of habitats for invertebrates and fish, and with increased recycling of nutrients within aquatic ecosystems. As the biomass of algae and TOC increases, the transparency of water and associated penetration of light to deeper water decline rapidly, resulting in reduced production of oxygen with increasing depth (reduced primary production in deeper water) and loss of oxygen through decomposition of sedimentary materials (Mazumder et al. 1990a). These may lead to disappearance of certain fish species, especially the cold-water species that require deeper and cooler waters of lakes with high levels of oxygen (Burgi and Stadelmann 2002; Lindholm et al. 1999; Ochumba 1990; Starling et al. 2003). Certain zooplankton species with low tolerance for depleted oxygen levels may be restricted to the upper water column and thus become more vulnerable to predation from fish and large invertebrates. However, the human health impacts of high algal biomass and TOC are related to disinfection of water for human consumption. When water with a high concentration of carbon is treated with chlorine or chloramine, high levels of disinfection by-products, such as various types of chlorinated hydrocarbons, are produced, many of which are carcinogenic to humans.

Other water quality indicators that have more significant human heath consequences are the concentrations of specific species of algae, bacteria, and protozoa. For example, certain species of algae and bacteria produce taste and odor compounds in drinking water (Davies, Roxborough, and Mazumder 2003; Furtula, Davies, and Mazumder 2004; Watson 2003; Watson and Stachwill 2003), thus making water less desirable to consumers. In eutrophic systems, several blue-green taxa (e.g., *Microcystis, Anabena, Aphanizomenon, Oscillatoria*) are the common sources of taste and odor in drinking water, while in oligotrophic systems, the chrysophyte taxa are the major sources of taste and odor.

One of the direct health consequences of changes in aquatic biodiversity is the dominance of blue-green algae (cyanobacteria) in freshwater and estuarine environments worldwide. The consequences of toxins produced by blue-green algae for humans through consumption of food and water are of increasing global concern as increased nutrient loading and associated eutrophication are increasing the extent and frequency of blue-green algal blooms. There have been many serious cases of human and livestock fatalities and other health consequences due to blue-green algal blooms during the last three decades in Australia, Brazil, and the United States (Burkholder et al. 1997; Carmichael 1994; reviewed in Cox et al. 2005; Glasgow et al. 1995; Jacoby et al. 2000; Yin et al. 1999; see also this volume, chapter 3). Blue-green algal blooms are observed in highly disturbed aquatic systems with high nutrient loading and reduced biodiversity (figures 8.1 and 8.3). Many records of human, livestock, and wildlife fatalities are caused by ingestion of water containing algal toxins in both freshwater and marine environments. In coastal marine ecosystems, increased anthropogenic loading of wastes and nutrients has been found to cause blooms of harmful organisms like *Pfiesteria*, *Chattonella*, and *Gymnodinium*, which cause severe health impacts to fish, shellfish, and humans (e.g., Burkholder et al. 1997; Glasgow, Burkholder, Mallin, et al. 2001; Glasgow, Burkholder, Schmechel, et al. 1995; Heil et al. 2001; Yin et al. 1999). In freshwater systems, some of the bloom-forming taxa, which also produce toxins harmful to humans and livestock, are *Microcystis, Anabaena, Aphanizomenon,* and *Oscillatoria*. While we know that excessive nutrients are one of the important causes of bloom formation by these algae, we still do not understand what types of lakes are more prone to bloom formation, and what factors trigger toxin production by these algae. More recently, Cox et al. (2005) identified beta-N-methylamino-L-alanine (BMAA) as a neurotoxic amino acid produced by diverse taxa of blue-green algae from freshwater and marine ecosystems, and as symbiotic blue-green algae in terrestrial flowers and fruits. This BMAA bioaccumulates along aquatic and terrestrial food chains and can produce neurodegenerative symptoms. One of the case studies exemplifying this linkage is that of the Chamorro people in Guam, who show an extremely high rate of neurodegenerative symptoms (Cox, Banack, and Murch 2003; Cox et al. 2005). In addition, blue-green algal blooms are often followed by severe oxygen depletion and massive fish kills (Burgi and Stadelmann 2002; Lindholm et al. 1999; Ochumba 1990; Starling et al. 2003).

Pathogen-related human health consequences are a serious global issue (Davies and Mazumder 2003; Epstein 2001), and it is estimated that over 5 million people, especially children and immunocompromised individuals, die every year from gastrointestinal disease worldwide. An emerging threat to global health is associated with intensive livestock farming along watercourses, which not only degrades biodiversity in the terrestrial ecosystem through grazing of vegetation but also increases loading of nutrients, contaminants, and pathogens to aquatic ecosystems (Buskey et al. 1997; Fisher et al. 2000; Fraser, Barten, and Pinney 1998; Tian et al. 2002). Loss of terrestrial biodiversity, especially of riparian habitats, leads to increasing loading of pathogenic bacteria (*E. coli*) and protozoa (*Giardia, Cryptosporidium*) to surface water, resulting in contaminated drinking water and shellfish. One clear linkage between aquatic biodiversity and distribution of pathogens in water is through the association between species of algae and zooplankton as a substrate for *Vibrio cholerae*, although this association and outbreaks of cholera need more extensive research (reviewed in Cottingham, Chiavelli, and Taylor 2003). Furthermore, far more research is needed to establish linkages and feedbacks among land use, aquatic biodiversity, water quality, and the distribution of waterborne pathogens.

Yet another significant health consequence of changes in aquatic biodiversity is through accumulation of persistent pollutants and heavy metals in fish and their consumption by humans. Among the persistent contaminants in aquatic systems, mercury (in methylated form) is considered to be a serious human health risk in terms of neurobehavioral deficits and motor performance (Mergler 2002). At least 85% of mercury in fish comes from their diet—that is, from feeding on lower trophic levels, and almost all (99%) mercury in fish is methyl mercury (MeHg). While the processes leading to the bioaccumulation of persistent organic pollutants (POPs) and heavy metals are quite different, the structure, composition, and thus the diversity of aquatic food webs are important determinants of these contaminants in higher trophic levels, which are important sources of diet for humans (deBruyn et al. 2007). As the exploitation of fisheries, invasion by new species, and loading of nutrients change the structure and function of aquatic food webs (figure 8.2; Mazumder 1994; Vander Zanden, Casselman and Rasmussen 1999), the concentrations of POPs and MeHg in predatory fish increase to levels threatening to human health (Cabana and Rasmussen 1994; Rasmussen et al. 1990). Others used zooplankton size to express planktonic food web structure, and found that MeHg in zooplankton increases with increasing body size (Kainz, Arts and Mazumder 2004). Existing studies have shown that the concentration of MeHg in zooplankton (Pickhardt et al. 2002) from oligotrophic lakes is higher than in zooplankton of an equivalent trophic level in eutrophic lakes. This reduced level of methyl mercury per unit biomass of zooplankton may be due to the dilution effect of the algal blooms (Pickhardt et al. 2002). A similar relationship between trophic status and accumulation of POPs in higher trophic levels has been shown by Ridal, Mazumder, and Lean (2001). One general observation is that the accumulation of contaminants in fish increases with increasing length of the food chain (Rasmussen et al. 1990). The lengthening of the aquatic food chain through invaded or introduced species, as it has been shown for the North American Great Lakes and smaller temperate lakes, causes the concentrations of MeHg and persistent organic pollutants in the top predators to increase significantly (Cabana and Rasmussen 1994, 1996; Rasmussen et al. 1990). Thus the existing research shows that the productivity of ecosystems and the structure and diversity of the food web are strongly linked to the concentrations of contaminants in fish and other aquatic resources, with significant consequences for human health.

Health Consequences of Changes in Biodiversity and Water Quality

A major challenge in the 21st century is to protect and manage the biological diversity and integrity of aquatic ecosystems for sustainable clean and healthy water and for healthy aquatic resources for global communities. This review of the current literature shows that only very limited research, terrestrial or aquatic, has focused on the implications of biodiversity changes for water quality, and that even less research has focused on the consequences of aquatic biodiversity for health. However, this synthesis of the available literature shows that there is significant feedback among anthropogenic changes to water and watershed; loading of nutrients, contaminants, and pathogens; water quality; and human and ecosystem health. Some of the emerging patterns of the association between changes in aquatic biodiversity, water quality, and health consequences are through increasing incidence of harmful algal blooms, waterborne pathogens, and contaminants in aquatic systems. As this is the first attempt to synthesize these linkages and feedbacks, and as not many quantitative results are available, the established patterns and associations are presented as conceptual frameworks, intended to generate further research questions and hypotheses relevant to ecosystem and human health.

Research Directions

Based on this review, the following emerged as the priority areas of research needed to establish more robust linkages and feedbacks among aquatic biodiversity, water quality, and their consequences for human health and well-being:

- Impacts of land and water use on nutrient loading, aquatic biodiversity, and water quality
- Consequences of changes in the structure and diversity of aquatic food webs through increased nutrient loading, exploitation of aquatic resources, and species invasion for water quality and human ecosystem health
- Factors and processes triggering toxin production by algae
- Consequences of changes in biodiversity in adjacent trophic levels for ecosystem function and water quality
- Quantitative relationships between aquatic biodiversity and the dynamics of pathogenic organisms in aquatic ecosystems

References

Barbour, C. D., and J. H. Brown. 1974. Fish species diversity in lakes. *American Naturalist* 108:473–89.

Burgi, H., and P. Stadelmann. 2002. Change of phytoplankton composition and biodiversity in Lake Sempech before and during restoration. *Hydrobiologia* 469:33–48.

Burkholder, J. M., M. A. Mallin, H. B. Glasgow Jr., L. M. Larsen, M. R. McIver, G. C. Shank, N. Deamer-Melia, et al. 1997. Impacts to a coastal river and estuary from rupture of a large swine waste holding lagoon. *Journal of Environmental Quality* 26:1451–66.

Buskey, E. J., P. A. Montagna, A. F. Amos, and T. E. Whitledge. 1997. Disruption of grazer populations as a contributing factor to the initiation of the Texas brown tide algal bloom. *Limnology and Oceanography* 42:1215–22.

Cabana, G., and J. B. Rasmussen. 1994. Modeling food-chain structure and contaminant bioaccumulation using stable nitrogen isotope. *Nature* 372:255–57.

———. 1996. Comparison of aquatic food chains using nitrogen isotopes. *Proceedings of the National Academy of Sciences* 93:10844–47.

Carmichael, W. W. 1994. The toxins of cyanobacteria. *Scientific American* 270:78–86.

Carpenter, S. R., N. F. Caraco, D. L. Correll, R. W. Howarth, A. N. Sharpley, and V. H. Smith. 1998. Nonpoint pollution of surface waters with phosphorus and nitrogen. *Ecological Applications* 8:559–68.

Carpenter, S. R., J. F. Kitchell, and J. R. Hodgson. 1985. Cascading trophic interactions and lake productivity. *BioScience* 35:634–39.

Chase, J. M., and M. A. Leibold. 2002. Spatial scale dictates the productivity-biodiversity relationship. *Nature* 416:427–30.

Chichilnisky, G., and G. Heal. 1998. Economic returns from the biosphere: Commentary. *Nature* 391:629–30.

Cottingham, K. L., D. A. Chiavelli, and R. K. Taylor. 2003. Environmental microbe and human pathogen: The ecology and microbiology of *Vibrio Cholerae*. *Frontiers in Ecology and Environment* 1:80–86.

Cox, P. A., S. A. Banack, and S. J. Murch. 2003. Biomagnification of cyanobacterial neurotoxins and neurodegenerative disease among the Chamorro people of Guam. *Proceedings of the National Academy of Sciences* 100:13380–83.

Cox, P. A., S. A. Banack, S. J. Murch, U. Rasmussed, G. Tien, R. R. Bidigare, J. S. Metcalf, et al. 2005. Diverse taxa of cyanobacteria produce -N-methylamino-L-alanine, a neurotoxic amino acid. *Proceedings of the National Academy of Sciences* 102:5074–78.

Davies, J.-M., and A. Mazumder. 2003. Health, and environmental policy issues in Canada: The role of watershed management in sustaining clean drinking water quality at surface sources. *Journal of Environmental Management* 98:399–414.

Davies, J.-M., M. Roxborough, and A. Mazumder. 2004. Origins and implications of drinking water odor in British Columbia lakes and reservoirs. *Water Research* 38:1900–1910.

deBruyn, A. M. H., M. Trudel, N. Eyding, J. Harding, H. McNally, R. Mountain, C. Orr, et al. Ecosystemic effects of salmon farming increase mercury contamination in fish. *Environmental Science and Technology* 40:3489–83.

Dodson, S. I., S. E. Arnott, and K. L. Cottingham. 2000. The relationship in lake communities between primary productivity and species richness. *Ecology* 81:2662–79.

Downing, A. L., and M. A. Leibold. 2002. Ecosystem consequences of species richness and composition in pond food webs. *Nature* 416:837–41.

Downing, J. A., and E. McCauley. 1992. The nitrogen-phosphorus relationship in lakes. *Limnology and Oceanography* 37:936–45.

Downing, J. A., and C. Plante. 1993. Production of fish populations in lakes. *Canadian Journal of Fisheries and Aquatic Sciences* 50:110–20.

Downing, J. A., S. B. Watson, and E. McCauley. 2001. Predicting cyanobacteria dominance in lakes. *Canadian Journal of Fisheries and Aquatic Sciences* 58:1905–8.

Epstein, P. R. 2001. Climate change and emerging infectious disceases. *Microbes and Infection* 3:747–54.

Firth, P. L. 1998. Fresh water: Perspectives on the integration of research, education and decision making. *Ecological Applications* 8:601–9.

Fisher, D. S., J. L. Steiner, D. M. Endale, J. A. Stuedemann, H. H. Schomberg, A. J. Franzluebbers, and S. R. Wilkinson. 2000. The relationship of land use practices to surface water quality in the Upper Oconee Watershed of Georgia. *Forest Ecology and Management* 128:39–48.

Fraser, R. H., P. K. Barten, and D. A. K. Pinney. 1998. Predicting stream pathogen loading from livestock using a geographical information system–based delivery model. *Journal of Environmental Quality* 27:935–45.

Furtula, V., J.-M. Davies, and A. Mazumder. 2004. An automated headspace SPME-GC-ITMS technique for taste and odor compound identification. *Water Quality Research Journal of Canada* 39:213–22.

Glasgow, H. B., J. M. Burkholder, M. A. Mallin, N. J. Deamer-Melia, and R. E. Reed. 2001. Field ecology of *Pfiesteria* complex species and a conservative analysis of their role in estuarine fish kills. *Environmental Health Perspectives* 109:715–30.

Glasgow, H. B., J. M. Burkholder, D. E. Schmechel, P. A. Tester, and P. A. Rublee. 1995. Insidious effects of a toxic estuarine dinoflagellate on fish survival and human health. *Journal of Toxicology and Environmental Health* 46:501–22.

Graham, J. L., J. R. Jones, S. B. Jones, J. A. Downing, and T. E Clevenger. 2004. Environmental factors influencing microcystin distribution and concentration in the midwestern United States. *Water Research* 38:4395–4404.

Heil, C. A., P. M. Gilbert, M. A. Al-Sarawl, M. Faraj, M. Behbehani, and M. Hussain. 2001. First record of a fish-killing *Gymnodinium* sp. bloom in Kuwait Bay, Arabian Sea: Chronology and potential causes. *Marine Ecology Progress Series* 214:15–23.

Horner-Devine, M. C., M. A. Leibold, V. H. Smith, and B. M. J. Bohannan. 2003. Bacterial diversity patterns along a gradient of primary productivity. *Ecology Letters* 6:613–22.

Huston, M. A. 1994. *Biological diversity: The coexistence of species in changing landscapes.* Cambridge: Cambridge University Press.

Jacoby, J. M., D. C. Collier, E. B. Welch, F. J. Hardy, and M. Crayton. 2000. Environmental factors associated with a toxic bloom of *Microcystis aeruginosa. Canadian Journal of Fisheries and Aquatic Sciences* 57:231–40.

Jeppesen, E., J. P. Jensen, M. Sondergaard, and T. Lauridsen. 2000. Trophic structure, species richness and biodiversity in Danish lakes: Changes along phosphorus gradient. *Freshwater Biology* 45:201–18.

Kainz, M., M. Arts, and A. Mazumder. 2004. Essential fatty acids in the planktonic food web and their ecological role for higher trophic levels. *Limnology and Oceanography* 49:1784–93.

Kainz, M., and A. Mazumder. 2005. The effects of algal and bacterial diet on methyl mercury concentrations in lake zooplankton. *Environmental Science and Technology* 39:1666–72.

Kennedy, T. A., S. Naeem, K. M. Howe, J. M. H. Knoops, D. Tilman, and P. Reich. 2002. Biodiversity as barrier to ecological invasion. *Nature* 417:636–38.

Lawton, L. A., and G. A. Codd. 1991. Cyanobacterial (blue-green algal) toxins and their significance in UK and European waters. *Journal of the Institute of Water and Environment Management* 5:460–65.

Leibold, M. A. 1989. Resource availability and the effects of predators and productivity on consumer-resource interactions. *American Naturalist* 134:922–49.

———. 1999. Biodiversity and nutrient enrichment in pond plankton communities. *Evolutionary Ecology Research* 1:73–95.

Lindholm, T., P. Ohman, K. Kurki-Helasmo, B. Kincaid, and J. Meriluoto. 1999. Toxic algae and fish mortality in a brackish-water lake in Angstrom Land, SW Finland. *Hydro-biologia* 397:109–20.

Martin, A., and G. D. Cooke. 1994. Health risks in eutrophic water supplies. *Lake Line* 14:24–26.

Mazumder, A. 1994. Phosphorus-chlorophyll relationships under contrasting herbivory and thermal stratification: Patterns and predictions. *Canadian Journal of Fisheries and Aquatic Sciences* 51:390–400.

Mazumder, A., D. J. McQueen, W. D. Taylor, and D. R. S. Lean. 1988. Effects of fertilization and planktivorous fish (yellow perch) predation on size distribution of particulate phosphorus and assimilated phosphate: Large enclosure experiments. *Limnology and Oceanography* 33:421–30.

———. 1990a. Pelagic food web interactions and hypolimnetic oxygen depletion: Results from experimental enclosures and lakes. *Aquatic Sciences* 52:143–55.

———. 1990b. Micro- and mesozooplankton grazing on natural pico- and nanoplankton in contrasting plankton communities produced by planktivore manipulation and fertilization. *Archiv für Hydrobiologie* 118:257–82.

McCauley, E., J. A. Downing, and S. Watson. 1989. Sigmoid relationships between nutrients and chlorophyll among lakes. *Canadian Journal of Fisheries and Aquatic Sciences* 46:1171–75.

Meng, F. R., C. P. A. Bourque, K. Jewett, D. Daugharty, and P. A. Arp. 1995. The Nash-waak Experimental Watershed Project: Analyzing effects of clearcutting on soil temperature, soil moisture, snowpack, snowmelt and stream flow. *Water, Air, and Soil Pollution* 82:363–74.

Mergler, D. 2002. Review of neurobehavioral deficits and river fish consumption from the Tapajo's (Brazil) and St. Lawrence (Canada). *Environmental Toxicology and Pharmacology* 12:93–99.

Naiman, R. J., J. J. Magnison, and P. L. Firth. 1998. Integrating culture, economic and environmental requirements for fresh water. *Ecological Applications* 8:569–70.

Ochumba, P. B. O. 1990. Massive fish kills within the Nyanza Gulf of Lake Victoria, Kenya. *Hydrobiologia* 208:93–99.

Policansky, D. 1998. Science and decision making for water resources. *Ecological Applications* 8:610–18.

Post, D. M., M. L. Pace, and N. G. Hairston. 2000. Ecosystem size determines food-chain length. *Nature* 405:1047–49.

Power, M. E., D. Tilman, J. A., Estes, B. A. Menge, W. J. Bond, L. S. Mills, G. Daily, J. C. Castilla, J. Lubchenco, and R. T. Paine. 1996. Challenges in quest for keystones. *BioScience* 46:609–20.

Pickhardt, P. C., C. L. Folt, C. Y. Chen, and J. D. Blum. 2005. Impacts of zooplankton composition and algal enrichment on the accumulation of mercury in an experimental freshwater food web. *Science of the Total Environment* 339:89–101.

Proulx, M., and A. Mazumder. 1998. Grazer reversal of plant species richness under contrasting nutrient richness. *Ecology* 79:2581–92.

Proulx, M., F. Pick, A. Mazumder, P. Hamilton, and D. R. S. Lean. 1996. Experimental evidence for interactive impacts of nutrients and herbivores on algal diversity. *Oikos* 76:191–95.

Rasmussen, J. B., D. J. Rowen, D. R. S. Lean, and J. H. Carey. 1990. Food-chain structure in Ontario lakes determines PCB levels in lake trout (*Silvenius namaycush*) and other pelagic fish. *Canadian Journal of Fisheries and Aquatic Sciences* 47:2030–38.

Ridal, J., A. Mazumder, and D. Lean. 2001. Effects of nutrient loading and planktivory on the accumulation of pesticides in aquatic foodchains. *Environmental Toxicology and Chemistry* 20:1312–19.

Ritchie, M. E., D. Tilman, and J. M. H. Knoops. 1998. Herbivore effects on plant and nitrogen dynamics in oak savanna. *Ecology* 79:165–77.

Rosenzwick, M. L. 1995. *Species diversity in space and time.* Cambridge: Cambridge University Press.

Sarnelle, O. 1993. Herbivore effects on phytoplankton succession in a eutrophic lake. *Ecological Monographs* 63:129–49.

Smith, V. H. 1983. Low nitrogen to phosphorus ratios favor dominance of blue-green algae in lake phytoplankton. *Science* 221:669–71.

———. 1998. Cultural eutrophication of inland, estuarine and coastal waters. In *Successes, limitations, and frontiers in ecosystem science,* ed. M. L. Pace and P. M. Goffman, 7–49. New York: Springer-Verlag.

Starling, F., X. Lazzaro, C. Cavalcanti, and R. Moreira. 2003. Contribution of omnivorous tilapia to eutrophication of a shallow tropical reservoir: Evidence from a fish kill. *Freshwater Biology* 47:2443–52.

Tian, Y. Q., P. Gong, J. D. Radke, and J. Scarborough. 2002. Spatial and temporal modeling of microbial contaminants on grazing farmlands. *Journal of Environmental Quality* 31:860–69.

Tilman, D. 1999. The ecological consequences of changes in diversity: A search for general principles. *Ecology* 80:1455–74.

Trimbee, A. M., and E. E. Prepas. 1987. Evaluation of total phosphorus as a predictor of the relative biomass of blue-green algae with emphasis on Alberta lakes. *Canadian Journal of Fisheries and Aquatic Sciences* 44:1337–42.

Vander Zanden, M. J., J. M. Casselman, J. B. Rasmussen. 1999. Stable isotope evidence for the food web consequences of species invasions in lakes. *Nature* 401:464–67.

Watson, S. B. 2003. Cyanobacterial and eukaryotic algal odor compounds: Signals or by-products? A review of their biological activity. *Phycologia* 42:332–50.

Watson, S. B., E. McCauley, and J. A. Downing. 1997. Patterns of phytoplankton taxonomic composition across temperate lakes of differing nutrient status. *Limnology and Oceanography* 42:139–49.

Watson, S. B., and T. Stachwill. 2003. Chrysophyte odor production: Resource-mediated changes at the cell and population levels. *Phycologia* 42:393–405.

Worm, B., H. K. Lotze, H. Hillebrand, and U. Sommer. 2002. Consumer versus resource control of species diversity and ecosystem functioning. *Nature* 41:848–51.

Yan, N. D., R. Girard, and S. Boudreau. 2002. An introduced invertebrate predator (Bythotrephes) reduces zooplankton species richness. *Ecology Letters* 5:481–85.

Yin, K. D., P. J. Harrison, J. Chen, W. Huang, and P. Y. Qian. 1999. Red tides during spring 1998 in Hong Kong: Is El Niño responsible? *Marine Ecology Progress Series* 187:289–94.

Zhu, Z., P. A. Arp, A. Mazumder, F. Meng, C. P. A. Borque, and N. W. Foster. 2005. Modeling stream water nutrient concentrations and loading in response to weather condition and forest harvesting. *Ecological Modeling* 185:231–43.

9

The Global Nitrogen Cycle, Biodiversity, and Human Health

Alan R. Townsend, Luiz A. Martinelli, and Robert W. Howarth

Natural patterns in global biodiversity and their current human-driven changes are intimately linked to the fundamental ecosystem processes of carbon, water, and nutrient cycling (Chapin et al. 1997, 2000; Hooper et al. 2005; Tilman 1996). The species composition of an ecosystem can affect, among other properties, hydrology and climate (Foley et al. 2003; Gordon and Rice 1992); the availability and fluxes of major nutrients (Hooper et al. 2005; Van Cleve et al. 1991; Vitousek et al. 1997); the structure and fertility of soils (Wardle et al. 2004); and total ecosystem carbon stocks (Christian and Wilson 1999). In turn, exogenous alteration of these variables, such as an increase in the supply of a limiting nutrient, can drive changes in species composition and distribution (Bobbink, Hornung, and Roelofs 1998; Stevens et al. 2004; Tartowski and Howarth 2000). Thus any evaluation of the ties between biodiversity and human health must also consider the role of ecosystem processes.

In particular, it is essential to understand the relationships between biodiversity and the nitrogen cycle, for two reasons. First, nitrogen is a limiting nutrient in a wide variety of Earth's ecosystems (Vitousek and Howarth 1991), and thus the structure and function of such ecosystems are regulated by nitrogen supply. Second, human activities are changing nitrogen availability at a pace and scale that is unprecedented in Earth's history (Galloway et al. 2004; Howarth 2003). Nitrogen fluxes through many regions of the world have increased by 10-fold or more over the past century, and these changes have already caused shifts in biodiversity in terrestrial, freshwater, and marine ecosystems (NRC 2000; Tartowski and Howarth 2000; Stevens et al. 2004; Vitousek et al. 1997). Thus this chapter will first discuss links between biodiversity and the changing nitrogen cycle, and then outline ways in which an altered nitrogen cycle can affect human health, with an emphasis on the ecological feedbacks that involve biodiversity.

Biodiversity and the Nitrogen Cycle

The majority of the atmosphere is nitrogen (N), but in the form of N_2 gas, which is available to only a taxonomically diverse group of microorganisms that are capable of "fixing" N_2 into reduced forms. Prior to the past few decades, such biological N fixation was the dominant source of new N inputs to Earth's ecosystems, a flux that is far below biotic demand (Cleveland et al. 1999; Vitousek et al. 2002). For this and other reasons, absent human disturbance, N limitation of primary production is widespread in terrestrial ecosystems, as well as in numerous freshwater and marine ecosystems (Vitousek and Howarth 1991). Low N availability not only can limit primary production but can also affect species composition; for example, in N-poor terrestrial systems, the dominant plant species often have relatively low tissue N concentrations. Low-N tissues tend to be low-quality food, and thus can affect the species composition of herbivores, with cascading consequences for the species that depend on those herbivores (Coley 1998; Throop and Lerdau 2004); in turn, herbivory itself can alter N cycling, with consequences for species composition (Belovsky and Slade 2000). As well, low-N plant tissues can retard decomposition (Melillo et al. 1989), creating feedbacks that help maintain the ecosystem in a nitrogen-poor state (Vitousek 1982). However, if nitrogen availability increases—for example, due to increasing N fixation—plant species with greater nitrogen requirements but also higher growth rates and reproductive output can gain a competitive advantage over those adapted to N-poor conditions. Once established, the more N-rich tissues of such plants can drive increased decomposition rates that further elevate N availability (Vitousek 1982). Thus any change to an N-limited ecosystem that improves N availability is not only likely to increase primary production but may also drive changes in species composition at multiple levels of organization. For example, in coastal marine ecosystems, increased nitrogen inputs lead predictably to both increased primary production and greatly reduced diversity (NRC 2000). In the United States, this increased nitrogen loading has now led to the degradation of two-thirds of coastal rivers and bays (NRC 2000).

Increases in N availability can and do occur naturally in any ecosystem type (Vitousek and Howarth 1991), but in the modern world, humans have become the dominant cause of such change across a substantial fraction of Earth's surface (Galloway et al. 2003, 2004; Vitousek et al. 1997). Human activities now fix atmospheric N_2 into biologically available forms on a scale comparable to all natural N_2 fixation in the planet's terrestrial ecosystems (figure 9.1; Galloway et al. 2004; Vitousek et al. 1997). The rate of change is incredible, and the past 15 years have seen the creation and use of half of the synthetic N fertilizer in Earth's history (Howarth 2004; Howarth et al. 2002). While synthetic fertilizer production is the single largest alteration of the nitrogen cycle by human activity, society also increases nitrogen availability through fossil fuel combustion and the widespread cultivation of leguminous crop species (Galloway et al. 2004) and redistributes nitrogen widely over the Earth's surface through shipments of foods and feedstocks

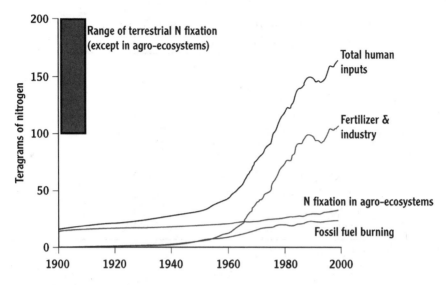

Figure 9.1. Global trends in the creation of reactive nitrogen from human activity. Units are in teragrams of nitrogen per year (TgN/yr). Adapted from Howarth et al. 2005.

(Howarth 2004; Howarth et al. 2002; Howarth, Sharpley, and Walker 2002). Taken as a whole, the global N cycle has been altered more rapidly, and to a greater extent, than any other major biogeochemical cycle (Howarth et al. 2002; Vitousek et. al. 1997); consider, for example, that anthropogenic N fixation has more than doubled in less than half a century, a time period over which atmospheric carbon dioxide (CO_2) increased less than 10%.

Moreover, unlike the rise in atmospheric CO_2, global changes in the N cycle are not evenly distributed but rather are concentrated in areas of intensive agricultural and industrial activity. It is here that large amounts of N are applied to agricultural fields, and the greatest quantities of N_2 are fixed during fossil fuel combustion. Much of the newly reactive N is transported in both air and water, causing substantial increases in N inputs to ecosystems across large regions. For example, riverine N inputs to the coastal oceans now exceed preindustrial fluxes by 10- to 15-fold in many parts of North America, Europe, and Asia (table 9.1); similar relative increases are documented for atmospheric N deposition (Galloway, Levy, and Kashibhatla 1994; Holland et al. 2005; Howarth et al. 1996).

The resultant environmental effects are widespread, are at times severe, and include a number of ways in which biodiversity can be directly or indirectly affected. First, as described above, N enrichment of traditionally N-limited ecosystems can lead to a cascade of changes that results in the loss of species adapted to low-N conditions (Bobbink et al. 1998; Bowman 2000; NRC 2000; Stevens et al. 2004), and that often appears to

Table 9.1. Human-driven increase in nitrogen fluxes from rivers to coastal oceans for several world regions

Region	Change
Labrador & Hudson's Bay	no change
Southwestern Europe	3.7-fold
Great Lakes/St. Lawrence basin	4.1-fold
Baltic Sea watersheds	5.0-fold
Mississippi River basin	5.7-fold
Yellow River basin	10-fold
Northeastern United States	11-fold
North Sea watersheds	15-fold
Republic of Korea	17-fold

Adapted from Boyer et al. 2006.

drive reductions in overall biodiversity (NRC 2000; Tilman and Lehman 2001). This response has been demonstrated experimentally (Wedin and Tilman 1996), as well as in long-term observations and/or comparative analyses of both terrestrial and aquatic ecosystems (Bobbink et al. 1998; Leibold 1999; Proulx and Mazumder 1998; Stevens et al. 2004; A. P. Wolfe, Baron, and Cornett 2001). Second, elevated N deposition contributes to soil and water acidification, changes that can in turn alter the species composition in affected ecosystems (Driscoll et al. 2003). Third, anthropogenic N_2-fixation fuels increased levels of tropospheric ozone (O_3; Chameides et al. 1994), a powerful oxidant that causes cellular damage in a wide variety of organisms, and may contribute to species shifts and declines (Bermejo et al. 2003). Fourth, elevated N inputs are causing increases in both O_3 and nitrous oxide (N_2O), gases that contribute significantly to the risks of future climate change (IPCC 2007). The relative importance of N_2O and O_3 to the greenhouse burden is predicted to increase in the future (IPCC 2007), and associated climate changes may drive substantial losses of biodiversity (Parmesan and Yohe 2003; Thomas et al. 2004).

Finally, the global trend toward intensification of agricultural practices (Matson et al. 1997), in which monospecific, fertilizer-intensive crops predominate, causes declines in biodiversity within already managed lands (Vandermeer et al. 2002), as well as from conversion of new land to support such agriculture (Sala et al. 2000). Much of this intensified agriculture supports a growing demand for meat in world diets, a practice that creates enormous needs for land to grow feed crops—a practice that is in part possible because of widespread, often relatively cheap N fertilizer in some parts of the world and which drives feedbacks to even greater inputs of new reactive N (Howarth et al. 2002; Smil 2001).

Until recently, the majority of changes to the global N cycle were concentrated in the temperate Northern Hemisphere (e.g., Howarth 1996), but some of the most rapid changes are now occurring in tropical and subtropical countries (Martinelli et al. 2006; McKenzie and Townsend 2007). This change is driven by explosive growth in N-intensive crop production and concentrated animal feeding operations, along with rapid industrialization leading to increased fossil fuel combustion (Martinelli et al. 2006).

Equatorial changes in the N cycle pose some unique and poorly understood threats to biodiversity. While bastions of species-rich ecosystems can be found throughout the globe, at least for terrestrial ecosystems, the tropics and subtropics have been called the "cradle of Earth's biodiversity," as these regions house the vast majority of the planet's terrestrial species (Wilson 1988). Twenty-two of the 34 "biodiversity hotspots" (e.g., Myers et al. 2000) identified by Conservation International are in tropical and subtropical regions (table 9.2). To date, loss of habitat via land conversion has been the major threat to tropical biodiversity, but additional threats are likely in the near future (Sala et al. 2000). These threats include climate change (Thomas et al. 2004), and though it has received little attention, may also include elevated N deposition (Krusche et al. 2003; Lara et al. 2001; Phoenix et al. 2006). Five- to 10-fold increases in atmospheric N deposition are possible in a substantial fraction of these tropical hotspots by the mid-21st century (table 9.2; Phoenix et al. 2006), with unknown effects on biodiversity.

Tropical regions, especially lowland rainforests, are unique in that N limitation does not appear to be as widespread as in temperate and high-latitude ecosystems (Martinelli et al. 1999; Vitousek and Sanford 1986). Instead, phosphorus (P), calcium (Ca), and/or potassium (K) limitation is thought to predominate in many lowland tropical forests (Cuevas and Medina 1988; Vitousek and Sanford 1986). This belief has understandably led some past evaluations of tropical biodiversity threats to discount N deposition as a major concern (Sala et al. 2000). However, a significant fraction of tropical forests may still be directly responsive to elevated N inputs (e.g., Tanner and Kapos 1992; Tanner, Vitousek, and Cuevas 1998), especially montane forests, which are often regions of high biodiversity (Pounds, Fogden, and Campbell 1999) and may also be highly sensitive to climate change (Still, Foster, and Schneider 1999). In addition, for those that are N-rich and limited by P, Ca, and/or K, additional N inputs may drive *declines* in the availability of these limiting elements, with unknown subsequent effects on community composition (Asner et al. 2001; Matson et al. 1999).

Moreover, many freshwater aquatic systems in the tropics appear to be N limited, and thus are likely to change in response to rising N inputs (Downing et al. 1999). For example, many tropical rivers and streams now receive substantial inputs of N-rich raw sewage, a disturbance that can lead to changes in aquatic biodiversity (Ometo et al. 2000; Singer and Battin 2007). Overall, the potential connections between a changing N cycle and biodiversity in the tropics clearly merit further study, especially given that rising N deposition will coincide with increasing pressure for land and a changing climate.

Table 9.2. Biodiversity hotspots by world region, as defined by Conservation International

These hotspots are defined by containing an unusually high number of endemic species, along with significant threats to their continued existence; each hotspot has lost 70% or more of its original vegetation. The total area of all the hotspot regions is 2.3% of Earth's land surface, yet more than half of all plant species and 42% of all terrestrial vertebrates are endemic to these regions. Regions in **boldface** type are projected to receive 5- to 10-fold increases in atmospheric nitrogen deposition by the mid-21st century.

North and Central America	Africa
California Floristic Province	**Cape Floristic Region**
Caribbean Islands	**Coastal Forests of Eastern Africa**
Madrean Pine-Oak Woodlands	**Eastern Afromontane**
Mesoamerica	**Guinean Forests of West Africa**
	Horn of Africa
	Madagascar and the Indian Ocean Islands
South America	**Maputaland-Pondoland-Albany**
Atlantic Forest	**Succulent Karoo**
Cerrado	
Chilean Winter Rainfall–Valdivian Forests	*Asia-Pacific*
Tumbes-Chocó-Magdalena	
Tropical Andes	East Melanesian Islands
	Himalaya
Europe and Central Asia	**Indo-Burma**
	Japan
Caucasus	Mountains of Southwest China
Irano-Anatolian	New Caledonia
Mediterranean Basin	New Zealand
Mountains of Central Asia	**Philippines**
	Polynesia-Micronesia
	Southwest Australia
	Sundaland
	Wallacea
	Western Ghats and Sri Lanka

Sources: Biodiversity hotspot listings based on www.biodiversityhotspots.org and Myers et al. 2000. Predictions of nitrogen deposition based on Galloway et al. 2004.

Finally, we note that while rising N inputs can clearly affect biodiversity, the species composition of an ecosystem can, in turn, regulate the N cycle (Hooper et al. 2005). Thus, for example, the diversity and composition of a landscape may affect losses of reactive N to atmospheric and aquatic realms. Experimental studies have shown that nitrogen losses can decrease with increasing species richness (e.g., Bigelow, Ewel, and Haggar 2004), although that relationship appears to saturate at low numbers of species (Hooper et al. 2005). As with many aspects of biodiversity's role in ecosystem function, the connections may be more dependent on species composition than on diversity per se. For example, the introduction of plant species capable of forming N-fixing symbioses can dramatically increase rates of N turnover, and overall ecosystem N stocks (Vitousek et al. 1987). In alpine tundra ecosystems, variation in N cycling rates between adjacent patches of *Acomastylis rossii* and *Deschampsia caespitosa*—species with very different tissue N content and overall tissue quality—can exceed the variation seen along substantial landscape-scale gradients in abiotic controls over N cycling (Bowman 2000; Steltzer and Bowman 1998). More generally, Hooper et al. (2005), among others, suggest that analyses of diversity-function relationships should focus on functional groups of species with similar responses to, or effects on, ecosystem properties, rather than on total species numbers.

Human Health and the Changing Global Nitrogen Cycle

As outlined earlier, human alteration of the global N cycle has widespread effects on the environment, but it also has diverse consequences for human health, the full scope of which have escaped significant attention (Townsend et al. 2003; A. H. Wolfe and Patz 2002). Many of the potential health effects arise because of ecological responses to increasing reactive N, and like those responses, links between the N cycle and human health range from well chronicled to theoretical. In addition, from a public health perspective, anthropogenic increases in fixed N can have both positive and negative outcomes. For example, the ability to produce synthetic N fertilizer on a large scale was an essential component of the green revolution, and allows human society to sustain much higher levels of food production than was possible several decades ago (Howarth et al. 2005; Smil 2001). Thus despite large population increases since the 1960s, N fertilizer has helped reduce starvation and malnutrition in Asia (Smil 2001), and even now, more fertilizer use in some world regions—most notably Africa—likely would have a significant public health benefit (Howarth et al. 2005; Sanchez 2002; Sanchez and Swaminathan 2005).

In a 2003 review, Townsend and colleagues presented a conceptual model that outlines the public health trade-offs inherent in the production of reactive N (figure 9.2; Townsend et al. 2003). The idea behind this model is that at low-to-moderate inputs of human-derived fixed N, the public health consequences are likely to be largely positive, because such inputs are associated with significant gains in food availability, nutrition, and the infrastructure that improves human welfare. However, as N inputs continue to rise, the relative benefits will decline, while the negative consequences begin to grow

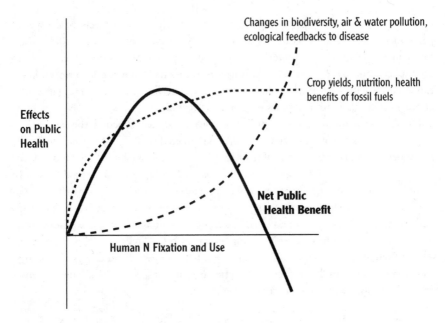

Figure 9.2. Conceptual model of the net public health benefits from increasing creation and use of reactive nitrogen. Adapted from Townsend et al. 2003.

rapidly; it is at these higher levels of N inputs, for example, that pronounced changes in biodiversity are most likely to occur. Thus, the model implies that while some creation and use of fixed N clearly improve public health, the net public health benefits must eventually peak, and as fixed N inputs continue to rise, those benefits will begin to decline (figure 9.2).

It is these negative health consequences of a changing N cycle that have received relatively little attention, and thus are often not well understood. In general, they may be grouped into three major categories (Townsend et al. 2003):

- Fertilizer, food, and nutrition
- The direct effects of N-related compounds on human health
- Indirect ecological feedbacks to human health

We will review the first two briefly before turning our attention to the last category in greater detail, as it is here that biodiversity may play an important role.

Fertilizer, Food, and Nutrition

As discussed earlier, the production and use of N fertilizers can and does confer major public health benefits. Beyond the obvious benefits of greater food production, a healthy diet with sufficient protein content is important to human immune responses against parasitic and infectious diseases (Nesheim 1993). Thus even small increases in the availability and use of fertilizer in impoverished regions can confer substantial health gains (Sanchez and Swaminathan 2005). However, despite unprecedented global fertilizer creation and use, malnutrition and hunger are still common in some parts of the world, a fact that now largely derives from inequitable food distribution rather than food shortages (Smil 2001). In large part, that distribution problem occurs because much of the world's grain is fed to animals, fueling the production of meat that is consumed mostly in wealthier regions (Howarth et al. 2002; Howarth et al. 2005), although the demand for grain-fed meat is increasingly rapidly in numerous world regions (Naylor et al. 2005). Feed crops such as corn and soybeans require large amounts of both fixed N and land per food calorie (Howarth et al. 2002); thus meat consumption, losses of biodiversity to land conversion, and an accelerated global N cycle are tightly linked.

In addition, high consumption of meat, often possible due to the growing prevalence of concentrated animal feeding operations that have their own suite of negative environmental and health consequences (Mallin 2000; Mallin and Cahoon 2003), has been linked to a range of health problems that plague developed nations, including heart disease and adult-onset diabetes (Hu and Willett 2002; Zimmet et al. 2001). Similarly, many of the crops used to produce animal feed, notably corn, are used far more to produce commercial sweeteners than for direct human consumption, and such sweeteners have their own ties to a range of human ailments (Hu and Willett 2002; Weisburger 2002; Zimmet et al. 2001). Thus even many wealthy nations have average diets that are far from optimal, while also typically being N-intensive. On a positive note, it is at least theoretically possible to improve nutrition in both impoverished and wealthy regions while simultaneously reducing the overall use and unwanted consequences of reactive N (Howarth et al. 2005; Townsend et al. 2003).

Direct Effects of N-related Compounds on Human Health

Nitrogen-containing compounds in air and water, and/or compounds formed due to N species, can directly affect human health. Concerns about drinking water nitrate caused the World Health Organization to adopt a 10 ppm nitrate-N standard, although this standard is often exceeded in many parts of the world. The initial worry over drinking water nitrate derived from its role in methemoglobinemia, or "blue baby" syndrome. While this disorder remains a valid concern in some regions (Gupta et al. 2000), and may be underdiagnosed (Kross, Ayebo, and Fuortes 1992), some recent evidence suggests that other health effects of drinking water nitrate may be more worrisome and pervasive. These effects include links to reproductive problems (Kramer 1996) and several cancers (De Roos et al. 2003; Ward et al. 2003; Ward et al. 2004), potentially even at levels below the World Health Organization (WHO) standard. Evidence for such links remains relatively sparse, and further study is clearly needed (Ward et al. 2005).

Next, increased reactive N contributes to harmful air pollution in several ways. The best known of these is the role of nitrogen oxides (NO_x) in the production of tropospheric ozone, an air pollutant that can trigger and/or worsen a variety of cardiopulmonary diseases (von Mutius 2000), increase pediatric emergency room visits (McConnell 2002), and perhaps increase overall mortality and morbidity in a wide range of urban centers (Cifuentes et al. 2001). Ironically, high levels of O_3 that can be partly fueled by fertilizer N also cause billions of dollars in annual crop damage (Chameides et al. 1994).

In addition, high levels of NO_x can have direct effects on human health: they can worsen viral infections and induce respiratory illnesses (Spannhake et al. 2002), and may contribute to substantial human mortality from indoor air pollution (A. H. Wolfe and Patz 2002). Finally, fine particulate air pollution is associated with a variety of cardiopulmonary diseases, increased mortality, and possibly several cancers (Allen, Cardoso, and da Rocha 2004; Arbex et al. 2000; Pope et al. 2002). Not all such pollution is related to elevated N, but it does play a significant, and at times dominant role (Allen, Cardoso, and da Rocha 2004).

Indirect Ecological Feedbacks to Human Health

As outlined earlier, rising nitrogen inputs can alter the structure and function of terrestrial, freshwater, and marine ecosystems, at times causing a cascade of ecological responses that include changes in species abundance and/or existence. Given that many human diseases are controlled by ecological interactions (Smith et al. 2005), it is nearly certain that ecological responses to a changing N cycle will include the dynamics of human diseases (McKenzie and Townsend 2007). In the case of infectious diseases, many of which depend on multiple hosts and/or environments during their life cycle, those dynamics often will be linked to changes in biodiversity (Dobson 2004). To date, very few studies have directly examined the connections between changing N inputs, community composition, and human disease, but some recent evidence is beginning to demonstrate the potential risks.

For example, starting in the early 1990s, several data sets showed positive correlations between the N content and/or algal productivity of N-limited freshwater ecosystems, and the larval abundance of malarial *Anopheles* mosquitoes (e.g., Rejmankova et al. 1991; Teng et al. 1998). Not all such correlations were positive, however (Gimnig et al. 2001), suggesting that mosquito responses to changing N were likely a complex function of how N levels affect mosquito breeding sites, parasites (e.g., Comiskey, Lowrie, and Wesson 1999), and predators. More recently, Rejmankova and colleagues have shed light on potential mechanisms by studying a range of marshes in Belize subject to different levels of N- and P-driven eutrophication (Rejmankova et al. 2005).

Prior work showed that habitat preferences of *Anopheles* mosquitoes were a function of the dominant plant species in a marsh (Rejmankova et al. 1993, 1998). In Belize, both

Anopheles vestitipennis and *A. albimanus* are carriers of the malarial parasite, but *A. vestitipennis* is a much more efficient vector that poses a greater malarial risk. As nutrient inputs to the Belizean marshes rise, overall plant biodiversity in the marshes appears to decline, and plant dominance shifts toward macrophytes from the genus *Typha*, which is the preferred habitat of *A. vestitipennis* mosquitoes (Rejmankova et al. 2005). Thus nutrient loading not only changes the species composition and biodiversity of the ecosystem, it also causes an ecological response that increases the risk of a serious human disease. Similar responses may occur for other mosquito-borne diseases; positive correlations between N levels and the larval abundances of *Culex* and *Aedes* carriers of several types of encephalitis, including the *Culex*-borne West Nile virus, have been demonstrated (Townsend et al. 2003). As with any disease subject to ecological controls, however, effects of N on mosquito-borne diseases are not likely to be simple or always in the same direction. Instead, the eventual response will depend on how changing nutrient availability affects the abundance of a variety of other species that in turn control mosquito populations.

Broadly similar feedbacks may exist for a variety of other vector-borne infectious diseases (McKenzie and Townsend 2007). For example, there is speculation that eutrophication could lead to an increase in the abundance of the snails that act as hosts for liver fluke, schistosomiasis, and other parasitic digenean trematodes of fish and livestock. Although we are aware of no long-term or experimental studies on the response of schistosomiasis or liver fluke to eutrophication, recent work in wildlife disease provides an important example from a related trematode species. In a mesocosm-based study of the trematode *Riberoia ondatrae*, which causes amphibian limb malformation, Johnson and colleagues (2007) demonstrated that increased nutrient levels substantially increased parasite prevalence in the planorbid snails that serve as an intermediate host of the parasite. This change occurred for two reasons: first, nutrient-stimulated algal growth led to greater food resources for the snails, and thus higher snail biomass, but second, the eutrophic conditions also caused an increase in the numbers of immature parasites produced and released to the environment, leading to higher rates of infection and higher infection loads per snail. This study has important implications for human disease, as the trematode diseases such as schistosomiasis have very similar life cycles to those of *Riberoia ondatrae*, including the intermediate snail hosts, and thus may demonstrate quite similar responses to eutrophication.

As well, the *Vibrio cholerae* bacteria responsible for cholera are harbored in a variety of marine organisms (Colwell and Huq 2001), and cholera outbreaks often occur following coastal algal blooms (Colwell and Huq 2001; Epstein 1993). Algal blooms can be fueled by elevated N inputs (NRC 2000); thus while a direct link between nutrient loading and cholera is difficult to demonstrate, several authors have suggested that the nutrient-rich conditions that favor algal blooms may pose elevated risks of cholera epidemics (Epstein 1993). Substantial evidence does exist that rising N inputs to both freshwater and coastal marine ecosystems can favor algal blooms that produce a range of human toxins, some of which can be fatal (Burkholder 1998; NRC 2000). For example, nitrogen-driven changes in the community composition of several Brazilian reservoirs led to blooms of toxic microcystin-producing cyanobacteria (Bittencourt-Oliveira 2003), a change that in one tragic instance, caused more than 50 deaths following the use of affected reservoir water for hemodialysis (Azevedo et al. 2002; Jochimsen et al. 1998).

Links between a changing N cycle, biodiversity, and human ailments have also been documented in terrestrial ecosystems, and are likely more pervasive than currently known. For example, allergic responses to pollen are a significant environmental health issue (NIH 1993), and pollen counts are rising in many regions (Wayne et al. 2002). Increasing N deposition is one potential driver of higher pollen production, both via the direct effects of N on a plant's ability to allocate resources to reproductive output, and via indirect changes that can favor the prevalence of weedy species (such as ragweed) that tend to produce large amounts of allergenic pollen (Townsend et al. 2003).

Other examples arise from the interactions between N-intensive agriculture and biodiversity. As described earlier, land conversion to N-intensive crops such as soy and corn has been rampant throughout much of South America over the past two decades (Martinelli et al. 2006). In at least one case, such conversion led to increased risks for Argentine hemorrhagic fever (AHF) (Daily and Ehrlich 1996). AHF is caused by the Junin virus, whose main reservoir is the mouse *Calomys musculinus*. The conversion of pampas grasslands in Argentina to maize production typically includes widespread use of weed-controlling herbicides; these, in turn, led to changes in grass species composition that favored higher populations of *C. musculinus*, thereby increasing human exposure to AHF. Likewise, conversion to N-intensive sugarcane is followed by repeated cycles of cane burning, a practice that leads to health problems arising from particulate air pollution (Arbex et al. 2000).

Data from studies of plant diseases also illustrate the potential risks a changing N cycle may pose. For example, Mitchell, Tilman, and Groth (2002) and Mitchell and colleagues (2003) studied plant fungal infections in a long-term experiment in Great Plains grasslands. Plots included treatments manipulating both N inputs and overall diversity in a full factorial design, and the results showed significant increases in fungal infection under higher N fertilization and in plots with lower species richness. Data from these and other experiments also show that increasing N can cause lower plant diversity, leading to a potential feedback in which pathogen loads could be substantially increased following rising N inputs. These results have implications both for low-diversity, N-intensive agriculture, and more broadly, for how N-driven declines in biodiversity may alter the dynamics of disease.

In general, both theoretical and experimental results suggest that declines in biodiversity may often favor increased disease transmission, and perhaps increased virulence (Dobson 2004). A central mechanism behind such links is the fact that greater species diversity can reduce the probability of disease outbreaks by "diluting" host availability (Schmid and Ostfeld 2001). As with the mosquito examples above, however, the complexity of ecological interactions ensures that not all diseases are likely to respond in this manner. For example, changes in the species composition of northeastern U.S. forests, including the near or complete loss of some species, have led to increased risk for Lyme disease in humans (Ostfeld and Keesing 2000). However, LoGiudice et al. (2003) showed that the relationship between diversity and Lyme disease risk was a function of the sequence of species loss; in some cases, reduced diversity could lead to reduced risk.

An Era of Overlapping Environmental Change

Finally, we wish to stress that the threats posed by a rapidly changing N cycle typically coincide with numerous other human-caused stresses to the environment, and therefore assessments of the risks for either ecological or human health must take a multifactorial perspective. Other chapters in this volume illustrate a number of ways in which biodiversity and human health are intimately related, and how the former may be affected by several global-scale environmental changes. In the case of the global N cycle, rising N inputs alone might not always trigger substantial changes in biodiversity, ecosystem function, and/or factors that affect human health. However, when they occur in combination with widespread land use change, rising atmospheric CO_2, a changing climate, and/or the stresses imposed by species invasions, the resultant effects on both ecosystem function and human welfare are likely to be much more pronounced.

Acknowledgments

We wish to thank Andy Dobson, Sybil Seitzinger, and Cory Cleveland for intellectual input that improved the content of this chapter. We also thank the organizers of the SCOPE-RAP meeting for their comments on this manuscript and for the meeting in general. A.R.T. was supported by funds from the Andrew Mellon Foundation, the Institute of Marine and Coastal Sciences at Rutgers University, and the University of Colorado.

References

Allen, A. G., A. A. Cardoso, and G. O. da Rocha. 2004. Influence of sugar cane burning on aerosol soluble ion composition in Southeastern Brazil. *Atmospheric Environment* 38 (30): 5025–38.

Arbex, M. A., A. L. Ferreira, J. E. D. Cancado, L. A. A. Pereira, P. H. Saldiva. 2000. Assessment of the effects of sugar cane plantation burning on daily counts of inhalation therapy. *Journal of the Air and Waste Management Association* 50 (10): 1745–49.

Asner, G. P., A. R. Townsend, W. J. Riley, P. A. Matson, J. C. Neff, and C. C. Cleveland. 2001. Physical and biogeochemical controls over terrestrial ecosystem responses to nitrogen deposition. *Biogeochemistry* 54 (1): 1–39.

Azevedo, S. M., W. W. Carmichael, E. M. Jochimsen, K. L. Rinehart, L. Lau, G. R. Shaw, and G. K. Eaglesman. 2002. Human intoxication by microcystins during renal dialysis treatment in Caruaru, Brazil. *Toxicology* 181:441–46.

Belovsky, G. E., and J. B. Slade. 2000. Insect herbivory accelerates nutrient cycling and increases plant production. *Proceedings of the National Academy of Sciences USA* 97 (26): 14412–17.

Bermejo, V., B. S. Gimeno, J. Sanz, D. de la Torre, and J. M. Gil. 2003. Assessment of the ozone sensitivity of 22 native plant species from Mediterranean annual pastures based on visible injury. *Atmospheric Environment* 37 (33): 4667–77.

Bigelow, S. W., J. J. Ewel, and J. P. Haggar. 2004. Enhancing nutrient retention in tropical tree plantations: No short cuts. *Ecological Applications* 14 (1): 28–46.

Bittencourt-Oliveira, M. D. 2003. Detection of potential microcystin-producing cyanobacteria in Brazilian reservoirs with a mcyB molecular marker. *Harmful Algae* 2 (1): 51–60.

Bobbink, R., M. Hornung, and J. G. M. Roelofs. 1998. The effects of air-borne nitrogen pollutants on species diversity in natural and semi-natural European vegetation. *Journal of Ecology* 86 (5): 717–38.

Bowman, W. D. 2000. Biotic controls over ecosystem response to environmental change in alpine tundra of the Rocky Mountains. *Ambio* 29 (7): 396–400.

Boyer, E. W., R. W. Howarth, J. N. Galloway, F. J. Dentener, P. A. Green, C. J. Vorosmarty. 2006. Riverine nitrogen export from the continents to the coasts. *Global Biogeochemical Cycles* 20 (1): GB1591 (1–9).

Burkholder, J. M. 1998. Implications of harmful microalgae and heterotrophic dinoflagellates in management of sustainable marine fisheries. *Ecological Applications* 8 (1): S37–S62.

Chameides, W. L., P. S. Kasibhatla, J. Yienger, and H. Levy. 1994. Growth of continental-scale metro-agro-plexes, regional ozone pollution, and world food-production. *Science* 264 (5155): 74.

Chapin, F. S., B. H. Walker, R. J. Hobbs, D. U. Hooper, J. H. Lawton, O. E. Sala, and D. Tilman. 1997. Biotic control over the functioning of ecosystems. *Science* 277 (5325): 500.

———. 2000. Consequences of changing biodiversity. *Nature* 405 (6783): 234–42.

Christian, J. M., and S. D. Wilson. 1999. Long-term ecosystem impacts of an introduced grass in the northern Great Plains. *Ecology* 80 (7): 2397–2407.

Cifuentes, L., V. H. Borja-Aburto, N. Gouveia, G. Thurston, and D. L. Davis. 2001. Climate change: Hidden health benefits of greenhouse gas mitigation. *Science* 293 (5533): 1257–59.

Cleveland, C. C., A. R. Townsend, D. S. Schimel, H. Fisher, R. W. Howarth, L. O. Hedin, S. S. Perakis, et al. 1999. Global patterns of terrestrial biological nitrogen (N-2) fixation in natural ecosystems. *Global Biogeochemical Cycles* 13 (2): 623–45.

Coley, P. D. 1998. Possible effects of climate change on plant/herbivore interactions in moist tropical forests. *Climatic Change* 39 (2–3): 455–72.

Colwell, R., and A. Huq. 2001. Marine ecosystems and cholera. *Hydrobiologia* 460:141–45.

Comiskey, N. M., R. C. Lowrie, and D. M. Wesson. 1999. Role of habitat components on the dynamics of *Aedes albopictus* (Diptera: Culicidae) from New Orleans. *Journal of Medical Entomology* 36 (3): 313–20.

Cuevas, E., and E. Medina 1988. Nutrient dynamics with Amazonian forests. II: Fine root growth, nutrient availability, and leaf litter decomposition. *Oecologia* 76:222–35.

Daily, G. C., and P. R. Ehrlich. 1996. Global change and human susceptibility to disease. *Annual Review of Energy and the Environment* 21:125–44.

De Roos, A. J., M. H. Ward, C. F. Lynch, and K. P. Cantor. 2003. Nitrate in public water supplies and the risk of colon and rectum cancers. *Epidemiology* 14 (6): 640–49.

Dobson, A. 2004. Population dynamics of pathogens with multiple host species. *American Naturalist* 164 (5): S64–S78.

Downing, J. A., M. Mcclain, R. Twilley, J. M. Melack, J. Elser, N. N. Rabalais, W. M. Lewis Jr., et al. 1999. The impact of accelerating land-use change on the N-cycle of tropical aquatic ecosystems: Current conditions and projected changes. *Biogeochemistry* 46 (1–3): 109–48.

Driscoll, C. T., D. R. Whitall, J. D. Aber, E. W. Boyer, M. Castro, C. Cronan, C. L. Goodale, et al. 2003. Nitrogen pollution: Sources and consequences in the U.S. northeast. *Environment* 45 (7): 8.

Epstein, P. R. 1993. Algal blooms in the spread and persistence of cholera. *Biosystems* 31 (2–3): 209–21.

Foley, J. A., M. H. Costa, C. Delire, N. Ramankutty, and P. Snyder. 2003. Green surprise? How terrestrial ecosystems could affect earth's climate. *Frontiers in Ecology and the Environment* 1 (1): 38–44.

Galloway, J. N., J. D. Aber, J. W. Erisman, S. P. Seitzinger, R. W. Howarth, E. B. Cowling, and B. J. Cosby. 2003. The nitrogen cascade. *Bioscience* 53 (4): 341–56.

Galloway, J. N., J. J. Dentener, D. G. Capone, E. W. Boyer, R. W. Howarth, S. P. Seitzinger, G. P. Asner, et al. 2004. Nitrogen cycles: Past, present, and future. *Biogeochemistry* 70 (2): 153–226.

Galloway, J. N., H. Levy, and P. S. Kashibhatla. 1994. Year 2020: Consequences of population-growth and development on deposition of oxidized nitrogen. *Ambio* 23 (2): 120–23.

Gimnig, J. E., M. Ombok, L. Kamau, and W. A. Hawley. 2001. Characteristics of larval anopheline (Diptera: Culicidae) habitats in western Kenya. *Journal of Medical Entomology* 38 (2): 282–88.

Gordon, D. R., and K. J. Rice. 1992. Partitioning of space and water between two California annual grassland species. *American Journal of Botany* 79 (9): 967–76.

Gupta, S. K., R. C. Gupta, A. K. Seth, A. B. Gupta, J. K. Bassin, and A. Gupta. 2000. Methaemoglobinaemia in areas with high nitrate concentration in drinking water. *National Medical Journal of India* 13 (2): 58–61.

Holland, E. A., B. H. Braswell, J. Sulzman, and J. F. Lamarque. 2005. Nitrogen deposition onto the United States and western Europe: Synthesis of observations and models. *Ecological Applications* 15 (1): 38–57.

Hooper, D. U., F. S. Chapin, J. J. Ewel, A. Hector, P. Inchausti, S. Lavorel, J. H. Lawton, et al. 2005. Effects of biodiversity on ecosystem functioning: A consensus of current knowledge. *Ecological Monographs* 75 (1): 3–35.

Howarth, R. W. 1996. Nitrogen cycling in the North Atlantic Ocean and its watersheds: Report of the International SCOPE Nitrogen Project. Special issue, *Biogeochemistry* 35 (1): 1.

———. 2003. Human acceleration of the nitrogen cycle: Drivers, consequences, and steps towards solutions. In *Proceedings of the Strong N and Agro 2003 IWA Specialty Symposium*, ed. E. Choi and Z. Yun, 3–12. Seoul: Korea University.

———. 2004. Human acceleration of the nitrogen cycle: Drivers, consequences, and steps toward solutions. *Water Science and Technology* 49 (5–6): 7–13.

Howarth, R. W., G. Billen, D. Swaney, A. Townsend, N. Jaworski, K. Lajtha, J. A. Downing, et al. 1996. Regional nitrogen budgets and riverine N and P fluxes for the drainages to the North Atlantic Ocean: Natural and human influences. *Biogeochemistry* 35 (1): 75–139.

Howarth, R. W., E. W. Boyer, W. J. Pabich, and J. N. Galloway. 2002. Nitrogen use in the United States from 1961–2000 and potential future trends. *Ambio* 31 (2): 88–96.

Howarth, R. W., K. Ramakrishna, E. Choi, R. Elmgren, L. Martinelli, A. Mendoza, W. Moomaw, et al. 2005. Nutrient management, responses assessment. In *Ecosystems and human well-being*. Vol. 3, *Policy responses*, 295–311. Millennium Ecosystem Assessment. Washington, DC: Island Press. See also http://www.millenniumassessment.org.

Howarth, R. W., A. Sharpley, and D. Walker. 2002. Sources of nutrient pollution to coastal waters in the United States: Implications for achieving coastal water quality goals. *Estuaries* 25 (4B): 656–76.

Hu, F. B., and W. C. Willett. 2002. Optimal diets for prevention of coronary heart disease. *JAMA: Journal of the American Medical Association* 288 (20): 2569–78.

IPCC (Intergovernmental Panel on Climate Change). 2007. Climate change 2007: Impacts, adaptation and vulnerability. Summary for policymakers. http://www.ipcc.ch.

Jochimsen, E. M., W. W. Carmichael, J. An, D. M. Cardo, S. T. Cookson, C. E. M. Holmes, M. B. Antunes, et al. 1998. Liver failure and death after exposure to microcystins at a hemodialysis center in Brazil. *New England Journal of Medicine* 338 (13): 873–78.

Johnson, P. T. J., J. M. Chase, K. L. Dosch, R. B. Hartson, J. A. Gross, D. J. Larson, D. R. Sutherland, and S. R. Carpenter. 2007. Aquatic eutrophication promotes pathogenic infection in amphibians. *Proceedings of the National Academy of Sciences* 104 (40): 15781–86.

Kramer, M. E. A. 1996. Surveillance of waterborne-disease outbreaks: United States, 1993–1994. *CDCP Surveillance Summaries, MMWR* 45:1–33.

Kross, B. C., A. D. Ayebo, and L. J. Fuortes. 1992. Methemoglobinemia: Nitrate toxicity in rural America. *American Family Physician* 46 (1): 183–88.

Krusche, A. V., P. B. de Camargo, C. E. Cerri, M. V. Ballester, L. B. L. S. Lara, R. L. Victoria, and L. A. Martinelli. 2003. Acid rain and nitrogen deposition in a sub-tropical watershed (Piracicaba): Ecosystem consequences. *Environmental Pollution* 121 (3): 389–99.

Lara, L. B. L. S., P. Artaxo, L. A. Martinelli, R. L. Victoria, P. B. Camargo, A. Krusche, G. P. Ayers, E. S. B. Ferraz, and M. V. Ballester. 2001. Chemical composition of rainwater and anthropogenic influences in the Piracicaba River Basin, Southeast Brazil. *Atmospheric Environment* 35 (29): 4937–45.

Leibold, M. A. 1999. Biodiversity and nutrient enrichment in pond plankton communities. *Evolutionary Ecology Research* 1 (1): 73–95.

LoGiudice, K., R. S. Ostfeld, K. A. Schmidt, and F. Keesing. 2003. The ecology of infectious disease: Effects of host diversity and community composition on Lyme disease risk. *Proceedings of the National Academy of Sciences USA* 1 (1): 73–95.

Mallin, M. A. 2000. Impacts of industrial animal production on rivers and estuaries. *American Scientist* 88 (1): 26–37.

Mallin, M. A., and L. B. Cahoon. 2003. Industrialized animal production: A major source of nutrient and microbial pollution to aquatic ecosystems. *Population and Environment* 24 (5): 369–85.

Martinelli, L. A., R. W. Howarth, E. Cuevas, S. Filoso, A. T. Austin, L. Donoso, V. Huszar, et al. 2006. Sources of reactive nitrogen affecting ecosystems in Latin America and the Caribbean: Current trends and future perspectives. *Biogeochemistry* 79 (1–2): 3–24.

Martinelli, L. A., M. C. Piccolo, A. R. Townsend, P. M. Vitousek, E. Cuevas, W. McDowell, G. P. Robertson, O. C. Santos, and K. Treseder. 1999. Nitrogen stable isotopic composition of leaves and soil: Tropical versus temperate forests. *Biogeochemistry* 46 (1–3): 45–65.

Matson, P. A., W. H. McDowell, A. R. Townsend, and P. M. Vitousek. 1999. The globalization of N deposition: Ecosystem consequences in tropical environments. *Biogeochemistry* 46:67–83.

Matson, P. A., W. J. Parton, A. G. Power, and M. J. Swift. 1997. Agricultural intensification and ecosystem properties. *Science* 277:504–9.

McConnell, R. 2002. Asthma in exercising children exposed to ozone. *Lancet* 359 (9309): 896.

McKenzie, V., and A. R. Townsend. 2007. Parasitic and infectious disease responses to changing global nutrient cycles. *EcoHealth* 4 (4): 384.

Melillo, J. M., J. D. Aber, A. E. Linkins, A. Ricca, B. Fry, and K. J. Nadelhoffer. 1989. Carbon and nitrogen dynamics along the decay continuum: Plant litter to soil organic matter. *Plant and Soil* 115:189–98.

Mitchell, C. E., P. B. Reich, D. Tilman, and J. V. Groth. 2003. Effects of elevated CO_2, nitrogen deposition, and decreased species diversity on foliar fungal plant disease. *Global Change Biology* 9 (3): 438–51.

Mitchell, C. E., D. Tilman, and J. V. Groth. 2002. Effects of grassland plant species diversity, abundance, and composition on foliar fungal disease. *Ecology* 83 (6): 1713–26.

Myers, N., R. A. Mittermeier, C. G. Mittermeier, G. A. B. da Fonseca, and J. Kent. 2000. Biodiversity hotspots for conservation priorities. *Nature* 403 (6772): 853–58.

Naylor, R., H. Steinfeid, W. Falcon, J. Galloways, V. Smil, E. Bradford, J. Alder, and H. Mooney. 2005. Losing the links between livestock and land. *Science* 310 (5754): 1621–22.

Nesheim, M. C. 1993. Human-nutrition needs and parasitic infections. *Parasitology* 107:S7–S18.

NIH (National Institutes of Health). 1993. *Something in the air: Airborne allergens.* Bethesda: National Institutes of Health.

NRC (National Research Council). 2000. *Clean coastal waters: Understanding and reducing the problems of nutrient pollution.* Washington, DC: National Academy Press.

Ometo, J. P. H. B., L. A. Martinelli, M. V. Ballester, A. Gessner, A. V. Krusche, R. L. Victoria, and M. Williams. 2000. Effects of land use on water chemistry and macroinvertebrates in two streams of the Piracicaba River Basin, south-east Brazil. *Freshwater Biology* 44:327–37.

Ostfeld, R. S., and F. Keesing. 2000. Biodiversity and disease risk: The case of Lyme disease. *Conservation Biology* 14 (3): 722–28.

Parmesan, C., and G. Yohe. 2003. A globally coherent fingerprint of climate change impacts across natural systems. *Nature* 421 (6918): 37–42.

Phoenix, G. K, W. K. Hicks, S. Cinderby, J. C. I. Kuylenstierna, W. D. Stock, F. J. Dentener, K. E. Giller, et al. 2006. Atmospheric nitrogen deposition in world biodiversity hotspots: The need for a greater global perspective in assessing N deposition impacts. *Global Change Biology* 12 (3): 470.

Pope, C. A., R. T. Burnett, M. J. Thun, E. E. Calle, D. Krewski, K. Ito, and G. D. Thurston. 2002. Lung cancer, cardiopulmonary mortality, and long-term exposure to fine particulate air pollution. *JAMA: Journal of the American Medical Association* 287 (9): 1132–41.

Pounds, J. A., M. P. L. Fogden, and J. H. Campbell. 1999. Biological response to climate change on a tropical mountain. *Nature* 398 (6728): 611–15.

Proulx, M., and A. Mazumder. 1998. Reversal of grazing impact on plant species richness in nutrient poor vs. nutrient-rich ecosystems. *Ecology* 79 (8): 2581–92.

Rejmankova, E., J. Grieco, N. Achee, P. Masuoka, K. Pope, D. Roberts, and R. M. Higashi. 2005. Freshwater community interactions and malaria. In *Disease ecology: Community structure and pathogen dynamics,* ed. S. K. Collinge and C. Ray. New York: Oxford University Press.

Rejmankova, E., K. O. Pope, D. R. Roberts, M. G. Lege, R. Andre, J. Greico, and Y. Alonzo. 1998. Characterization and detection of *Anopheles vestitipennis* and *Anopheles punctimacula* (Diptera: Culicidae) larval habitats in Belize with field survey and SPOT satellite imagery. *Journal of Vector Ecology* 23 (1): 74–88.

Rejmankova, E., D. R. Roberts, R. E. Harbach, J. Pecor, E. L. Peyton, S. Manguin, R. Krieg, J. Polanco, and L. Legters. 1993. Environmental and regional determinants of *Anopheles* (Diptera: Culicidae) larval distribution in Belize, Central America. *Environmental Entomology* 22 (5): 978–92.

Rejmankova, E., H. M. Savage, M. Rejmanek, J. I. Arredondojimenez, and D. R. Roberts. 1991. Multivariate-analysis of relationships between habitats, environmental-factors and occurrence of anopheline mosquito larvae *Anopheles albimanus* and *A. pseudopunctipennis* in southern Chiapas, Mexico. *Journal of Applied Ecology* 28 (3): 827–41.

Sala, O. E., F. S. Chapin, J. J. Armesto, E. Berlow, J. Bloomfield, R. Dirzo, E. Huber-Sanwald, L. F. Huenneke, et al. 2000. Biodiversity: Global biodiversity scenarios for the year 2100. *Science* 287 (5459): 1770.

Sanchez, P. A. 2002. Ecology: Soil fertility and hunger in Africa. *Science* 295 (5562): 2019–20.

Sanchez, P. A., and M. S. Swaminathan. 2005. Hunger in Africa: The link between unhealthy people and unhealthy soils. *Lancet* 365 (9457): 442–44.

Schmid, K. A., and R. S. Ostfeld. 2001. Biodiversity and the dilution effect in disease ecology. *Ecology* 82 (3): 609–19.

Singer, G. A, and T. J. Battin. 2007. Anthropogenic subsidies alter stream consumer-resource stoichiometry, biodiversity, and food chains. *Ecological Applications* 17 (2): 376–89.

Smil, V. 2001. *Enriching the earth*. Cambridge, MA: MIT Press.

Smith, K. F., A. P. Dobson, F. E. McKenzie, L. A. Real, D. L. Smith, and M. L Wilson. 2005. Ecological theory to enhance infectious disease control and public health policy. *Frontiers in Ecology and the Environment* 3 (1): 29–37.

Spannhake, E. W., S. P. M. Reddy, D. B. Jacoby, X.-Y. Yu, B. Saatian, and J. Tian. 2002. Synergism between rhinovirus infection and oxidant pollutant exposure enhances airway epithelial cell cytokine production. *Environmental Health Perspectives* 110 (7): 665–70.

Steltzer, H., and W. D. Bowman. 1998. Differential influence of plant species on soil nitrogen transformations within moist meadow alpine tundra. *Ecosystems* 110 (7): 665–70.

Stevens, C. J., N. B. Dise, J. O. Mountford, and D. J. Gowing. 2004. Impact of nitrogen deposition on the species richness of grasslands. *Science* 303 (5665): 1876–79.

Still, C. J., P. N. Foster, and S. H. Schneider. 1999. Simulating the effects of climate change on tropical montane cloud forests. *Nature* 398 (6728): 608–10.

Tanner, E. V. J., and V. Kapos. 1992. Nitrogen and phosphorus fertilization effects on Venezuelan montane forest growth and litterfall. *Ecology* 73:78–86.

Tanner, E. V. J., P. M. Vitousek, and E. Quevas. 1998. Experimental investigation of the role of nutrient supplies in the limitation of forest growth and stature on wet tropical mountains. *Ecology* 79:10–22.

Tartowski, S., and R. W. Howarth. 2000. Nitrogen, nitrogen cycling. *Encyclopedia of Biodiversity* 4:377–88.

Teng, H. J., Y. L. Wu, S. J. Wang, and C. Lin. 1998. Effects of environmental factors on abundance of *Anopheles minimus* (Diptera: Culicidae) larvae and their seasonal fluctuation in Taiwan. *Environmental Entomology* 27 (2): 324–28.

Thomas, C. D., A. Cameron, R. E. Green, M. Bakkenes, L. J. Beaumont, Y. C. Collingham, B. R. N. Erasmus, et al. 2004. Extinction risk from climate change. *Nature* 427 (6970): 145–48.

Throop, H. L., and M. T. Lerdau. 2004. Effects of nitrogen deposition on insect herbivory: Implications for community and ecosystem processes. *Ecosystems* 7 (2): 109–33.

Tilman, D. 1996. Biodiversity: Population versus ecosystem stability. *Ecology* 77:350–63.

Tilman, D., and C. Lehman. 2001. Human-caused environmental change: Impacts on plant diversity and evolution. *Proceedings of the National Academy of Sciences USA* 98 (10): 5433–40.

Townsend, A. R., R. W. Howarth, F. A. Bazzaz, M. S. Booth, C. C. Cleveland, S. K. Collinge, A. P. Dobson, et al. 2003. Human health effects of a changing global nitrogen cycle. *Frontiers in Ecology and the Environment* 1 (5): 240–46.

Van Cleve, K., F. S. Chapin, C. T. Dyrness, and L. A. Viereck. 1991. Element cycling in taiga forests: State-factor control. *Bioscience* 41 (2): 78.

Vandermeer, J., D. Lawrence, A. Symstad, and S. Hobbie. 2002. Effect of biodiversity on ecosystem functioning in managed ecosystems. In *Biodiversity and ecosystem functioning: Synthesis and perspectives*, ed. M. Loreau, N. Shahid, and P. Inchausti, 221–33. Oxford: Oxford University Press.

Vitousek, P. M. 1982. Nutrient cycling and nutrient use efficiency. *American Naturalist* 119:553–72.

Vitousek, P. M., J. D. Aber, R. W. Howarth, G. E. Likens, P. A. Matson, D. W. Schindler, W. H. Schlesinger, and D. G. Tilman. 1997. Human alteration of the global nitrogen cycle: Sources and consequences. *Ecological Applications* 7 (3): 737.

Vitousek, P. M., K. Cassmand, C. Cleveland, T. Crews, C. B. Fields, N. B. Grimm, R. W. Howarth, R. Marino, et al. 2002. Towards an ecological understanding of biological nitrogen fixation. *Biogeochemistry* 57 (1): 1–45.

Vitousek, P. M., and R. W. Howarth. 1991. Nitrogen limitation on land and in the sea: How can it occur? *Biogeochemistry* 13:87–115.

Vitousek, P. M., and R. L. Sanford Jr. 1986. Nutrient cycling in moist tropical forest. *Annual Review of Ecology and Systematics* 17:137–67.

Vitousek, P. M., L. R. Walker, L. D. Whiteaker, D. Muellerdombois, and P. A. Matson. 1987. Biological invasion by *Myrica faya* alters ecosystem development in Hawaii. *Science* 238 (4828): 802–4.

von Mutius, E. 2000. The environmental predictors of allergic disease. *Journal of Allergy and Clinical Immunology* 105 (1): 9–19.

Ward, M. H., K. P. Cantor, J. Cerhan, C. F. Lynch, and P. Hartge. 2004. Nitrate in public water supplies and risk of cancer: Results from recent studies in the midwestern United States. *Epidemiology* 15 (4): S214.

Ward, M. H., K. P. Cantor, D. Riley, S. Merkle, and C. F. Lynch. 2003. Nitrate in public water supplies and risk of bladder cancer. *Epidemiology* 14 (2): 183–90.

Ward, M. H., T. M. deKok, P. Levallois, J. Brender, G. Gulis, B. T. Nolan, and J. Van Derslice. 2005. Workgroup report: Drinking-water nitrate and health—recent findings and research needs. *Environmental Health Perspectives* 113 (11): 1607–14.

Wardle, D. A., R. D. Bardgett, J. N. Klironomos, H. Setala, W. H. van der Putten, and D. H. Wall. 2004. Ecological linkages between aboveground and belowground biota. *Science* 304 (5677): 1629–33.

Wayne, P., S. Foster, J. Connolly, F. Bazzaz, and P. Epstein. 2002. Production of allergenic pollen by ragweed (*Ambrosia artemisiifolia* L.) is increased in CO_2-enriched atmospheres. *Annals of Allergy, Asthma & Immunology* 88 (3): 279–82.

Wedin, D. A., and D. Tilman. 1996. Influence of nitrogen loading and species composition on the carbon balance of grasslands. *Science* 274:1720–23.

Weisburger, J. H. 2002. Lifestyle, health and disease prevention: The underlying mechanisms. *European Journal of Cancer Prevention* 11:S1–S7.

Wilson, E. O., ed. 1988. *Biodiversity*. Washington, DC: National Academy Press.

Wolfe, A. H., and J. A. Patz. 2002. Reactive nitrogen and human health: Acute and long-term implications. *Ambio* 31 (2): 120–25.

Wolfe, A. P., J. S. Baron, and R. J. Cornett. 2001. Anthropogenic nitrogen deposition induces rapid ecological changes in alpine lakes of the Colorado Front Range (USA). *Journal of Paleolimnology* 25 (1): 1–7.

Zimmet, P., K. Alberti, and J. Shaw. 2001. Global and societal implications of the diabetes epidemic. *Nature* 414 (6865): 782–87.

10

Links between Microbial Diversity, Stressors, and Detoxification Capacity

Kerstin Hund-Rinke and Anne Winding

The numerous functions accomplished by soil microorganisms require a large and diverse microbial community. Microorganisms are different from organisms at higher trophic levels (plants and animals) in that only a minor fraction of the total number of species is known. When estimated by direct counting, the abundance of bacteria in soil is up to 10^9–10^{10} per gram of soil (dry weight) in agricultural, forest, and meadow soils. This number is very dependent on the organic matter content, whereas latitude is of less importance. When genetic approaches are used, the number of bacterial types is estimated to be 10^4 per gram of soil (Torsvik et al. 1996). Therefore, environmental microbial communities are expected to consist of known species and a large group of unknown species.

This lack of knowledge is a constant challenge for microbiologists and is caused by several features of microorganisms. Bacteria have very few distinctive morphological features, and therefore a taxonomy is not possible. Rather, bacteria have traditionally been identified on the basis of physiological features, especially their ability to grow at various conditions (e.g., temperature, redox potential) and on different carbon substrates. Only limited morphological features, such as the shape of the cells, cell membrane construction, and mobility, have been used as part of identification. This approach to identification requires cells to grow under laboratory conditions. However, only 0.1% to a maximum of 30% of the total number of cells in an environmental sample can be cultured in the laboratory. This leaves up to 99.9% of the bacteria unidentified. Fungi are usually identified by the morphology of the reproductive structures. However, a large proportion of the fungi are only known in their nonreproductive stage.

The availability of molecular techniques in environmental microbiology led to a revolution in bacterial taxonomy. Various definitions of differences between bacterial species were brought into play—for example, DNA-DNA hybridization of more than 70% and similarity of 16S rDNA sequences of more than 97%. Also, the biochemical technique of phospholipid fatty analysis of the cellular membrane is in use for species identification. The current rather pragmatic definition of a species is that it should be based on molecular as well as physiological test results. Microorganisms show an ever-changing gene composition, as they are able to take up and discard genes at a much higher rate than any other organism. The species concept is hence not as stable for microorganisms as for plants and animals.

These challenges in defining and determining microbial species and the high number of species make it practically impossible to determine the total number of species and the number of individuals of each species in an environmental sample. It would be a great achievement for microbial ecology if the diversity of microorganisms could be clearly determined. Toward this end, alternative approaches have been developed to estimate microbial diversity, and a large number of techniques are presently in use. Generally, the techniques can be divided into those estimating the functional diversity, the biochemical diversity, and the genetic diversity of microbes. Estimation of the functional diversity is based on the metabolic capabilities of the living microorganisms—for example, the carbon substrates the microbial community can mineralize. Measurement of the biochemical diversity is based on the composition of the phospholipid fatty acids that are components of bacterial and fungal cell membranes. Finally, estimates of genetic diversity are based on the base composition of DNA and ribosomal RNA. In the near future, the composition of messenger RNA will probably be able to elucidate the functional diversity of microbial communities. Already a vast spectrum of metabolic capabilities has been found by metagenomic analysis of soil communities (Handelsman 2004). This technique finds functional genes by sequencing the DNA and aligning it to known genes. Metagenomic analysis is therefore limited by our knowledge of functional genes, and thus the importance of these bacteria for the soil ecosystem can only be assumed. A comprehensive overview of the various techniques used to determine microbial diversity has been published (Winding, Hund-Rinke, and Rutgers 2005).

General Aspects of Effects of Stressors

Several principal and different hypotheses exist that describe biodiversity as well as its relationship to stressors. These hypotheses address nutrient cycling and ecosystem processes, the link between resilience toward toxic substances and biodiversity, and the link between concentration of toxic substances and microbial diversity.

Loreau (2001) developed a model predicting microbial diversity to have a positive effect on the efficiency of nutrient cycling and ecosystem processes through either a greater intensity of microbial exploitation of organic compounds or functional niche complementation. The insurance hypothesis (Yachi and Loreau 1999) describes a general principle of biological reactions—that is, that high diversity of microorganisms as well as other organisms increases resilience and provides insurance against large changes in eco-

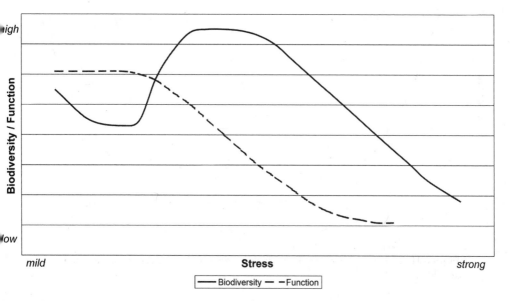

Figure 10.1. Interrelationship between stress and microbial diversity (according to the hypothesis of Giller, Witter, and McGrath 1998), and between stress and function. The position of the curves to one another may vary.

system processes. Giller, Witter, and McGrath (1998) hypothesized that, as a result of mild stress, competitive microbial species may predominate, causing a decline of diversity, whereas with increased stress, these individuals lose their competitive advantage, and more types may proliferate. At high levels of stress, progressive extinction of organisms and species leads to a loss of diversity and function (figure 10.1). As long as competitive species substitute for the function fulfilled by another species, the functionality of the microflora may not be disturbed. With greater stress, microbial activities will decrease, and the damage becomes obvious.

Besides the concentration of a chemical stressor, its properties may influence microbial functions and microbial diversity. It should be considered whether stressors are principally essential (e.g., trace elements like manganese, zinc, or iron, or nutrients such as ammonium and phosphate); degradable (e.g., polycyclic aromatic hydrocarbons [PAHs], mineral oil) or nondegradable (e.g., heavy metals); or intrinsically toxic.

If a stressor is essential, very low concentrations result in stimulated reactions, an effect known as hormesis. The dose-response curve of these elements is U shaped (figure 10.2). The bottom of the curve indicates the region of essentiality; the left-hand and right-hand portions define deficiency and excess (toxicity) (Baker et al. 2003). An area of no toxicity is expected to be within the bottom segment of the U shape, in the area of essentiality.

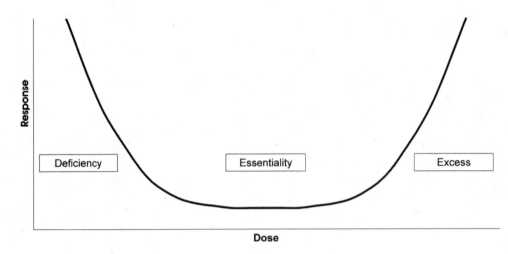

Figure 10.2. Dose-response construct for essential trace elements (according to Baker et al. 2003).

Concerning degradability and toxicity, several scenarios can be distinguished:

The stressor can be degraded, and no significant toxicity at the considered concentration exists. Microbial communities residing in contaminated ecosystems will be dominated by organisms that are capable of degrading and utilizing the stressor. The diversity might be influenced by the complexity of chemical mixtures present and the time period during which the populations have been exposed. On termination of the degradation process, the community changes again. Depending on the recovery potential of the system, the microbial diversity will bounce back to a condition similar to the original functional state.

The stressor is degradable, and a toxic potential at the considered concentration exists. The sensitive species will decline, and competing species will increase until concentration and toxicity reach a level that makes them decline as well. Furthermore, at concentrations also toxic for degraders, it is expected that increasing stressor concentrations will result in a decreasing rate of degradation. If a microorganism can perform the complete degradation of a stressor, remaining metabolites (dead-end metabolites) are not expected to occur. If an interaction of several microorganisms is needed for complete degradation, as every microorganism is able to perform only specific steps necessary for degradation, transformation will break down as soon as one of the relevant groups of microorganisms is inhibited and the transformation step cannot be performed. Metabolites will accumulate.

The stressor is not biodegradable. Microbial communities of contaminated ecosystems will be dominated by organisms less sensitive to the stressor; the complexity of the chemical mixtures present and the time period of exposure influence the diversity. This situation will pose the highest long-term risk to human health.

Case Studies of Microbial Diversity and Toxicity/Detoxification

The following section presents some examples from the literature as an overview of present knowledge on the issue of "microbial diversity and stressors." Clearly, knowledge is still limited. In all studies, only specific aspects are considered. The examples focus on natural diversity and the effects of chemical pollution. Effects caused by climatic changes and by species introduction are beyond the scope of this chapter.

When comparing the effects of stressors in various soils, it is important to consider the influence of different physicochemical soil properties. Physicochemical soil properties have an influence on (1) the composition of microbial communities and (2) the bioavailability of contaminants. Multiple factors, such as organic carbon (C_{org}), clay, acidity/alkalinity (pH), or ionic strength, may influence bioavailability and toxicity and, consequently, the effects of contaminants (Baker et al. 2003).

Furthermore, the pollution history, the introduction, and the concentration of a stressor in the soil are important factors influencing its effects (Giller, Witter, and McGrath 1998). In real-life contamination of soils, stressor concentrations may build up gradually over many years due to atmospheric deposition, the use of pesticides, or land application of contaminated wastes. In contrast, artificially contaminated soils (experiments) or soils polluted by environmental spills are rapidly exposed to single large doses.

Inorganic Stressors (Heavy Metals)

In laboratory experiments, Frostegård et al. (1993, 1996) investigated the effect of different concentrations of several heavy metals in a forest soil and an arable soil. The soils differed in their organic matter content and pH. The experiments revealed that, in both soils, relatively mild concentrations of stressors caused changes in biochemical diversity. Effects on total microbial biomass (determined as total amount of phospholipid fatty acids) and functional parameters (adenosine triphosphate, soil respiration) were obvious at higher concentrations. The results indicate that (1) different soils can be affected to a different extent and (2) the microbial community can be modified without obviously influencing overall microbial activity.

The results obtained for the artificially contaminated soils are principally confirmed by investigations of brownfields. For example, Müller et al. (2001) studied the genetic diversity of microbial communities of a 30-year-old mercury-contaminated soil. Experimental results showed that the genetic diversity decreased with increasing stressor (mercury) concentration, as also predicted in figure 10.1 at mild stressor concentrations.

Organic stressors

In laboratory experiments, the presence of 50 g/kg crude oil in a sandy soil resulted in variations of the functional, biochemical, and morphological diversity of microbial communities (Peressutti, Alvarez, and Pucci 2003). A selective enrichment of hydrocarbon-degrading bacteria was observed from 0.028% at the beginning of the assay to almost 100% two months after spillage, indicating the advantage of microorganisms capable of utilizing the new carbon source. Besides the overall increase of the hydrocarbon-utilizing microorganisms, an adaptation of the microbial community to the different components present in the hydrocarbon mixture of the crude oil could be observed. At the beginning, the microorganisms utilizing the more easily degradable hydrocarbons dominated (defined according to the structure of their cell wall as "gram-negative" bacteria). Then a shift in the bacterial community occurred, and the number of gram-positive bacteria increased. These organisms display broad metabolic capabilities and can utilize complex hydrocarbons as well as a variety of oxidation products, including cyclohexanone, salicylate, phenol, phenanthrene, and benzopyrene. Other experiments showed that, depending on the contamination, different groups of microorganisms dominated, indicating their selective advantage. The increase in the relative abundance of gram-negative bacteria in the presence of oil is assumed to be due to the growth of a limited number of species that belong to the alpha subgroup of the Proteobacteria and a less pronounced increase in species of the Flexibacter-Cytophaga-Bacteroides phylum (Macnaughton et al. 1999).

The *Acinetobacter* community was studied in soils with different oil contaminations, as earlier studies had indicated this community to be important for the degradation of the contaminants. Soils with mineral oil or kerosene contaminations contained relatively simple communities of *Acinetobacter*, whereas soils with contaminations of methyl tertiary butyl ether (MTBE; a gasoline component) or of the aromates benzene, toluene, ethylbenzene, and xylene (BTEX; gasoline components) had no detectable *Acinetobacter* populations (Vanbroekhoven et al. 2004).

Long-term experiments with the explosive TNT (trinitrotoluene) revealed that this substance favors a special group of microorganisms. A TNT-resistant community was formed in the presence of 29 g/kg TNT, and the biomass decreased, showing a low or no resistance toward TNT (Gong et al. 2000). The alteration of the microbial community by TNT can also be demonstrated and characterized by changes in the utilization spectrum of specific carbon substrates (Siciliano at al. 2000). In contrast, changes observed in genetic diversity were difficult to interpret: changes were observed, but whether they are linked to a modification in functional diversity remains unknown.

At a site contaminated with polyaromatic hydrocarbons (PAHs) and pentachlorophenol (PCP), it was shown that functional and biochemical diversity can differ. When comparing two bioremediation techniques, the extent of contaminant reduction was comparable (Hansen et al. 2004). In contrast, the formation of the microbial biomass and the diversity differed significantly. The concentration of PCP was in a range showing toxicity in other studies (Hund-Rinke and Simon 2004). The degradation of PCP during bioremediation clearly indicates the formation of a PCP-resistant microbial biomass.

A dose-response relationship was achieved for bacterial communities in three soils differing in soil properties and hydrocarbon pollution history. The size and functional diversity of the microbial community was higher in the slightly contaminated soil. In contrast, the genetic diversity was highest in soil with intermediate contamination and lower in uncontaminated and heavily long-term-contaminated soils (Andreoni et al. 2004). These results support the hypothesis described in figure 10.2.

Griffiths et al. (2000) found a decrease in diversity as an effect of fumigation. Later stresses due to copper addition or heat treatment showed that the more disturbed communities displayed less resilience (less resistance toward stressors). This indicates that high diversity increases resilience and protects against large changes in ecosystem processes, as suggested in the insurance hypothesis (Yachi and Loreau 1999).

Aspects of Relationships between Contamination and Diversity

The Role of Contamination Level and Biodegradability

Theoretically, optimal microbial diversity depends on the history as well as the present conditions in the soil. For example, an intact, undisturbed ecosystem is characterized by a relatively high diversity. However, as the number and concentrations of biodegradable pollutants increase, the diversity will first increase as microorganisms capable of mineralizing the organic contaminants adapt and proliferate. This increase in diversity is very beneficial for bioremediation. As the concentration of organic contaminants increases further, the diversity will decrease as sensitive organisms die off. Therefore, the soils contaminated with biodegradable pollutants are inhabited by a specialized microbial community, which is dominated by species capable of degradation. These changing diversities are very beneficial for bioremediation and have been modified from the indigenous diversity by adaptation and gene modification. These changes are the prerequisite for a disappearance of the contaminant and the reestablishment of a microbial community that resembles the undisturbed diversity.

In soils contaminated with nonbiodegradable toxic substances, a special microbial community dominated by insensitive species will establish. As the contaminant is not biodegradable, the adapted microflora will be present until the contaminant has disappeared as a result of, for example, leaching, sequestering, or accumulation in plant biomass.

If the soils are so heavily contaminated with nonbiodegradable toxic substances, or if the environmental conditions exclude degradation processes, the microbial diversity as well as activity will be low.

The Aspect of Structural and Functional Biodiversity

There are numerous data indicating effects on functional, biochemical, and genetic diversity in the presence of stressors. Addition of a stressor or modifications of the stressors (e.g., during bioremediation) results in changes of the microbial community. A number of studies (Chapin et al. 1997; Loreau 2001; Symstad et al. 1998; Tilman 1996) indicate that the relevance of changes in biochemical and genetic diversity is unclear without referring to functionality. A specific decrease in the diversity of critical soil functions may be significant (Giller, Witter, and McGrath 1998). On the other hand, if the function is still carried out but by a less diverse microbial community, the function is more sensitive to further stress. Studying the relation between the microbial diversity behind a function and the genetic or biochemical diversity should be given high priority.

In theory, dramatic scenarios may be created. For example, when one function is reduced by a stressor, which in turn affects further transformations, it could result in an amplification of the pollution effect and consequently have a severe impact on human health. Amplification of a pollution may be (1) the incomplete degradation of a pollutant with the formation of stable metabolites (dead-end metabolites) that are more toxic than the original pollution or (2) imbalanced natural nutrient cycles (e.g., carbon cycle, nitrogen cycle) with the increased formation of undesired compounds such as CO_2 or NO_x. For example, Fuller and Manning (1998) suggested that the explosive TNT causes a decrease in the rate of denitrification (step of the nitrogen cycle) and decomposition of soil organic matter.

Usually the impact of a stressor on soil microbial functions is assessed by studying selected functional aspects, such as respiration, nitrification, and degradation of selected carbon sources (e.g., according to ISO 1997, 2002, 2004; OECD 2000a, 2000b). These selected aspects are considered as isolated individual effects, and only rarely are interactions between various functions taken into account. Therefore, the knowledge concerning interactions is comparably limited. However, it is known that in soil ecosystems, microbial transformations may regulate each other as well as their predators and plants growing on the minerals recycled by the microorganisms. If, for example, the oxidation of nitrate to nitrite is inhibited and results in an accumulation of nitrite, the formation of nitrite from ammonium will be stopped. Comprehensive information on all direct and indirect effects of a stressor can only be achieved by complex long-term studies.

The Aspect of Availability

The number of microorganisms has been found to be high, even in highly polluted soils. This observation is probably due either to the microorganisms' being in a dormant state with a high tolerance to contaminants; the bioavailability of the contaminant, which again depends on the chemical properties of the contaminant; or the microorganisms' hiding in protective niches within the soil structure—for example, in very small micropores. By one or several of these strategies, microorganisms can avoid the toxic effects of the contaminant even though they are physically very close. In addition, the distribution of contaminants in soil is usually heterogeneous, depending on their chemical capabilities (e.g., hydrophobicity) and the soil structure—for example, the content of clay and organic matter. The microbial diversity in different soil aggregates is known to be different (Winding 1994).

Outlook on Research Needs

Currently, there is a gap in knowledge about the correlation of microbial diversity and related functions, ecosystem stability, resilience, and the respective effects of environmental impacts (Loreau et al. 2001). Systematic investigations into the potential consequences of a modified microbial structure on specific microbial functions and the resultant consequences for the environment have not yet been performed.

For a comprehensive understanding of the links between the composition of a microbial community, its function and activity, and the effects of multiple different stressors, numerous topics must be studied in more detail. Open questions are:

- What are the consequences resulting from a modified—that is, slightly affected—microbial community causing no detectable effects in functionality? To what extent will the resilience of a microbial community be decreased so that a further stressor will cause a stronger effect compared with its effect on a noncontaminated soil?
- Which concentration of a stressor will result in a structural change, and which concentration is needed to achieve functional modifications? In this context, the interactions of various functions and the impact for the ecosystem must be considered.
- Is pollution amplification a realistic scenario?
- To what extent can a decrease in microbial diversity and function be tolerated?
- Are there general principles concerning the correlation of stressors and the loss of specific microbial activities or transformations? Will the same functions always be deleted first, or is there a dependence on the stressor?
- Can genetically identified genomes or specific phospholipid fatty acids be linked to special functions?
- Can the microhabitats in which microorganisms survive the exposure to a stressor be characterized?
- Can the recolonization of the soil be characterized?
- What are the consequences for different soil types?

References

Andreoni, V., L. Cavalca, M. A. Rao, G. Nocerino, S. Bernasconi, E. Dell'Amico, M. Colombo, and L. Gianfreda. 2004. Bacterial communities and enzyme activities of PAHs polluted soils. *Chemosphere* 57:401–12.

Baker, S., M. Herrchen, K. Hund-Rinke, W. Klein, W. Kördel, W. Peijnenburg, and C. Rensing. 2003. Underlying issues including approaches and information needs in risk assessment. *Ecotoxicological and Environmental Safety* 56:6–19.

Chapin, F. S., B. J. Walker, R. J. Hobbs, D. U. Hooper, J. H. Lawton, O. E. Sasla, and D. Tilman. 1997. Biotic control over the functioning of ecosystems. *Science* 277:500–504.

Frostegård, A., A. Tunlid, and E. Bååth. 1993. Phospholipid fatty acid composition, biomass and activity of microbial communities from two soil types experimentally exposed to different heavy metals. *Applied and Environmental Microbiology* 59:3605–17.

———. 1996. Changes in microbial community structure during long-term incubation in two soils experimentally contaminated with metals. *Soil Biology and Biochemistry* 28:55–63.

Fuller, M. E., and J. F. Manning. 1998. Evidence for differential effects of 2,4,6-trinitrotoluene and other munitions compounds on specific subpopulations of soil microbial communities. *Environmental Toxicology and Chemistry* 17:2185–95.

Giller, K. E., E. Witter, and S. P. McGrath. 1998. Toxicity of heavy metals to microorganisms and microbial processes in agricultural soils: A review. *Soil Biology and Biochemistry* 30:1389–1414.

Gong, P., P. Gasparrine, D. Rho, J. Hawari, S. Thiboutot, G. Ampleman, and G. I. Sunahara. 2000. An in situ respirometric technique to measure pollution-induced microbial community tolerance in soils contaminated with 2,4,6-trinitrotoluene. *Ecotoxicological and Environmental Safety* 47:96–103.

Griffiths, B., K. Ritz, R. D. Bardgett, R. Cook, S. Christensen, F. Ekelund, S. J. Sørensen, et al. 2000. Ecosystem response of pasture soil communities to fumigation-induced microbial diversity reductions: An examination of the biodiversity–ecosystem function relationship. *Oikos* 90: 279–94.

Handelsman, J. 2004. Metagenomics: Application of genomics to uncultured microorganisms. *Microbiology and Molecular Biology Reviews* 68:669–85.

Hansen, L. D., C. Nestler, D. Ringelberg, and R. Bajpai. 2004. Extended bioremediation of PAH/PCP contaminated soil from the POPILE wood treatment facility. *Chemosphere* 54:1481–93.

Hund-Rinke, K., and M. Simon. 2004. Terrestrial ecotoxicity of eight chemicals in a systematic approach. *Journal of Soils and Sediments* 5:59–65.

ISO (International Organization for Standardization). 1997. Soil quality—Biological methods — Determination of nitrogen mineralization and nitrification in soils and the influence of chemicals on these processes. ISO 14238: 1997. International Organization for Standardization. http://www.iso.org/iso/home.htm.

———. 2002. Soil quality—Determination of abundance and activity of soil microflora using respiration curves. ISO 17155: 2002. International Organization for Standardization. http://www.iso.org/iso/home.htm.

———. 2004. Soil quality—Determination of potential nitrification and inhibition of nitrification—Rapid test by ammonium oxidation. ISO 15685: 2004. International Organization for Standardization. http://www.iso.org/iso/home.htm.

Loreau, M. 2001. Microbial diversity, producer-decomposer interactions and ecosystem processes: A theoretical model. *Proceedings of the Royal Society of London* B 268:303–9.

Loreau, M., S. Naeem, P. Inchausti, J. Bengtsson, J. P. Grime, A. Hector, D. U. Hooper, et al. 2001. Biodiversity and ecosystem functioning: Current knowledge and future challenges. *Science* 294:804–8.

Macnaughton, S. J., J. R. Stephen, A. D. Venosa, G. A. Davis, Y.-J. Chang, and D. C. White. 1999. Microbial population changes during bioremediation of an experimental oil spill. *Applied and Environmental Microbiology* 65:3566–74.

Müller, A. K., K. Westergaard, S. Christensen, and S. J. Sørensen. 2001. The effect of long-term mercury pollution on the soil microbial community. *FEMS Microbiology Ecology* 36:11–19.

OECD (Organisation for Economic Co-operation and Development). 2000a. Guidelines for the Testing of Chemicals. Test no. 216. Soil microorganisms: Nitrogen transformation test. http://www.oecd.org/document/62/0,3343,en_2649_34377_2348862_1_1_1_1,00.html.

———. 2000b. Guidelines for the testing of chemicals. Test no. 217: Soil microorganisms: Carbon transformation test. http://www.oecd.org/document/62/0,3343,en_2649_343 77_2348862_1_1_1_1,00.html.

Peressutti, S. R., H. M. Alvarez, and O. H. Pucci. 2003. Dynamics of hydrocarbon-degrading bacteriocenosis of an experimental oil pollution in Patagonian soil. *International Biodeterioration and Biodegradation* 51:299–302.

Siciliano, S. D., P. Gong, G. I. Sunahara, and C. W. Greer. 2000. Assessment of 2,4,6-trinitrotoluene toxicity in field soils by pollution-induced community tolerance, denaturing gradient gel electrophoresis and seed germination assay. *Environmental Toxicology and Chemistry* 19:2154–60.

Symstad, A. J., D. Tilman, J. Wilson, and J. M. H. Knops. 1998. Species loss and ecosystem functioning: Effects of species identity and community composition. *Oikos* 81:389–97.

Tilman, D. 1996. Biodiversity: Population versus ecosystem stability. *Ecology* 77:350–63.

Torsvik, V., R. Sørheim, and J. Goksøyr. 1996. Total bacterial diversity in soil and sediment communities: A review. *Journal of Industrial Microbiology and Biotechnology* 17:170–78.

Vanbroekhoven, K., A. Ryngaert, P. Wattiau, R. De Mot, and D. Springael. 2004. *Acinetobacter* diversity in environmental samples assessed by 16S rRNA gene PCR-DGGE fingerprinting. *FEMS Microbiology Ecology* 50:37–50.

Winding, A. 1994. Fingerprinting bacterial soil communities using Biolog microtitre plates. In *Beyond the biomass: Compositional and functional analysis of soil microbial communities,* ed. K. Ritz, J. Dighton, and K. E. Giller, 85–94. New York: Wiley.

Winding, A., K. Hund-Rinke, and M. Rutgers. 2005. The use of microorganisms in ecological soil classification and assessment. *Ecotoxicological and Environmental Safety* 62 (2): 230–48.

Yachi, S., and M. Loreau. 1999. Biodiversity and ecosystem productivity in a fluctuating environment: The insurance hypothesis. *Proceedings of the National Academy of Science USA* 96:1463–68.

PART IV
Biodiversity Change and the Spread of Infectious Diseases

A number of human and ecological factors have been identified with the emergence or spread of disease organisms, such as development and agricultural conversion of land, introductions of invasive species, and changes in host, vector, or prey populations. Certain types of habitat alteration, such as irrigation and removal of forests or other native vegetation, often increase the amount of standing water, thereby creating favorable conditions for disease-carrying vector populations. With an ever-expanding human population, people are penetrating farther into animal habitat and increasing human encounters with wildlife. The consequences can be twofold: higher rates of zoonotic disease transmission and overcrowding of increasingly smaller wildlife habitat, which facilitates the transmission of pathogens. In addition, global climate change is expected to exacerbate and even to facilitate the spread of some pathogens and their associated vectors.

Part IV of this book (chapters 11–14) focuses on the role of biodiversity in the spread of infectious diseases. Parasitic and infectious diseases are caused by living organisms and therefore constitute a part of global biodiversity. Microbial diversity reflects more than 3.8 billion years of adaptation to major environmental changes as well as to their own ever-evolving community. Microbes are found everywhere in soil, air, water, rock, and even in or on hosts. Some microorganisms are harmless while others are even beneficial, living as component parts of ecosystems and host bodies, where they help to decay and digest matter (e.g., detoxifying microbes, as discussed in chapter 10), or to destroy disease-causing microbes. To properly place microbial systems in the context of human health and sustainable development, we must first understand the basic ecology and natural history of microbes, and how that information relates to well-studied phenomena from macroplant and animal ecology. This section is provocative, because it includes the point of view that, with respect to microbial species, there may be times when humans would benefit from reduced, rather than increased, biological diversity.

In chapter 11, Jean-François Guégan and Hélène Broutin review what is known about microbial global diversity patterns and explore how these patterns relate to other well-known ecological patterns, such as the latitudinal species gradient, species/area relationships, productivity/diversity relationships, stability/diversity relationships, and resource/prey dynamics. The extent to which microbial groups obey ecological rules similar to those of better-studied groups has obvious ramifications for understanding global patterns of disease vulnerability in human populations. Guégan and Broutin shed light on these questions, revealing instances where data support basic ecological tenets and where data are sorely insufficient to interpret.

In chapter 12, Alain Froment explores the history of disease in the context of human and cultural evolution. A relatively small portion of parasitic and infectious diseases is restricted to humans, most having an animal reservoir. In the former case, eradication is possible, as was achieved with smallpox. However, Froment uses a suite of real-world examples to point out that in the latter case, eradication is almost out of reach without mass destruction of the natural world. Arguing from a historical framework, Froment proposes that disease should not be considered in isolation from a larger epidemiological context, and that an ecological approach to human health must occur within a holistic context, taking into account the pathogenicity of disease agents under conditions of early humans, preindustrial traditional societies, and modern industrial development.

In chapter 13, Matthew Thomas, Kevin Lafferty, and Carolyn Friedman review mechanisms by which biodiversity can affect the risk and spread of infectious disease. Their view of diversity is broad, including pathogen/parasite diversity, host species richness and species composition, and habitat and ecosystem diversity. Likewise, these authors delve into observed and hypothesized consequences of biodiversity loss on disease prevalence and transmission across a broad range of global drivers, including agricultural conversion, water development, loss of top predators, and invasive species. They extend the scope of their review to some of the current theoretical work in disease ecology and transmission, in concert with empirical evidence in support (or not) of those theories.

Camille Parmesan and Pim Martens in chapter 14 focus their review on a single global driver—anthropogenic climate change. Microbes exist as a part of nature, subject to the same climatic, and therefore geographic, constraints on their global distributions (as discussed in chapter 11), and just as susceptible to alterations of global climate as much more visible plants and animals. This chapter discusses the difficulties in attributing disease emergence or outbreaks to gradual climate change, given that most of the data are correlational, rather than experimental, in nature, and offers suggestions for improvement of study designs under these constraints. Even with these caveats, however, the authors find sufficient evidence to conclude that climate change has already affected human health. So far, much of the impact has been from direct effects of floods, storms, and heat waves, and hard evidence for three-way linkages between climate change, biodiversity change, and human health impacts is rare. Nevertheless, the few relatively clear examples that are highlighted in this chapter offer useful, if sobering, insight.

11

Microbial Communities: Patterns and Processes

Jean-François Guégan and Hélène Broutin

Microbes are essential components for Earth's global functioning. They cover a vast array of species from highly divergent groups, including the Archaea, bacteria, viruses, fungi, and Protista. Their diversity reflects more than 3.8 billion years of adaptation to major environmental changes as well as to their own ever-evolving community. Microbes are found everywhere in soil, air, water, and rock, and even in or on hosts, where most are harmless or even beneficial. They have adapted to a broad range of conditions, including extreme environments—for example, hot springs or freezing cold—where no other living forms can survive. They can be invisible invaders, creating diseases or becoming amazing allies for their hosts. Microbes are not just killers that have caused millions of deaths in humans. Some microorganisms are harmless, while others are even beneficial, living as component parts of ecosystems and host bodies, where they help to decay and digest matter or destroy disease-causing microbes. What is commonly considered the "true microbial world" is in reality a small fraction: only around 5% of bacteria and viruses may cause diseases in humans, animals, and plants (see this volume, chapter 12). The world of microbes is poorly understood because traditional studies by microbiologists and medical doctors focus only on the tip of the iceberg of microbial diversity. Microbes make up the majority of living biomass on Earth, forming a lost world of fascinating and unbelievable diversity. One common property is their capacity to cope with challenging environmental conditions, developing ingenious new survival tactics, making them what they are!

Nasty Pathogens versus Good Friends

Microbes are divided into nonliving substances—that is, viruses and other viruslike agents (e.g., prions: infectious proteins and viroids) that cannot self-replicate—and living organisms constituted by four domains: the prokaryotic Archaea and (Eu)bacteria, and the eukaryotic Fungi and Protista. Viruses cause numerous diseases of plants and animals (including humans), and even of bacteria and fungi. Prions have recently caused transmissible spongiform encephalopathies (TSEs), such as bovine spongiform encephalopathy (BSE, or "mad cow disease") and "kuru" in native populations of Papua New Guinea. The Archaea, recently discovered as a separate phylogenetic domain from bacteria, are tiny (usually less than one micron long) organisms found virtually everywhere. Research has shown that Archaea are abundant in plankton communities, and are unique in that they can live in extreme habitats.

Bacteria form an important group of microorganisms that both directly and indirectly provide human services. Some bacteria do invade host organisms, derive nutrition from host cells, and cause diseases in hosts, most notably by producing toxins. However, this is a small minority of this group. About 90% of known bacteria are beneficial microbes, serving as nitrogen-fixers that increase available nitrogen in soils and assist in the decomposition of dead plants and animals in ground litter, or living as internal mutualists—such as gut bacteria supplying vitamins, digesting food, and protecting against invaders. Moreover, many bacteria perform important functions such as fermentation of milk into cheese, or of apple juice into cider. Nowadays, genetically engineered bacteria may produce insulin, interferon, and other important products for medical concerns.

Fungi include single-celled organisms like yeasts and filamentous-like molds (which we are including in our discussion of microbes in this chapter), but also multicellular bunches like mushrooms. They feed on nutrients from living or dead organic materials, thus serving an important role in ecosystem services. Some fungi are useful to humans in that they produce natural antibiotics to fight against harmful bacteria, and digestive enzymes to brew beer. Pathogen fungi may cause disease in hosts, a trait that has been exploited by industries that produce biopesticides marketed for controlling target pests.

The last major group, Protista, includes primitive algae, protozoa, and slime molds. Slime molds, long classified within the fungi, are now known to form a separate group of nonpathogenic protists. Microscopic algae are found in fresh- and saltwater systems, as well as terrestrial systems, growing on rocks, tree trunks, or soil. Algae can even be found on animals, such as on the South American sloth (genera *Bradypus* and *Choloepus*). Photosynthetic algae produce a substantial amount of the oxygen we breathe. Conversely, aquatic algae may sometimes produce blooms or red tides that negatively impact ecosystems by depleting dissolved oxygen in the water. Dinoflagellates like *Pfiesteria* may produce toxic substances that cause bleeding sores in fish. Protozoans are classified into three groups:

- The ciliates, a diverse group that can act as predators (e.g., aquatic *Paramecium* and *Vorticella*); parasites (e.g., aquatic *Balantidium coli*); commensals (e.g., *Buetschelia* in horses); or symbiotes (e.g., *Entodinium* in ruminant guts of many mammals)

- The amoebas, subdivided into testate and naked categories, both of which may cause diseases
- The flagellates, with the majority causing no disease, but with notable exceptions like the protozoans *Cryptosporidium parvum* and *Plasmodium falciparum,* the parasite that causes malaria in human populations

How Many Species of Microbes Are There?

Despite their importance for human services and well-being, little is really known about the microbial world. There is general agreement that microbes house a high proportion of the Earth's species biodiversity, but there has been little attempt to fully describe individual species. The low percentage of described microbe species contrasts strongly with the relatively high percentage of described plants and vertebrates, though even in these latter groups, major discoveries are regularly reported. It has been estimated that microbial species make up about 60% of Earth's biomass, and up to 90% of all living things in the oceans. They also represent a large portion of life's genetic diversity (Whitman, Coleman, and Wiebe, 1998).

Though large in total biomass, microbes are often microscopic as individuals. Large viruses are approximately 100 nm (nanometers) in diameter, small bacteria like *Mycoplasma* are 150–250 nm, *E. coli* is 2 μm (micrometers), and large amoebas are up to 800 μm. By comparison, a ribosome in a typical human cell is 11 nm long.

Cases and de Lorenzo (2002) have indicated that we know of (have described) about 1.5 million species of animals, 0.25 to 0.3 million species of plants, and only around 0.2 million types of microorganisms (Margulis and Schwartz 1998). Current estimates indicate that more than 90% of the microorganisms remain unknown (Rondon et al. 2000; Torsvik, Goksøyr, and Daae 2002).

An indirect estimation only of bacteria indicates that there might be up to 0.3 million bacteria species (with estimates up to 5×10^{30}), but only 3,000 to 4,000 bacteria species have been described (Hawksworth and Colwell 1992). A more recent estimate is up to 5,000 described bacteria species (Pace 1999). Meeûs, Durand, and Renaud (2001) claim there are 9,280 known bacterial species (see also Lecointre and Le Guyader 2001). The number of viral taxonomic units (genetic isolates) has been estimated to be 3,500–4,000 by the International Committee on Taxonomy of Viruses (Kommedahl and Robinson 2001), with exactly 1,938 known species. A more recent council estimated that there are 6,000 taxonomic units (Fauquet et al. 2004).

The vast majority of microorganisms in the environment remain unculturable; as a consequence, our understanding of their diversity is limited, and good estimates are elusive to calculate. We may be moving from an extreme underestimation of species diversity to possibly a severe overestimation. Microbiologists and microbial ecologists are unfortunately faced with an identification nightmare, so the answer to the question, How many microbes are there? led to interest in Who are they? and What are they doing? (Ward 2002). We are just beginning to appreciate that these often ultramicroscopic entities provide essential ecological services.

Microbial "Species," You Said!

The species concept is very controversial in the microbial world (Weinbauer and Rassoulzadegan 2004). In contrast to higher taxonomic groups, in microbes it is difficult to discern clear-cut species. The biological species concept was articulated by the ornithologist Ernst Mayr, but its use has been seriously criticized for many other taxa. Microorganisms in particular possess traits that do not fit the biological species concept (Meeûs, Durand, and F. Renaud 2001). Prokaryotes, like bacteria, may exchange genetic materials between and across "species" (Ochman, Lawrence, and Groisman 2000), rendering this definition inappropriate. For essentially asexual haploid organisms such as prokaryotes, mutations are also a major source of genetic diversity and one of the essential factors in the formation of novel forms (Whitman, Coleman, and Wiebe 1998). Reconsideration of the species concept has been possible partially because of important technological advances in the field (Weinbauer and Rassoulzadegan 2004). For bacteria, the operational definition of a species is "a monophyletic and genomically coherent cluster of individual organisms that show a high degree of overall similarity in many characteristics, and is diagnosable by a discriminative phenotypic property" (Rosselló-Mora and Amann 2001). Some authors have also argued that a more "natural view" of microorganisms should be adopted that would define microbial species in the context of their natural ecological and evolutionary framework, and not just from a high-tech, genome-focused laboratory perspective (Ward 1998). "Metagenomics," which would simultaneously incorporate both genomic and environmental information, is the logical next step to be employed as a means of systematically investigating and classifying microbial diversity in the wild (Weinbauer and Rassoulzadegan 2004).

Quantitative Measures of Microbial Diversity

Simply counting microbial species is a herculean, endless task. But, then, what would be a more appropriate strategy to describe microbial diversity? The answer could come from the development of more and more sophisticated molecular biological methods in environmental microbiology (Ward 1998; Weinbauer and Rassoulzadegan 2004). DNA-DNA hybridization is a crucial tool for characterization of microbe isolates, with a 70% hybridization corresponding to around 96% sequence similarity, used as delineation between "species" (Stackebrandt and Goebel 1994; Weinbauer and Rassoulzadegan 2004). The widely used method by microbial ecologists for evaluating genetic diversity in environmental samples is 16S ribosomal DNA gene analysis, which uses a threshold of 97% similarity of the 16S rRNA gene that roughly matches the concept of 70% hybridization (Stackebrandt and Goebel 1994). Amann, Hawksworth, and Colwell

(1995), Dykhuizen (1998), and Girvan et al. (2005) for bacteria, and Weinbauer and Rassoulzadegan (2004) for viruses, respectively, have provided some reviews on the different molecular techniques having great potential for microbial diversity analysis.

In the Sargasso Sea, water samples from 1 mm up to 10 liters contained up to 160 bacterial species, as defined using metagenomics (Torsvik, Øvreä, and Thingstad 2002). This value matches well with that calculated on the basis of lognormal distribution curves, of 163 species (Curtis, Sloan, and Scannell 2002). But these estimates were questioned by Venter et al. (2004), who found 148 new phylotypes in this same well-surveyed area, which led them to estimate true bacterial diversity as being from 1,800 to 45,000 taxa. Investigations that have targeted algal or cyanophages (using different approaches, from molecular to statistical techniques) estimated up to 36 distinct viral groups in similar-sized seawater samples (from few to tens of liters) (Steward, Montiel, and Azam 2000). When assessing the metagenome of viral communities in two marine samples of 200 liters from coastal waters, the estimated number of viral genotypes was evaluated to be between 374 and 7,114 (Breitbart et al. 2002). On assumption that every prokaryotic species should harbor at least one virus species type, these findings clearly show that viral diversity is higher than that of prokaryotes on the order of 10 specific viruses per bacterial species (Weinbauer and Rassoulzadegan 2004; Wommack and Colwell 2000). This value also matches well with data on isolated bacteriophages (Weinbauer 2004). Weinbauer and Rassoulzadegan (2004) have recently provided a review of microbial diversity estimates using different methods of calculation.

A reconsideration of data in Whitman, Coleman, and Wiebe (1998) on prokaryotic cells in soils from different ecosystems shows that their number is a rough linear function of the total surveyed surface area across the different samplings (see also Venter et al. 2004). This trend, a phenomenon well known in community ecology, clearly indicates the strong necessity of considering sampling effects in any microorganism biodiversity estimates, notably because of their relevance to species-area relationships (Rosenzweig 1995). But, then, how to cope with the challenge of measuring microorganism biodiversity? Given the economic cost of molecular techniques and the difficulty of measuring microbial species from the field, one recent solution has been to use available statistical methods inherited from traditional community ecology (Curtis, Sloan, and Scannell 2002; Lunn, Sloan, and Curtis 2004). These methods are based on the simple observation that for large communities of free-living organisms, the frequency distribution of species abundances is a bell-shaped lognormal curve, with most species having an intermediate number of individuals and few species having very small or very large populations (Begon, Harper, and Townsend 1996). This trend can be used to estimate the expected number of microorganism species at both small and large scales from limited data (figure 11.1).

Using a community ecology approach, Mullins et al. (1995) estimated that the number of bacterial species in water samples from the Sargasso Sea was about 163 taxa. This value appears to be consistent with the value estimated by Torsvik, Øvreä, and Thingstad

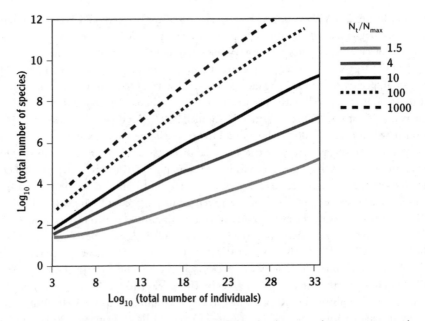

Figure 11.1. The double-logarithmic positive relationship between the maximum number of bacterial species and the total number of individual bacterial cells (in log) in the environment, under the basic assumption made by Curtis et al. (2002) that $N_{min} = 1$. The maximum bacterial species richness depends on N_T/N_{max} ratios—that is, the proportional abundance of the most abundant taxon, with a ratio of 4 corresponding to water systems, and of 1,000– 100 corresponding closer to soils and sediments. Adapted from Curtis et al. 2002.

(2002), but see recent discoveries by Venter et al. (2004). Using the same approach, Mullins et al. (1995) also estimated the number of bacterial species in a gram of soil at 6,300 taxa, an estimate identical to the value proposed by Torsvik, Goksøyr, and Daae (1990) (but see Dykhuizen 1998 for contradictory results). However, such parametric methods are highly dependent on the assumption of a lognormal species abundance distribution and a particular minimum abundance (typically 1 species at an abundance of 1), which can lead to an overestimate of community species diversity when this assumption is not met (Lunn, Sloan, and Curtis 2004). Hughes et al. (2001) and Ward (2002) indicated that there was insufficient information available today to assume that microbial populations are lognormally distributed.

Other statistical methods such as nonparametric approaches and community phylogenetics developed by community ecologists and evolutionary biologists exist to analyze microbial biodiversity data, reviewed by Bohannan and Hughes (2003). Contrary to parametric methods, nonparametric approaches do not assume any particular species abundance distribution model, and most of them are based on the probability of detecting a taxon in a given environment based on capture-recapture methods inherited from population biology (see Bohannan and Hughes 2003). For instance, Hughes et al. (2001) obtained estimates of around 467 to 590 taxa using clone libraries from two grazed grassland soils, whereas Curtis, Sloan, and Scannell (2002), analyzing the same data under the assumption of a bell-shaped species abundance distribution, reached an estimate of 6,300 species.

Interestingly, to understand microbial systems, we will have to correctly consider patterns of biodiversity that are observed to occur, or are likely to occur (Curtis and Sloan 2004). Curtis, Sloan, and Scannell (2002) have undoubtedly developed a useful starting point from which to build a new quantitative theory of microbial ecology and microbial biodiversity measurements. Examples that follow show that many other patterns of microbial biodiversity might still exist that, to our minds, should be a prerequisite for testing whether microbial species richness and relative abundances can be tied into a more general theoretical framework (Finlay 2002).

Important Microbial Community Species Patterns to Explore

One of the most consistent large-scale trends in biogeography and macroecology is the latitudinal gradient in species richness, with a tendency toward higher diversity in the tropics (Chown et al. 2004; Hawkins et al. 2003; Hillebrand 2004; Willig, Kaufman, and Stevens 2003). This pattern holds for most large plants and animals, although there are a few exceptions (Brown 1995; Gaston and Blackburn 2000; Rosenzweig 1995). According to Finlay (2002), however, [all] microbial species do not show biogeographic patterns, because they are sufficiently abundant to have worldwide distribution. But detailed field studies for some tiny microbes like parasitic and infectious disease organisms give evidence for a latitudinal diversity gradient similar to that found for larger organisms (Guernier, Hochberg, and Guégan 2004; Nunn et al. 2005), although, as with macroecology, exceptions do exist (Poulin 2001; Poulin and Mouritsen 2003).

According to Hillebrand and Azovsky (2001) and Azovksy (2002), the latitudinal gradient of diversity is most pronounced for large organisms and decreases in strength with decreasing body size. Does this apparently fundamental characteristic of biodiversity extend to microscopic organisms like bacteria, viruses, or fungi? The extent to which there is a tendency toward greater microbial diversity in the tropics seems particularly important to know, especially because the tropics generally support more complex and/ or more productive food webs than do higher latitudes. For instance, the production of planktonic heterotrophic bacteria represents a large fraction of the flow of energy and matter in aquatic systems, and prokaryotic photoautotrophs, including cyanobacteria and prochlorophytes, are very important in primary production. We need to know more about the biogeography of these organisms if we want to make an important step toward understanding the processes that are behind observed patterns of microbial diversity (Horner-Devine et al. 2003).

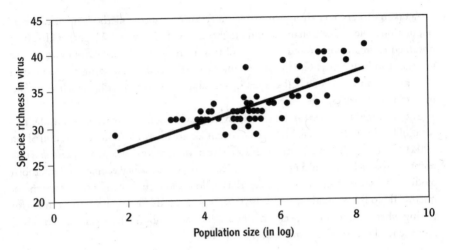

Figure 11.2. Linear relationship between human population size and species diversity in viruses across 71 human populations on islands. The linear regression is $y = 1.67x + 23.97$, $r^2 = 0.551$, $p<0.0001$. The population size variable is log-transformed. From J. F. Guégan, personal data.

Host Population Size and Microbial Diversity

Many animal and plant species contain dense populations of microorganisms that play important roles in nutrition—as the prokaryotes do in the gastrointestinal tracts of animals—as well as in disease, as in the case of viruses impacting host populations (Whitman, Kaufman, and Stevens 1998). Unfortunately, values for host population sizes are known only for a small number of domestic animals and human populations. Many recent studies have clearly shown the importance of population size for an infectious disease agent to persist within human communities, and studies on measles or whooping cough are now classical illustrations of the importance of host density in microbial success (Broutin et al. 2004; Grenfell and Harwood 1997; Rohani, Earn, and Grenfell 1999). Intuitively, larger host population size should also favor higher microbe species diversity, but there are no clear actual demonstrations that this hypothesis is correct (Guégan, Morand, and Poulin 2005; Morand and Guégan 2000). Human pathogens may provide a clue to test this hypothesis since humans are the best-documented animal species on Earth. Figure 11.2 illustrates the positive relationship observed between human population size and species diversity of viruses across 71 different islands all over the world. This observation is in keeping with the "critical community size" concept (Grenfell and Harwood 1997). But to what extent might this pattern apply to other tiny associated microbes? Because of empirical evidence that host resources strongly influence microbial ecology, research on microorganisms should give better consideration to their associated hosts, including local host characteristics.

Surface Area and Microbial Diversity

As pointed out by Horner-Devine et al. (2004), species-area relationships have not been examined explicitly for many microbial taxonomic categories. The absence of such a pattern is in contradiction with the many observations that free-living bacteria are heterogeneously distributed in space and that environmental heterogeneity can alter their distribution. Empirically resolving species-area relationships can be informative in that differences in characteristics such as dispersal ability, survival, or habitat conditions might result in a different species-area relationship within and between some taxonomic groups of microorganisms (see Finlay 2002; Horner-Devine et al. 2004; K. Smith et al., 2007). The general equation of cumulative increase in species number with increasing surface area is of the form $S = cA^z$, with S = number of species, A = surface area, and c and z = two constants. Empirically derived estimates for the slope coefficient z range from values of around 0.043 for ciliate species to 0.31 for insect species (Finlay 2002). According to Finlay, a high local/global species ratio will produce relatively flat species-area curves.

Horner-Devine et al. (2004) studied the bacteria species–area relationship in a salt marsh of New England in the United States, and found among the lowest z-values, with z being the slope of the species-area curve, reported for any organisms in the world. They went one step further by demonstrating that this bacterial species–area pattern is driven primarily by environmental heterogeneity and not by geographic distance or by plant species composition. Smith et al. (2005) have also reported a species-area curve for phytoplankton microalgae in different water systems, demonstrating that the z-exponent value was invariant across aquatic ecosystems from small-scale experimental ecosystems to lakes and oceans, then confirming that this pattern for microbial diversity was consistent with those reported for macroorganisms, whatever the magnitude of spatial scale. This result suggests that species-area curves might serve to inform us about microbial diversity at the whole-ecosystem level based on small-scale outdoor mesocosms and laboratory microcosm experiments.

The Regional-to-Local Microbial Diversity Relationship

Even if area size may control the maximum microbial species attainable at any site, local communities represent subsets of regional species pools (Guégan, Morand, and Poulin 2005). Dispersal limitation and thus regional control of biological diversity has become a major topic in ecology over the past decade (Ricklefs and Schluter 1993; Tilman and Kareiva 1997), and it is becoming an important domain of research in parasitology and epidemiology (Guégan, Morand, and Poulin 2005; Holt and Boulinier 2005). It is now implicit from many ecological studies that regional and historical processes may profoundly affect local community diversity and composition in free-living organisms (Gaston and Blackburn 2000; Lawton 1999, 2000). Questions of spatial scale have been addressed only very recently for free-living microorganisms (Fenschel, Esteban, and Finlay 1997; Finlay, Maberly, and Cooper 1997; Smith et al. 2005; Wardle, Bonner, and Nicholson 1997). For instance, Finlay, Maberly, and Cooper (1997) found very high local species richness of free-living ciliates compared with their regional or even global species richness, and they suggested that high dispersal capacity of these unicellular organisms ("everything is everywhere") should mean that microbial diversity does not play an important role in variation in ecosystem functioning.

The paper by Hillebrand et al. (2001) is really one of the first studies to analyze the importance of the spatial dimension in potentially shaping local species richness assemblages for different taxonomic groups of unicellular taxa (e.g., microalgae, protozoans, and ciliates) using appropriate statistical methods—that is, meta-analyses. The relative absence of consideration of space in microorganism studies, for both free-living and associated forms, is a common tradition in bacteriology, microbiology, parasitology, and epidemiology. These traditional fields tend to focus only on very fine scales, considering de facto that biogeographical processes of microbe diversity do not matter (Guégan, Morand, and Poulin 2005). However, recent evidence suggests that, depending on the microbe taxa, biogeographical patterns might emerge (see the recent review by Horner-Devine et al. 2004). The work by Smith et al. (2005), for instance, clearly emphasizes the need to consider space, such as how distance affects emigration-colonization balances, since generating observed local patterns of microbial diversity (Havel and Shurin 2004), as well as other local influences—for example, resource limitation, disturbance frequency, and strength of interaction between species (Hillebrand and Blenckner 2002).

The Productivity-Microbial Diversity Relationship

The relationship between macroorganism (especially plant) diversity and productivity has received much attention in ecology, but this relationship has been little explored for microbial communities. Responses of species and communities to local habitat productivity has been observed for many organisms (Chase and Leibold 2002; Mittlebach et al. 2001), and local factors known to influence distributions of species include area size, chemical environment, supply of limiting nutrients, competition, and predation-consumption (Havel and Shurin 2004). Horner-Devine et al. (2003) reviewed studies of the distribution of free-living bacterial diversity in aquatic mesocosms, and compared those findings with what we know of patterns in plants and animals. They showed that bacterial diversity may vary along a gradient of primary productivity—that is, based on chlorophyll-a concentration, in experimental mesocosms mimicking natural small ponds, but that different taxonomic groups of bacteria exhibit different responses to primary productivity variation, from being hump shaped (Cytophaga-Flavobacteria-Bacteroides group) to U shaped (a-Proteobacteria) to no relationship (b-Proteobacteria). These patterns were consistent across many of the reviewed studies. Generally, total bacterial abundance rose with an increase of primary productivity across aquatic mesocosms, suggesting that, in general, energy availability is a fundamental constraint to microbial diversity, at least for aquatic systems. Concerning soil microbial communities that use carbon resources derived from plants, microbial diversity may also respond to changes in plant diversity or productivity (Broughton and Gross 2000). This finding would be analogous to observed results from host/pathogen dynamics, in which host abundance may be considered as a measure of energy supply for pathogens (figure 11.2). While many factors might be at work, the supply of energy availability has been demonstrated to be determinant in shaping species diversity in a region (Mittlebach et al. 2001). Understanding observed variation in productivity–microbial diversity relationship is thus crucial for future work on microorganisms.

Microbial Diversity and Ecosystem Processes

To what extent might members of the tiny microbial world be keystone species within ecosystems? Microorganisms are involved in the making of several food products, such as yogurt, cheese, sauerkraut, wine, and beer. They have been the primary source of several antibiotics, fungicides, and insecticides. They are keys in the remediation of polluted sites, and are essential in wastewater treatment processes and in the rehabilitation of our natural water streams (see this volume, chapters 8 and 10). Moreover, they have adapted to several higher organisms, including humans, and some have coevolved with these organisms. Ironically, given that we have clearly profited from their socioeconomic benefits, it is surprising that their importance in ecosystem functioning has clearly been neglected until recently (Allsopp, Hawksworth, and Colwell 1995; Mitchell 2003). The study of "laboratory adapted" microorganisms—necessary because the vast majority of microbial species cannot be cultivated—certainly has had a strong impact on our understanding of their functional role within ecosystems. In this final section, we will focus on the potential for microbes to control major ecosystem functioning.

Microbial Diversity and the Element Cycle of Ecosystems

One major property of ecosystems is their capacity to generate and to circulate substances, such as carbohydrates and nutrients, and some groups of microorganisms may be keystone species, or have keystone effects within this framework (see this volume, chapters 8–10). Microbes range from organisms that decompose macromolecules to organisms that are responsible for the breakdown of specific chemical compounds. They may exhibit a large variety of functional properties within ecosystems—for example, metabolic pathways—and this diversity of microbial taxa is essential to ecosystem maintenance and functioning. Using a simple ecosystem model linking primary plant producers and microorganisms decomposers through a limiting nutrient factor, Loreau (2001) showed that the efficiency of nutrient recycling from organic compounds to decomposers was a key parameter controlling ecosystem productivity and biomass. Microbial diversity had a positive effect on nutrient cycling efficiency, and contributed to increased ecosystem processes.

More generally, the functional role of microorganisms is highly apparent in the following ways:

- They are primary producers that synthesize organic matter through photosynthetic and chemosynthetic processes.
- They decompose organic matter to release inorganic (e.g., mineralization) or organic compounds (as in the case of lignin decomposers).
- They produce nutrient-rich food sources for other microorganisms and animal predators.
- They change the amounts of material in soluble and gaseous forms.
- They make inhibitory compounds that limit the survival and functioning of other organisms.
- They contribute to the functioning of plants and animals through positive and negative symbiotic interactions.

Microorganisms have a major importance in nutrient cycling interactions involving both biological and chemical processes, most notably the carbon, sulfur, nitrogen, and iron cycles. In every situation, microbes affect the balance between inputs and outputs, which in turn determines the whole ecosystem response. The relationship between microbial diversity and ecosystem processes might not be linear because many processes are carried out by a consortium of microorganisms playing in concert within the system (Torsvik and Øvreäs 2002), thus certainly complicating the diversity-functioning relationships.

Microbial Diversity, Community Stability, and Ecosystem Resilience

Biodiversity has been assumed to influence ecosystem stability, productivity, and resilience toward environmental stress and disturbance, diversity tending to stabilize community and ecosystem processes in plant systems (Tilman 1996). The relationship between microbial diversity and function is largely unknown, and recent insights into soil ecology have addressed this issue. Girvan et al. (2005) investigated the effect of severe stress and disturbance on the diversity and resilience of different soil microbial communities. They perturbed different soil (from mineral to more diverse organomineral) systems using copper sulfate and benzene doses of equivalent toxicity. Environmental pollution caused by benzene yielded important decreases in total bacterial numbers and biomass in all soils but, in contrast to copper treatment, led to larger shifts in bacterial community structure. The mineral soil appeared to be less stable than the organomineral soil to environmental perturbations, the latter exhibiting greater resistance and functional resilience. The organomineral soil recovered its mineralization function by the end of the experimentation, but the mineral soil did not.

These findings on soil bacterial communities support the view that decreased diversity could lead to decreased stability following environmental perturbations, a scenario reflecting adaptation by the surviving microbial communities to stress. Similar results have been reached in manipulated aquatic systems (McGrady-Steed, Harris, and Morin 1997) and other experimentations on soil communities (Griffiths et al. 2000). Bardgett (2002) has reached different conclusions in that soil diversity is better attributed to the nature of the soil environment. There is now increasing evidence showing the existence of aboveground-belowground feedbacks in controlling ecosystem processes and properties. Plants provide both the organic carbon and the resources for the belowground-life kingdom, which in turn breaks down dead plant material and regulates the aboveground life by determining the supply of available soil nutrients (Wardle et al. 2004). Mutualist symbiots in soil may also influence plant community structure and composition. As such, aboveground and belowground communities interact with each other, and they should be studied in concert in the future since both positive and negative feedbacks may happen.

Microorganisms and Ecosystem Functioning: Implications for Global Environmental Change

Despite their acknowledged importance in ecology, most research on microorganisms has focused on molecular genetics, and on individual and population investigations. Exploration of the interface between population and ecosystem-level ecology has been too seldom envisaged (Mitchell 2003; Wardle et al. 2004), and this relationship should be thoroughly considered in the near future. From the limited evidence available, it appears that the microbial world may play an important role in ecosystem performance. Current patterns of alteration of climate through atmospheric CO_2 enrichment and ozone depletion, nitrogen deposition, invasion of alien host and pathogen species into new territories, land use changes, and use of pesticides, insecticides, or other drugs will have consequences on the interactions between ecosystems and microorganisms (Townsend et al. 2003; see also this volume, chapter 9). Numerous human-induced disturbances impinge on microbial functioning—for instance, in the interactions between plant communities and arbuscular mycorrhizae (Rillig 2004). Because of the variety of ways in which ecosystem processes and microorganisms are interconnected with each other, direct and indirect effects on global change may occur in a cascade-type phenomenon. A growing number of studies point to how anthropogenic changes can affect soil microorganisms, which in turn may affect plant biota (Wardle et al. 2004), or can impact infectious disease distribution and their vectors through changes in azote availability, thus impacting human health (Townsend et al. 2003). New insights from studies on the role of microorganisms in ecosystem maintenance and functioning will improve our understanding of the effects of anthropogenic changes on biodiversity and ecosystem properties.

Acknowledgments

This work was supported by the Institut de Recherche pour le Développement and the Centre National de la Recherche Scientifique.

References

Allsopp, D., D. L. Hawksworth, and R. R. Colwell. 1995. *Microbial diversity and ecosystem function*. Wallingford, UK: CABI Publishing.

Amann, R. I., W. L. Ludwig, and K. H. Schleifer. 1995. Phylogenetic identification and in situ detection of individual microbial cells without cultivation. *Microbiological Reviews* 59:143–69.

Azovsky, A. I. 2002. Size-dependent species-area relationships in benthos: Is the world more diverse for microbes? *Ecography* 25:273–82.

Bardgett, R. D. 2002. Causes and consequences of biological diversity in soil. *Zoology* 105:367–74.

Begon, M., J. L. Harper, and C. R. Townsend. 1996. *Ecology: Individuals, populations and communities*. 3rd ed. Oxford: Blackwell Science.

Bohannan, B. J. M., and J. Hughes. 2003. New approaches to analyzing microbial biodiversity data. *Current Opinion in Microbiology* 6:282–87.

Breitbart, M., P. Salamon, B. Andresen, J. M. Mahaffy, A. M. Segall, D. Mead, F. Azam, and F. Rohwer. 2002. Genomic analysis of uncultured marine viral communities. *Proceedings of the National Academy of Sciences USA* 99:14250–55.

Broughton, L. C., and K. L. Gross. 2000. Patterns of diversity in plant and soil microbial communities along a productivity gradient in a Michigan old-field. *Oecologia* 125:420–27.

Broutin, H., E. Elguero, F. Simondon, and J. F. Guégan. 2004. Spatial dynamics behaviour of pertussis in a small area in Senegal. *Proceedings of the Royal Society of London* B 217:2091–98.

Brown, J. H. 1995. *Macroecology*. Chicago: University of Chicago Press.

Cases, I., and V. de Lorenzo. 2002. The grammar of (micro)biological diversity. *Environmental Microbiology* 4:623–27.

Chase, J. M., and M. A. Leibold. 2002. Spatial scale dictates the productivity-biodiversity relationship. *Nature* 416:427–30.

Chown, S. L., B. J. Sinclair, H. P. Leinass, and K. J. Gaston. 2004. Hemispheric asymmetries in biodiversity: A serious matter for ecology. *PLoS Biology* 2:1701–7.

Curtis, T. P., and W. T. Sloan. 2004. Prokaryotic diversity and its limits: Microbial community structure in nature and implications for microbial ecology. *Current Opinion in Microbiology* 7:221–26.

Curtis, T. P., W. T. Sloan, and J. W. Scannell. 2002. Estimating prokaryotic diversity and its limits. *Proceedings of the National Academy of Sciences USA* 99:10494–99.

Dykhuizen, D. E. 1998. Santa Rosalia revisited: Why are there so many species of bacteria? *Antonie van Leeuwenhoek* 73:25–33.

Fauquet, C., M. Mayo, J. Maniloff, and L. Ball. 2004. *Virus taxonomy: The eighth report of the International Committee on Taxonomy of Viruses*. New York: Academic Press.

Fenchel, T., G. F. Esteban, and B. J. Finlay. 1997. Local versus global diversity of microorganisms: Cryptic diversity of ciliated protozoa. *Oikos* 80:220–25.

Finlay, B. J. 2002. Global dispersal of free-living microbial eukaryote species. *Science* 296:1061–63.

Finlay, B. J., S. C. Maberly, and J. I. Cooper. 1997. Microbial diversity and ecosystem function. *Oikos* 80:209–13.

Gaston, K. J., and T. M. Blackburn. 2000. *Pattern and process in macroecology.* Oxford: Blackwell Science.

Girvan, M. S., C. D. Campbell, K. Killham, J. I. Prosser, and L. A. Glover. 2005. Bacterial diversity promotes community stability and functional resilience after perturbation. *Environmental Microbiology* 7:301–13.

Grenfell, B. T., and J. Harwood. 1997. (Meta)population dynamics of infectious diseases. *Trends in Ecology and Evolution* 12:395–99.

Griffiths, B. S., K. Ritz, R. D. Bardgett, R. Cook, S. Christensen, F. Ekelund, S. J. Sørensen, et al. 2000. Ecosystem response of pasture soil communities to fumigation-induced microbial diversity reductions: An examination of the biodiversity-ecosystem functioning relationship. *Oikos* 90:279–94.

Guégan, J. F., S. Morand, and R. Poulin. 2005. Are there general laws in parasite community ecology? The emergence of spatial parasitology and epidemiology. In *Parasitism and ecosystems*, ed. F. Thomas, F. Renaud, and J. F. Guégan, 22–42. Oxford: Oxford University Press.

Guernier, V., M. E. Hochberg, and J. F. Guégan. 2004. Ecology drives the worldwide distribution of human infectious diseases. *PLoS Biology* 2:740–46.

Havel, J. E., and J. B. Shurin. 2004. Mechanisms, effects, and scales of dispersal in freshwater zooplankton. *Limnology and Oceanography* 49:1229–38.

Hawkins, B. A., R. Field, H. V. Cornell, D. J. Curie, J. F. Guégan, D. M. Kaufman, J. T. Kerr, et al. 2003. Energy, water, and broad-scale geographic patterns of species richness. *Ecology* 84:305–17.

Hawksworth, D., and R. R. Colwell. 1992. Biodiversity amongst microorganisms and its relevance. *Biodiversity and Conservation* 1:219–345.

Hillebrand, H. 2004. On the generality of the latitudinal diversity gradient. *American Naturalist* 163:192–211.

Hillebrand, H., and A. I. Azovsky. 2001. Body size determines the strength of the latitudinal diversity gradient. *Ecography* 24:251–56.

Hillebrand, H., and T. Blenckner. 2002. Regional and local impact on species diversity: From pattern to processes. *Oecologia* 132:479–91.

Hillebrand, H., F. Watermann, R. Karez, and U.-G. Berninger. 2001. Differences in species richness patterns between unicellular and multicellular organisms. *Oecologia* 126:114–24.

Holt, R. D., and T. Boulinier. 2005. Ecosystems and parasitism: the spatial dimension. In *Parasitism and ecosystems*, ed. F. Thomas, F. Renaud, and J. F. Guégan, 68–84. Oxford: Oxford University Press.

Horner-Devine, M. C., M. Lage, J. B. Hughes, and J. M. Bohannan. 2004. A taxa-area relationship for bacteria. *Nature* 432:750–53.

Horner-Devine, M. C., M. A. Leibold, V. H. Smith, and B. J. M. Bohannan. 2003. Bacterial diversity patterns along a gradient of primary productivity. *Ecology Letters* 6:613–22.

Hughes, J. B., J. J. Hellmann, T. H. Ricketts, and B. J. M. Bohannan. 2001. Counting the uncountable: Statistical approaches to estimating microbial diversity. *Applied Environmental Microbiology* 67:4399–4406.

Kommedahl, T., and P. Robinson. 2001. Updates on nomenclatures of viruses, plants, and animals. *Science* 24:86–87.

Lawton, J. H. 1999. Are there general laws in ecology? *Oikos* 84:177–92.

———. 2000. *Community ecology in a changing world*. Oldendorf-Kuhe, Germany: Ecology Institute.

Lecointre, G., and H. Le Guyader. 2001. *Classification phylogénétique du vivant*. Paris: Éditions Belin.

Loreau, M. 2001. Microbial diversity, producer-decomposer interactions and ecosystem processes: A theoretical model. *Proceedings of the Royal Society of London* B 268:303–9.

Lunn, M., W. T. Sloan, and T. P. Curtis. 2004. Estimating bacterial diversity from clone libraries with flat rank abundance distributions. *Environmental Microbiology* 6:1081–85.

Margulis, L., and K. V. Schwartz. 1998. *Five kingdoms: An illustrated guide to the phyla of life on earth*. New York: Freeman.

McGrady-Steed, J., P. M. Harris, and P. J. Morin. 1997. Biodiversity regulates ecosystem predictability. *Nature* 390:162–65.

Meeûs, T. de, P. Durand, and F. Renaud. 2001. Species concept: What for? *Trends in Parasitology* 19:425–27.

Mitchell, C. E. 2003. Trophic control of grassland production and biomass by pathogens. *Ecology Letters* 6:147–55.

Mittlebach, G. G., C. F. Steiner, S. M. Scheiner, K. L. Gross, H. L. Reynolds, R. B. Waide, M. R. Willig, S. I. Dodson, and L. Gough. 2001. What is the observed relationship between species richness and productivity? *Ecology* 82:2381–96.

Morand, S., and J. F. Guégan. 2000. Patterns of endemism in host-parasite associations: Lessons from epidemiological models and comparative tests. *Belgian Journal of Entomology* 2:135–47.

Mullins, T. D., T. B. Bristschgi, R. L. Krest, and S. J. Giovannoni. 1995. Genetic comparisons reveal the same unknown bacterial lineages in Atlantic and Pacific bacterioplankton communities. *Limnology and Oceanography* 40:148–58.

Nunn, C. L., S. M. Altizer, W. E. S. Sechrest, and A. A. Cunningham. 2005. Latitudinal gradients of parasite species richness in primates. *Diversity and Distribution* 11 (3): 249–56.

Ochman, H., J. Lawrence, and E. Groisman. 2000. Lateral gene transfer and the nature of bacterial innovation. *Nature* 405:299–304.

Pace, N. R. 1999. Microbial ecology and diversity. *ASM News* 65:328–33.

Poulin, R. 2001. Another look at the richness of helminth communities in tropical freshwater fish. *Journal of Biogeography* 28:737–43.

Poulin, R., and K. N. Mouritsen. 2003. Large-scale determinants of trematode infections in intertidal gastropods. *Marine Ecology Progress Series* 254:187–98.

Ricklefs, R. E., and D. Schluter, eds. 1993. *Species diversity in ecological communities: Historical and geographical perspectives*. Chicago: University of Chicago Press.

Rillig, M. C. 2004. Arbuscular mycorrhizae and terrestrial ecosystem processes. *Ecology Letters* 7:740–44.

Rohani, P., D. J. Earn, and B. T. Grenfell. 1999. Opposite patterns of synchrony in sympatric disease metapopulations. *Science* 286:968–71.

Rondon, M. R., P. R. August, A. D. Bettermann, S. F. Brady, T. H. Grossman, M. R. Liles, K. A. Loiacono, et al. 2000. Cloning the soil metagenome: A strategy for accessing the genetic and functional diversity of uncultured microorganisms. *Applied Environmental Microbiology* 66:2541–47.

Rosenzweig, M. L. 1995. *Species diversity in space and time.* Cambridge: Cambridge University Press.

Rosselló-Mora, R., and R. Amann. 2001. The species concept for prokaryotes. *FEMS Microbiology Reviews* 25:39–67.

Smith, K., S. Gaines, D. Sax, V. Guernier, and J. F. Guégan. 2007. Global homogenization of human diseases. *Ecology* 88:1903–10.

Smith, V., B. L. Foster, J. P. Grover, R. D. Holt, M. A. Leibold, and F. deNoyelles. 2005. Phytoplankton species richness scales consistently from laboratory microcosms to the world's oceans. *Proceedings of the National Academy of Sciences USA* 102:4393–96.

Stackebrandt, E., and B. M. Goebel. 1994. Taxonomic note: A place for DNA-DNA reassociation and 16S rRNA sequence analysis in the present species definition in bacteriology. *Applied Environmental Microbiology* 44:846–49.

Steward, G., J. Montiel, and F. Azam. 2000. Genome size distributions indicate variability and similarity among marine viral assemblages from diverse environments. *Limnology and Oceanography* 45:1697–1706.

Tilman, D. 1996. Biodiversity: Population versus ecosystem stability. *Ecology* 77:350–63.

Tilman, D., and P. Kareiva. 1997. *Spatial ecology. The role of space in population dynamics and interspecific interactions.* Princeton, NJ: Princeton University Press.

Torsvik, V., J. Goksøyr, and F. L. Daae. 1990. High diversity in DNA of soil bacteria. *Applied Environmental Microbiology* 56:782–87.

Torsvik, V., and L. Øvreås. 2002. Microbial diversity and function in soil: From genes to ecosystems. *Current Opinion in Microbiology* 5:240–45.

Torsvik, V., L. Øvreä, and T. F. Thingstad. 2002. Prokaryotic diversity: Magnitude, dynamics and controlling factors. *Science* 296:1064–66.

Townsend, A. R., R. W. Howarth, F. A. Bazzaz, M. S. Booth, C. C. Cleveland, S. K. Collinge, A. P. Dobson, et al. 2003. Human health effects of a changing global nitrogen cycle. *Frontiers in Ecology and Environment* 1:240–46.

Venter, J., K. Remmington, J. Heidelberg, A. Halpern, D. Ruscch, J. Eisen, D. Wu, et al. 2004. Environmental genome shotgun sequencing of the Sargasso Sea. *Science* 304:66–74.

Ward, B. B. 2002. How many species of prokaryotes are there? *Proceedings of the National Academy of Sciences USA* 99:10234–36.

Ward, D. M. 1998. A natural species concept for prokaryotes. *Current Opinions in Microbiology* 1:271–77.

Wardle, D. A., R. D. Bardgett, J. N. Klironomos, H. Setälä, W. H. van der Putten, and D. H. Wall. 2004. Ecological linkages between aboveground and belowground biota. *Science* 304:1629–33.

Wardle, D. A., K. I. Bonner, and K. S. Nicholson. 1997. Biodiversity and plant litter: Experimental evidence which does not support the view that enhanced species richness improves ecosystem function. *Oikos* 79:247–58.

Weinbauer, M. G. 2004. Ecology of prokaryotic viruses. *FEMS Microbiology Reviews* 28:127–81.

Weinbauer, M. G., and F. Rassoulzadegan. 2004. Are viruses driving microbial diversification and diversity? *Environmental Microbiology* 6:1–11.

Whitman, W. B., D. C. Coleman, and W. J. Wiebe. 1998. Prokaryotes: The unseen majority. *Proceedings of the National Academy of Sciences USA* 95:6578–83.

Willig, M. R., D. M. Kaufman, and R. D. Stevens. 2003. Latitudinal gradients of biodiversity: Pattern, process, scale and synthesis. *Annual Review of Ecology and Systematics* 34:273–309.

Wommack, K. E., and R. R. Colwell. 2000. Virioplankton: Viruses in aquatic ecosystems. *Microbiology and Molecular Biology Reviews* 64:69–114.

12

Biodiversity and Health: The Place of Parasitic and Infectious Diseases

Alain Froment

The relation between diseases and environment is one of the oldest objects of medical thinking. Hippocrates (ca. 460–377 BC), the founding father of medicine, devoted a whole treatise ("On Air, Waters, and Places") to this question, stating that "the appearance of disease in human populations is influenced by the quality of air, water, and food; the topography of the land, and general living habits." The link between environment and disease has long been acknowledged—for example, in the ancient names of diseases like malaria (from the Italian *mala aria,* "bad air"), also called paludism (from the ancient French *palud,* "swamp"). Thanks to Louis Pasteur (1822–1895), who demonstrated unequivocally the relationship between one category of germ and one disease and stressed the role of the host (terrain, or ground), medicine entered the modern era. The input of ecological thinking led to the shift from a linear paradigm (one germ gives one disease) to a more complex systemic perspective, which marked the start of an ecological, evolutionary approach to medicine (Froment 1997a, 1997b; Lappé 1994; Stearns 1998; Trevethan, McKenna, and Smith 1998).

A major effect of degraded environments, and one of the most studied, is the increase of noninfectious diseases that are related to pollution. Pollution has major effects on biodiversity and on chronic diseases (e.g., cancers, poisoning, endocrine disruption, genetic mutations). However, these important aspects of health will not be discussed here. Conversely, parasitic and infectious diseases (PIDs) are caused by living organisms and therefore constitute a part of global biodiversity. According to World Health Organization (WHO) vital statistics, of 140,000 daily deaths, 30,000 are children under age five, one-third of whom are killed by diarrheas and dehydration, one-third by measles and respiratory infections, and one-third by other diseases, mainly infectious, of which malaria ranks first. A total of 50,000 persons die of infections (at least 5,000 of human immunodeficiency virus/acquired immune deficiency syndrome [HIV/AIDS], and

3,000 of malaria), while 40,000 die of cardiovascular diseases, 15,000 of cancers, and 10,000 of violent deaths (WHO 2006). These figures emphasize the dominant role of PIDs in mortality. In this chapter I shall mainly focus on the direct role of this specific aspect of biodiversity on human health, which seems more complex than expected.

The Perception of Ecological Changes

The ability to provide good-quality air and water and a rich biodiversity, in this volume termed "ecosystem services," is globally encompassed in the concept of ecosystem health. It then seems obvious that human health is somewhat correlated with ecosystem health (Grifo and Rosenthal 1997; Soskolne 2003), even if the concept of ecosystem health itself has been questioned, as the term *health* is difficult to apply to objects that are not living organisms. As will be examined in further detail, there is no obvious link between "natural" environment and human health. For example, the health of a population living in a remote, virtually untouched area of a tropical rainforest is by far much worse in terms of mortality, morbidity, and biological parameters than the health of a group living in a totally controlled and artificial environment like a large city, or, in the future, a submarine city, or a Moon or Mars station.[1] In this chapter, then, I shall review some links between general biodiversity, pathogen biodiversity, and human health.

Clearly, biodiversity is severely threatened by several factors, which can be summarized by the acronym HIPPO: habitat loss, introduced species, pollution, population growth, and overconsumption—all responsible for a "biotic holocaust" (Myers 1999). Though figures regarding the number of extinct species are often contradictory,[2] human activities have severely reduced natural habitats. Fragmentation of ecosystems by agricultural encroachment, logging activities, road construction, wetland modification, mining activities that poison the rivers, the expansion of urban environments and its related pollution, and coastal zone degradation can have health consequences, but no systematic evaluation of these factors is currently available. Beside these direct modifications linked to economic development, social problems related to poverty, uncontrolled demography, transmigration, wars, and refugees make up another component of ecosystem degradation. In addition, political unrest has disrupted former medical care and vaccination and prevention campaigns, an explanation for the resurgence of old scourges like yellow fever and sleeping sickness. Modern climate change is also perceived as another threat to living species.

It must be remembered that, historically, the planet has experienced constant climatic variations, the most spectacular of which in geologically recent times was the change observed at the end of the last glacial period, 10,000 years ago, when the Earth's temperature rose approximately 7°C in probably less than a century. Very important changes in fauna and flora occurred, one direct consequence being the start of the Neolithic Transition and its related demographic explosion. At that time, all human societies were hunter-gatherer bands having no recorded impact on the Earth's climate.[3]

Ecologists recognize, however, that environmental changes can affect biological evolution. According to Gould and Eldredge's (1993) theory of punctuated equilibrium, evolution does not progress at a regular pace, but rather displays phases of stagnation and acceleration. The reason given for this phasal pattern is that crises offer new opportunities for emerging species. It seems that these abrupt changes may have had, for the human species also, the important effect of accelerating evolution (Calvin 2002). Some people argue that rapid changes do not give time for species to adapt, but, in fact, opportunistic variants preexist a change and are immediately given a chance when a catastrophe occurs.

Ironically, global warming is favorable to the expansion of tropical forests (of course, only where logging companies do not destroy them). This phenomenon is very well documented on the forest-savannah ecotone in Cameroon, Congo, and Gabon, where tropical forests began to expand about 500 years ago, and is observed on satellite images as progressing at a pace of 1 to 2 meters a year (Froment and Guffroy 2003). On the other hand, moderately opening the forest does not reduce biodiversity but, conversely, increases it, because moderate opening gives room to both open-field species and those that live on ecotone borders, especially birds and small mammals.

Psychologically, the contemplation of a "natural" landscape is recognized as excellent for mental health, but biodiversity is not a factor here. First, most of the landscape, such as a garden or the countryside, is not "natural," but humanized. Second, viewing an environment poor in biodiversity, such as bears on the Arctic Circle, may be more mentally beneficial than a rich environment like a jungle, which may cause anguish in some. For the "civilized"[4] world, forests (from the Latin *foris,* "outside") are savage (from the Latin *silva,* "forest") and wild jungles (from the Hindi *jangal,* "uninhabited space") can generate anxiety. There is, then, no direct correlation between the relaxing role and comfort provided by nature, and wealth of biodiversity.

Cultural Background

One of the main concerns with the destruction of ecosystems is that traditional societies using the resources of these ecosystems are collapsing. Some observers (e.g., Maffi 2001) have noted a parallel between biodiversity and cultural diversity.[5] Loss of traditional knowledge is among the risks to these small societies. Some ecological activists believe that the traditional way of life of tribal groups should be conserved—for example, that the absence of technology would be good for Africa, allowing it to stay "close to nature," as in the Pygmy lifestyle. This view could be extended to mean that, among other things, these children should not go to "modern" schools or universities, because it is well known that instruction and Western life lead to new needs and irreversible shifts from traditions. The idea that native societies have managed their environment with wisdom is a modern myth (e.g., the myths of the Noble Savage and the Ecological Indian) (Abruzzi 2000), but it is only the industrial societies that have pushed this disturbance beyond good management of natural ecosystems.

There is no doubt that many plants, and a lot of animals, have the potential to cure disease: the research of such substances is the field of bioprospection. It has been estimated (see this volume, chapter 5) that more than 50,000 species are used for medicinal resources; however, only 121 Western drugs are derived from plants (and many of them, like the anticancer substances, are not used in traditional medicine for that purpose). If 50,000 plants, each containing dozens of active molecules, had to be tested for, say, 100 diseases, hundred of millions of tests and several centuries of research would be required—a time long enough for many plants to disappear. It could further be argued that such extinctions would have no significant impact on medicinal resources, as all those plants could be stored in botanical collections. In addition, some traditional remedies, such as rhinoceros horn or tiger bones, used in Chinese medicine, have seriously threatened the survival of the corresponding species. Last, many herbal remedies like ginseng (*Panax quinquefolium*), gingko (*Gingko biloba*), valerian (*Valeriana officinalis*), kava (*Piper methysticum*), or St. John's wort (*Hypericum perforatum*), very popular in the West, are more toxic than previously believed, and present dangerous interactions with prescription drugs, including antidepressants and some antiviral and blood-thinning drugs.

Considering that traditional healers in developing countries are about 100 times more numerous than physicians, the idea of associating them with disease control is interesting, and has been considered by WHO. Unfortunately, many abuse the trust of their patients and supply no effective service. Dr. Kapunda Kambale, a physician from Goma, Democratic Republic of Congo, recently complained that "for 2003 only, over 100 patients died of AIDS in Goma hospitals because they had lost too much time with herborists" (Kambale 2001).[6] AIDS specialists often have to cope with a certain Afrocentric view about the disease that can deny its cause, instead offering alternative treatments or preventatives that give false hope to people who, believing that accessible treatments are available, do not protect themselves. Traditional practices like scarifications, tattooing, or shaving several people with the same blade spread viral blood diseases very efficiently. In Namibia, which has one of the highest rates of infection with the human immunodeficiency virus (HIV) in the world, a rise in child rape has been observed, related to the recommendation of traditional healers that HIV-positive people have sex with virgin minors to "cure" their infection (Ahmad 2001).[7] However, some initiatives associating traditional healers and modern practitioners, such as THETA in Uganda,[8] seem promising (King 2002).

Ironically, countries in which traditional healers are most active also have populations with the shortest life spans, illustrating the limited effectiveness of such therapies. In all likelihood, the situation would be even worse without these healers, but, in fact, patients do not visit traditional healers and hospitals for the same purpose—for reasons of culture as well as cost. Witch doctors are excellent in psychiatry—for example, in treating possession and other beliefs—but are weak in treating infectious and chronic diseases, as well as in surgery. They cannot deal effectively with many of the causes of mortality listed in the WHO vital statistics given in the introduction to this chapter.

And when it comes to traditional explanations of disease etiology involving witchcraft, little room is left for managing appropriate prevention and care (Riach 2004). Instead of according witch doctors too much importance, then, a better choice would be to convince international medical authorities to provide better and cheaper medical care to everyone. The example of anti-AIDS therapies, for example—the cost of which has been divided by 20 during the last few years, and which are now affordable for many, but unfortunately not all, patients—is encouraging.

Destruction of Ecosystems and Emergence of Diseases

A persisting myth says that tropical forests are pristine, virgin, natural ecosystems that have only recently been penetrated by humans, whose presence is causing new opportunities for disease emergence. In fact, the development of archaeological research in rainforests shows that they have been inhabited for millennia. Early inhabitants were not only small bands of hunter-gatherers, for the use of agriculture for at least three millennia has been documented in Asia, Africa, and Amazonia (Mercader 2003). Tropical forests can therefore be viewed as elaborate, cultivated anthroposystems somewhat comparable to gardens (Balée 1989). Humans have enriched natural biodiversity by selecting and creating appropriate varieties of crops or animals, which are now dependent on humans for their survival. Though thousands of tropical plants are potentially edible, and rich in proteins and vitamins, our studies in Cameroon show that they are not used, because people have enough calories with cassava, bananas, maize, and cocoyams—all of which are imported crops—and cultivated yams.

Some PIDs depend on a vector or an intermediary host, while others have a direct transmission. We can therefore define an ecosensitivity of diseases: infectious diseases are either "eco-dependent"—meaning that they are restricted, directly or through the need of their vector, to specific ecosystems and act at a local level—or they are ecoindependent, with the potential to spread to the global level, the extreme form being pandemics. A portion of PIDs (16% of the 270 identified) is restricted to the human species, while others have an animal reservoir. In the former case, eradication is possible, as was achieved with smallpox, while in the latter, eradication is almost out of reach.

The link between environment and disease has been explored for the last decades by medical geographers, who defined the concepts of pathogenic complex landscape epidemiology, geomedicine, and ecoepidemiogenic systems (Rémy 1992). More globally, it would be useful to popularize the concept of pathocenosis (Grmek 1983, 15), referring to the complex interdependence between all the diseases observed within a certain population at a certain time, which appears to be a very powerful tool. Pathocenosis is defined by three propositions: (1) it is a system, with peculiar structural properties; (2) the distribution of each disease is influenced by all the others; and (3) pathocenosis tends toward equilibrium in a stable ecological situation. This concept means that a disease can no longer be considered in isolation from a larger epidemiological context, and that the ecological approach to human health must occur within a holistic perspective, taking into account the competition between agents—in particular, PIDs.

Of course, human activities change the pattern of diseases. For example, deforestation and mining works created water collections offering new opportunities for some disease vectors, like *Anopheles darlingi*, inducing expansion of malaria, dengue, and leishmaniasis in Amazonia (Coimbra 1991). But deforestation did not create those diseases, which for millions of years have been present in savannahs and have coevolved with humankind.[9] Dam building, irrigation, and extension of rice culture also have this perverse effect—the stimulation of vector multiplications in particular molluscs, vectors of bilharziasis (schistosomiasis), and in arthropods, vectors of many diseases, such as malaria and arboviruses (the prefix *arbo* referring to *ar*thropod-*bo*rne). But most of the time, it is not local water surface modifications, but an increased pluviometry, that boosts biodiversity, and directly generates more mosquitoes. Japanese encephalitis and Rift Valley fever epidemics have been related to exceptional rains, amplified by the development of domesticated animal farms—pigs in the case of Japanese encephalitis, and cattle in the case of Rift Valley fever. Raising huge number of pigs or poultry in the close vicinity of huge numbers of people is a very sure way to get outbreaks of viral infections like flu: pigs can be infected by both avian and human viruses, giving an opportunity for those viruses to recombine.

Indirectly, better rains, which mean more abundant harvests, can result in an explosion of rodents that act as vectors for many viruses, such as the hantavirus epidemics among the Navajo in the United States in 1993. This virus, first described as Korean hemorrhagic fever, has been found in ancient Chinese medicine texts, and is not a new one. Dengue is an old disease too, described in Philadelphia in 1780. But hemorrhagic dengue forms are recent and are linked to new strains, probably due to mutations. These mutations, and recombinations, explain the virulence of some new aspects of old diseases, such as avian flu, Crimea-Congo fever, the Egyptian strain of Rift Valley fever, and the U.S. variant of West Nile virus, probably brought by migratory birds. The most threatening epidemics of today—severe acute respiratory syndrome (SARS), avian flu, HIV/AIDS, tuberculosis, measles, and cholera—are in no way the result of a threat on biodiversity, but are problems linked to increased interactions with animals or overcrowded households. The mechanism is not an ecological disruption in itself, but an increasing economic pressure. Conversely, the extension of Lyme disease[10] since its appearance in Lyme, Connecticut, in 1975, seems linked to the restoration of an ecosystem, the forest, after cultivated lands have been abandoned. The extension of suburbs, or leisure activities in "nature," into wooded areas has also increased the contact between humans and ticks living on large and small wild mammals (e.g., deer and rodents, respectively). Lyme disease has received much publicity and has become emblematic of emergent disease caused by ecological disruption, though no fatal cases have been reported.

In Amazonia, where biodiversity is extreme, huge surfaces of forest have been transformed into pastures, and no emergence has occurred—that is, no unknown virus has surfaced since then; the forest clearing has only increased the contact between *Aedes*—the yellow fever mosquito, which usually feeds on monkeys—and humans, and has triggered epidemics of some hemorrhagic fevers due to the Oropouche and Machupo ("black typhus") viruses (Confalonieri 2000). In deep forest, mosquitoes are active year-round, and their bites are "diluted" on more animal targets. Surprisingly, yellow fever, a huge killer disease in Africa and Latin America, is absent in Asia, despite great deforestation, the presence of the *Aedes* mosquito vector, and some intense population exchanges—for example, the Indian communities settled in eastern Africa and visiting India (Rogers et al. 2006). No simple explanation has been found, but such an outbreak is the number-one fear for WHO.

While there is still a debate about the animal reservoir of Ebola virus, evidence has been found that small rodents and bats living in forest-savannah ecotones are involved (Morvan et al. 2000). One of the early epidemics, in 1977, was observed in South Sudan, in an open landscape. This finding contradicts the idea that Ebola virus emergence is linked to forest penetration. Epidemics of Marburg and Lassa hemorrhagic fever have always occurred in savannah areas of Africa. Major ecosystem attacks are worse in Indonesia and Amazonia, but it is in Africa that new viruses were identified in the wild. The fact that humankind originated there—thus African apes are genetically closer to humans than any other animal species—and that bushmeat consumption is linked to some of the most severe infectious hazards—such as HIVs, Ebola and Marburg fever, HTLVs (human T-cell lymphotropic viruses), and anthrax—may be an explanation (Peeters et al. 2002; Wolfe et al. 2004).

An Evolutionary View

What is emergence? First, it can be the appearance of entirely new diseases by mutation, a purely stochastic and unpredictable process, mainly independent of human activities. Or it can be a virus existing in animals but breaking the species barrier and infecting humans. Or it can be the discovery of the agent of an old disease, such as herpesvirus-8 in Kaposi's sarcoma. Or it can be the spread of a well-known disease in previously unknown zones. Though new pathogens always appear, a distinction must be made between the appearance of the germ and the appearance of the disease. To be surprised by the emergence of new diseases would be a fixist view. New diseases seemed to appear in the last years of the 20th century, but this event could well be an illusion. The French bacteriologist Charles Nicolle, who in 1928 was awarded the Nobel Prize in Medicine for his identification of lice as the vector of epidemic typhus, warned long ago that infectious diseases constantly appear and disappear (Nicolle 1933). The Athens "plague" (possibly typhus, 430 BC) was not comparable to the Black Plague of the 14th century. Syphilis existed in Europe in Roman times, but exploded only after America's discovery (Froment 1994). During the same time, leprosy disappeared in Europe without clear explanation.

Due to the absence of records, we know nothing of the diseases prevalent, say, in Central African rainforests before European penetration a little more than a century ago. A compilation of radiocarbon dates (Delneuf, Essomba, and Froment 1998; Oslisly 2001) indicates a two-thirds drop of sites dated between 600 and 1,400 BP in equatorial forest, compared with earlier and more recent periods. Such a deficit is not observed in the northern savannah region of the continent. Though it is impossible at the moment to explain this phenomenon, a severe depopulation due to a medical scourge is one of the plausible hypotheses. Sociological or political disruption could be considered, but the dates are too early for Western slave trade, while Muslim trade did not go that far in West-Central Africa. The fact that Gabon is still underpopulated could be a consequence of this unexplained event.

We do not know if climatic evolution will lead to an impoverishment of ecosystems (McMichael 2001), a restriction of natural resources (soils, fish, etc,), desertification, a drop in agriculture production, mass south-north migrations, or wars for water access or other ecological problems. For the very reason that PIDs are more diverse in low latitudes, the consequence of rising temperature would be a spread of tropical diseases like dengue, yellow fever, Rift Valley fever, and so forth. Bacterial growth is stimulated by temperature, as observed for *Salmonella,* responsible for typhoid fever (WHO 2003), or the relationship of cholera to ocean surface temperature and El Niño episodes (Pascual et al. 2000). Excessive summer heat killed 15,000 elderly persons in France in August 2003; however, a deficit of 14,000 deaths was observed in France in the first six months of 2004 (Valleron and Boumendil 2004), suggesting that the effect of heat was to shorten the lives of people already at the end of their lives by an average of nine months.

However, in the case of recent extensions of viral infections like West Nile virus or hemorrhagic dengue, not to mention malaria and so many other arthropod-borne diseases, one may question the role of the worldwide ban on use of the insecticide DDT.[11] This molecule (dichloro-diphenyl-trichloroethane), a kind of magic bullet against mosquitoes, was abandoned in fear of its destructive effects on the food chain. Whether this decision was ecologically right or wrong belongs to specialists, but in relation to biodiversity and human health, the question is, Did this decision allow the expansion of mosquitoes, and mosquito-borne diseases, and what has been the cost in human lives? More generally, may some policies targeting environmental protection increase human mortality? Or, reciprocally, to what extent is it acceptable to degrade environments to save human lives? For example, in past centuries, malaria has been eliminated at the expense of drying the marshes; draining the marshes, however, has created ecological disequilibria, and ecological concerns now demand their restoration.

The answer to the preceding questions is that a higher medical technology may reconcile these needs. With better antimalarial drugs, and the hope of a vaccine, it is now possible to restore humid zones. In regard to human activities like dam building, a balance must be considered between the health cost (e.g., knowing that bilharziasis is cured by a single dose of drug), and the benefits, such as irrigation and food production. Intestinal worms each have specific and cheap treatments; the same is true for onchocerciasis (river blindness) and most of the parasitic diseases. Dracunculiasis (Guinea worm disease) is now nearly eradicated, owing to very simple hygiene rules. Sleeping sickness was controlled between 1920 and 1930 by fairly simple epidemiological methods under the direction of Dr. Eugène Jamot (1879–1937), who invented the modern, mobile strategies of survey; the present resurgence of sleeping sickness is due only to political unrest, not to ecosystem disruption.

Presently the world is divided between people from developed countries, who pollute much but are not numerous, and people from developing countries,[12] who are many, and who legitimately aspire to enjoy the same conditions. When the concept of sustainable development is used, a clear definition of "development" is needed. There are many definitions of sustainable development, but, from a medical perspective, there is only one: the optimization of human life. Optimization of human life means both an increase in life quality, especially for elderly people, and an increase in life span. In this respect, Japan and France are more developed than the United States. However, a successful medical strategy initially leads to an increase of population size, and therefore an increased pressure on natural resources. The demographic explosion in underdeveloped countries is mainly due to the influence of modern medical practices. This very success is the "population bomb" threat, and many ecologists think that only a decrease in world population is compatible with sustainability. A solution is an observation that has been noted repeatedly—namely, that increasing development provokes a reduction in fertility (called the demographic transition), leading to a decrease of population size: many developed countries currently have negative growth, balanced only by foreign immigration.

Development, causing a major increase in international travel by boats and planes, by refugees and tourists, is also the cause of an increase in spreading of disease (globalization and "microbial unification" of the planet). Examples of this new "pathogen pollution" (Daszak, Cunningham, and Hyatt 2000) are diseases transmitted through pet trade (e.g., monkeypox, in which conservationists and health specialists share the same interest) and food-borne diseases transmitted through exportation of fruits. Medical waste management also causes increasing problems, as iatrogenic practices, such as blood transfusions and nonsterile vaccinations and injections, can transmit several blood viruses, especially HIV/AIDS and hepatitis B and C. Regarding viruses, most of those that are transmitted directly, and that have no animal reservoir—from varicella (chicken pox) to flu, and from measles to HIV/AIDS—are driven by social parameters, such as human densities, poverty, and behavior. None of these factors is related to a biodiversity issue.

Environmental Biodiversity and Diversity of Transmissible Diseases

In an Italian cartoon published in 2003, a smiling SARS virus asks a puzzled Friend of the Animals: "And me, in which protected species shall I be registered?" This cartoon reminds us that there is a "good" and a "bad" biodiversity. We are ready to admit that, along with elephants and dolphins, some ugly animals like toads and insects deserve protection. But PIDs are rarely associated with the idea of biodiversity, though their mass is probably larger than the rest of living organisms. This mass, however, is not evenly distributed. An ecological rule called Rapoport's rule states that the biodiversity of plants and animals increases as the distance to the equator decreases. Recently, it was demonstrated that this rule also applies to PIDs (Guernier, Hochberg, and Guégan 2004; see also this volume, chapter 11). It is then necessary to explore the link between the complexity of ecosystems, and both the number and intensity of pathogens.

As mentioned earlier, Amazonia is the richest area in biodiversity. Noting the misery of the Amazonian *caboclos* (peasants of mixed ancestry), Bennett (1996) asked if rainforests were adapted for modern populations. One might expect that indigenous societies who have lived in this ecosystem for millennia would be more adapted than poor *caboclos*. Such is not the case. Bacterial diseases like yaws are plaguing the forest people: 80% in Eastern Pygmies, 37% among Bantus nowadays; 48% of Mbuti Pygmies have enlarged spleen, a sign of chronic malaria, and 85% have hookworms (Mann et al. 1962). Of course, one reason why present-day indigenous people seem in bad health could be the fact that they have been marginalized and impoverished in the recent past. Were those people healthier in earlier times? There are not many archaeological studies to prove it (Klepinger 1992), but one done in Australia did not show in a convincing manner that health was better in the past (Webb 1995). What we are able to say is that there was a major epidemiological shift, or transition, between precolonial and postcontact times, from infectious diseases to chronic ones (Froment 2001), which resulted globally in a gross improvement in life span.

No study ever tried to make a census of all the infections met by a single adult. Not only would it be tedious to test hundreds of antigens, but many germs are still unknown, so there is still no test to identify them. For the category of arenaviruses only, a family of hemorrhagic fever agents, one new virus is discovered every two years. Saluzzo, Vidal, and Gonzalez (2004) report that in a systematic survey of fauna (rodents, birds, arthropods) conducted in Central African Republic, 919 viruses (39 new) belonging to 84 species were observed; on average, there was 1 new virus for each 100 birds captured. Virological studies (isolation, culture, tests) are not easy to perform in the field, but our knowledge about this invisible world is growing. The late Stephen Jay Gould used to say that the truly successful organisms on the planet are not humans, but bacteria, which are found in the most diverse conditions. This tremendous success and diversity cannot be ignored, and is genuinely part of the total biodiversity.

For practical reasons, because they are much bigger than viruses or bacteria, and easier to identify, human parasites are virtually all known. Their epidemiology confirms that they are much more numerous in rich biodiversity areas. Dunn (1977) noted that the average number of parasites is 1 in the Central Australian desert, 3 among the San Bushmen of the Kalahari Desert, but 20 among the Pygmies of Central Africa and 22 among the Semang of Malaysia. Also, the prevalence of parasites, or percentage of people carrying a parasitic disease, is proportional to heat and humidity. A study by Ratard et al. (1991), completed by our own surveys, shows that in Cameroon, a country stretching between the Sahara Desert and the equator, *Ascaris* and pinworms affect about 2% of people in the driest area, and 95–98% in the most humid, with a cline following climate gradient. An excellent diet rich in protein, energy, and iron[13] does not prevent short stature among hunter-gatherers,[14] who show low skin fold thickness, especially in women, and even anemia. The life expectancy of hunter-gatherers is lower than that of anyone else (Pennington 2001).

The greatest burden to human health in tropical countries is diarrheal diseases linked to water pollution. This pollution, however, is rarely chemical, except in areas where industries, intensive agriculture, and gold mining release poisons like mercury or pesticides. The hazard is mainly organic, caused first by viruses (such as rotavirus, calicivirus), and second by bacteria (cholera, salmonellas, the pathogen *Escherichia coli*, and other fecal germs). Here again, those germs are the result of a microbial biodiversity, not related to ecosystem modification, but belonging to the local epidemiological landscape and favored by lack of hygiene.

Research Directions

The purpose of this chapter has been not to discuss environment and health in general, but only to focus on the relationship between general biodiversity and infectious and parasitic diseases. There is no doubt that a growing environmental crisis is in process, and that maximal efforts to protect natural ecosystems are required to maintain and improve ecoservices. The industrial lobby would be all too happy to hear that ecological wounds to the planet are not harmful to human health. Such is not the case, however, as organic and chemical pollution, the greenhouse effect, and the role of chlorofluorocarbons (CFCs) have all been clearly identified as health threats. We have just addressed a couple of questions: Does reduction of biodiversity induce an increase of human infections? How should we deal with the hidden face of biodiversity—that is, harmful animals, like arthropods, that are vectors of diseases and which display a huge biodiversity, and parasitic and infectious diseases themselves, most of them still unknown? As an example, smallpox was eradicated around 1979; the virus is kept in some labs and is now a subject of fear, related to bioterrorism. However, the smallpox virus is also a part of biodiversity: should it then be kept or destroyed?

Let us be reasonably pessimistic: solutions will not come from traditional medicine; threatened animals and plants are not the solution to public health problems; restored ecosystems will not reduce emerging diseases; and, to be blunt, the extinction of the blue whale or the tiger will have zero effect on human health. In public health, biodiversity, most of which is invisible, is more the problem than the solution. The spots of highest biodiversity are the equatorial rainforests. These ecosystems are not friendly to humans; they can tolerate an enormous vegetal biomass and 1 ton of mammals per square kilometer, but only 30 to 100 kilograms of humans per square kilometer—that is, a density of 1 human/sq km for hunters, or 8 humans/sq km for agriculturists. Out of 6 billion humans, only 200 million, or 3.3%, live there, and only 12 million live directly from the products of the tropical forest itself (Bahuchet and De Maret 2000).

Now, as we have shown, the link of human health with ecosystem disruption is far from clear. Conservation medicine—bridging human, wildlife, and ecosystem health and disease emergence—is trying to reconcile environmental protection and human health. This approach contradicts Western medical strategy, which targets, first and directly, the biodiversity of pathogen germs and, second and more indirectly, the biodiversity of vectors. The effects of the first strategy on ecosystems are unknown. As for the second strategy, the reduction of some populations of vectors (e.g., mosquitoes, tsetse flies) may have consequences for food chains (e.g., birds), but this effect seems limited. Attempts to avoid chemical products by releasing irradiated sterile flies, or genetically modified *Anopheles* mosquitoes unable to transmit malaria, have failed, because these insects are always less competitive than their wild counterparts. The problem is that when we change one component of the environment, because of the complex interrelationships of the whole system, the consequences can be far reaching.

When reading reports dealing with biodiversity and health (Chivian 2002), one gets the impression that infectious disease hazards are increasing with ecosystem destructions. However, the picture seems distorted. Thanks to the development of drugs and vaccines, there has been a constant decrease in mortality rates from infectious diseases, even in countries where modern medicine is still incompletely available. A painful exception, HIV/AIDS, is not such a recent disease,[15] but its spread is due solely to sociological factors. There is no reason to expect that restoration of ecosystems will bring relief to human diseases. On the contrary, human health has improved constantly during recent decades, a time when ecosystems have been constantly degraded. The common factor in these two phenomena is technological development. Even traditional hunter-gatherers have seen their population size augmented, thanks to modern medical care, to the extent that they are no longer able to live from wild resources alone and are obliged to turn to agriculture.

Development was viewed until the middle of the 20th century as a blind exploitation of Earth's resources. Humankind ransacked the environment to such an extent that an awareness of irreversible danger finally arose. In regard to biodiversity, we have seen that temperature increase can boost biodiversity. However, usually, human activities generate ecological destruction and related species loss. Species loss is considered unacceptable—the usual reason given being that disappearance of a species may mean the potential loss of a therapeutic miracle molecule. However, as discussed earlier, considering the number of tests that would be necessary, this reason for protecting biodiversity is fairly unrealistic. In fact, most endangered species, especially among animals like birds and mammals, will not produce any usable substance, and the idea of "utility" is therefore a bit surprising in the mouths of conservationists. The true reason why we must protect biodiversity is that we are responsible for the optimal management of the Earth, and we should protect species just because they are here.

Traditional knowledge and local communities should be involved in conservation policies. Very often, however, indigenous populations are kicked out of protected parks and are no longer allowed to use the plants or animals. On the periphery of those parks, destruction of fields by elephants or apes is not compensated, so villagers prefer to kill the animals rather than to stand those costs. In the same vein, it is difficult to convince peasants to protect venomous snakes, tigers, and other dangerous animals. Ebola virus happens to be a greater killer for large African apes (chimpanzees *Pan paniscus* and *Pan troglodytes*, and gorillas *Gorilla gorilla*) than all poaching activities (Leroy et al. 2004). That is why, in Congo, in areas affected by the outbreak, and in Cameroon, in areas still untouched, villagers aware of the risk of epidemics are threatening to slaughter all the apes preventively.[16]

Parasitic and infectious diseases (PIDs) have been the major burden of humankind since its very beginning; they are still the first cause of mortality in most of the world, especially in developing countries. People living traditional life in ecosystems rich in biodiversity are the most exposed to this mortality, a factor neglected by the advocates of natural ecosystems, who usually do not take into account the microscopic world. PIDs are part of the global biodiversity, and both their prevalence and their number increase in parallel with other components of the biosphere. If pathogens are eradicated or controlled, what can be the consequences on biocenosis? Is it ecologically acceptable to eliminate a virus, a bacterium, or a parasite? How then can such destruction be compatible with medical ethics? A major question cannot be avoided in the implementation of conservation medicine: what is the ecological price to pay for wiping out pathogens?

Notes

1. "In the United States in the 19th and early 20th centuries, there was a substantial mortality penalty to living in urban places. This circumstance was shared with other nations. By around 1940, this penalty had been largely eliminated, and it was healthier, in many cases, to reside in the city than in the countryside" (Haines 2001).

2. It is not uncommon to read that "thousands" of species disappear each year, yet no lists of these thousands are ever provided. Others say that the rate of extinction is now 1,000 to 10,000 times above normal, though we have no definition of what is normal.

3. It is probable that early faunal extinctions were due to overexploitation by Paleolithic hunters, but these extinctions, and other human activities, had no visible impact on the global climate change.

4. The term *civilized* is here taken in its etymological sense—the development of cities (from the Latin *civitas*).

5. See http://www.terralingua.org/. The correlation between biodiversity and cultural diversity is not very surprising, as remote ecosystems were occupied by small, segmentary groups, who did not exchange much, while empires developed mainly in savannahs. But there are numerous exceptions to this apparent correlation.

6. http://www.ql.umontreal.ca/volume11/numero12/mondev11n12d.html.

7. A similar phenomenon has been noted in Zambia and South Africa.

8. http://www.aidsuganda.org/response/govt_sectors/cso_programs/theta.htm.

9. The coevolution of germs and man is such that the polymorphism of some viruses, such as human T-cell lymphotrophic viruses (HTLVs) and bacteria (*Helicobacter pylori*) reveals the history of human migrations.

10. Lyme disease is caused by the bacterium *Borrelia burgdorferi* and is transmitted by tick bites (which may also transmit several viruses and other germs like *Rickettsia* and *Ehrlichia*, and parasites like *Babesia*). The estimated 20,000 cases a year in the United States are treated with antibiotics.

11. DDT was banned because of the impact of Rachel Carson's 1962 best seller, *A Silent Spring*, which announced that its use was causing an "ecological genocide." Recent attacks on the DDT ban, however, could be political attempts to discredit ecological movements by journalists close to industry and by conservative-right capitalists.

12. Note that, in tropical countries, traditional activities such as burning the landscape may be as polluting as industrial activities; see http://www.asb.cgiar.org/PDFwebdocs/ASBPolicyBriefs4.pdf.

13. Among Cameroon Pygmies, the daily mean consumption of meat is 200 g, 285 g for an adult male (Koppert et al., 1993).

14. Tropical hunters are generally short. For Pygmies, short stature is because of a genetic mutation, but for others, it is the result of stunting. A higher stature is attained when children are raised in a safer environment (Froment 2001).

15. Molecular studies (Rambaud et al. 2001) show that HIV/AIDS has been present for several decades, beginning no later than the 1930s.

16. Information for Congo was given by Dr. Alain Epelboin, medical doctor and anthropologist, a member of the WHO team during the outbreak. Information on Cameroon is from a personal survey conducted on the border of the Jane Goodall Foundation's gorilla sanctuary in Oveng.

References

Abruzzi, W. S. 2000. The myth of Chief Seattle. *Human Ecology Review* 7:72–75.

Aguirre, A. A., R. S. Ostfeld, G. M. Tabor, C. House, and M. C. Pearl. 2002. *Conservation medicine: Ecological health in practice.* New York: Oxford University Press.

Ahmad, K. 2001. Namibian government to prosecute healers. *Lancet* 357 (9253): 371.

Bahuchet S., De Maret P., eds. 2000. *Les peuples des forêts tropicales aujourd'hui : 3. Région Afrique centrale.* Bruxelles: APFT (Avenir des Peuples Forestiers Tropicaux), ULB (Université Libre de Bruxelles), 45–66.

Balée, W. 1989. The culture of the Amazonian forest. *Advances in Economic Botany* 7:1–21.

Bennett, C. F. 1996. Les forêts tropicales humides constituent-elles un habitat adapté à l'homme du XXIe siècle? In *L'alimentation en forêt tropicale: Interactions bioculturelles et perspectives de développement,* ed. C. M. Hladik, A. Hladik, H. Pagezy, O. F. Linares, G. J. A. Koppert, and A. Froment, 1303–8. Paris: UNESCO.

Calvin, W. H. 2002. *A brain for all seasons: Human evolution and abrupt climate change.* Chicago: University of Chicago Press.

Chivian, E., and A. Bernstein, eds. 2008. *Sustaining life: How human health depends on biodiversity.* Oxford, UK: Oxford University Press.

Coimbra, C. E. A. 1991. Environmental changes and human disease: A view from Amazonia. *Journal of Human Ecology* 2:15–21.

Confalonieri, U. 2000. Environmental change and human health in the Brazilian Amazon. *Global Change & Human Health* 1:174–83.

Daszak, P., A. A. Cunningham, and A. D. Hyatt. 2000. Infectious diseases of wildlife: Threats to biodiversity and human health. *Science* 287:443–49.

Delneuf, M., J. M. Essomba, and A. Froment, eds. 1998. *Paléo-anthropologie en Afrique centrale: Un bilan de l'archéologie au Cameroun,* 349–52. Paris: L'Harmattan.

Dunn, F. L. 1977. Health and disease in hunter-gatherers: Epidemiological factors. In *Culture, disease, and healing: Studies in medical anthropology,* ed. D. Landy, 99–114. New York: Macmillan.

Froment, A. 1994. Les tréponématoses: Une perspective historique. In *L'origine de la syphilis en Europe,* ed. O. Dutour, G. Pàlfi, J. Berato, and J.-P. Brun, 260–68. Paris: Éditions Errance.

———. 1997a. Écologie humaine et médecine tropicale. *Bulletin de la Société de Pathologie Exotique* 90:131–38.

———. 1997b. Une approche écoanthropologique de la santé publique. *Nature, Sciences, Sociétés* 5:5–11.

———. 2001. Evolutionary biology and health of hunter-gatherer populations. In *Hunter-gatherers: An interdisciplinary perspective,* ed. C. Panter-Brick, R. Layton, and P. Rowley-Conwy, 239–66. Cambridge: Cambridge University Press.

Froment, A., and J. Guffroy, eds. 2003. *Peuplements anciens et actuels des forêts tropicales,* 157–68, 211–18. Paris: Éditions IRD.

Froment, A., and G. Koppert. 1999. Malnutrition chronique et gradient climatique en milieu tropical. In *L'homme et la forêt tropicale,* ed. S. Bahuchet, D. Bley, H. Pagezy, and N. Vernazza-Licht, 639–59. Société d'Écologie Humaine-APFT. Marseille: Éditions de Bergier.

Gould, S. J., and N. Eldredge. 1993. Punctuated equilibrium comes of age. *Nature* 366 (6452): 223–27.

Grifo, F., and J. Rosenthal., eds. 1997. *Biodiversity and human health.* Washington, DC: Island Press.

Grmek, M. 1983. *Les maladies à l'aube de la civilisation occidentale.* Paris: Payot.

Guernier, V., M. E. Hochberg, and J. F. Guégan. 2004. Ecology drives the worldwide distribution of human diseases. *PLoS Biology* 2:740–46.

Haines, M. R. 2001. The urban mortality transition in the United States, 1800–1940. *Annales de Démographie Historique* 1:33–64.

Kambale, L. K., and J. P. Lumbila Musongela. Problématique du VIH/Sida dans le contexte de la République Démocratique du Congo. *Congo medical* 3/3 (Sept. 2001): 195–97.

King, R. 2002. *Ancient remedies, new disease: Involving traditional healers in increasing access to AIDS care and prevention in East Africa.* UNAIDS Best Practice Collection. Geneva: Joint United Nations Programme on HIV/AIDS (UNAIDS). http://data. unaids.org/Publications/IRC-pub02/jc761-ancientremedies_en.pdf.

Klepinger, L. L. 1992. Innovative approaches to the study of past human health and subsistence strategies. In *Skeletal biology of past peoples: Research methods*, ed. S. R. Saunders and M.A. Katzenberg, 121–30. New York: Wiley-Liss.

Koppert, G., E. Dounias, A. Froment, and P. Pasquet. 1993. Food consumption in three forest populations of the southern coastal Cameroon. In *Tropical forests, people and food: Biocultural interactions and applications to development*, ed. C. M. Hladik, A. Hladik, O. Linares, H. Pagezy, A. Semple, and M. Hadley, 295–311. Man and the Biosphere. London: Parthenon-UNESCO.

Lappé, M. 1994. *Evolutionary medicine: Rethinking the origin of disease*. San Francisco: Sierra Club Books.

Leroy, E., P. Rouquet, P. Formenty, S. Souquière, A. Kilbourne, J. M. Froment, M. Bermejo, et al. 2004. Multiple Ebola virus transmission events and rapid decline of central African wildlife. *Science* 303:387–90.

Maffi, L. (Ed.) 2001. *On biocultural diversity: Linking language, knowledge and the environment*. Washington D.C., USA: Smithsonian Institution Press.

Mann, G. V., A. Roels, D. L. Price, and J. M. Merrill. 1962. Cardiovascular disease in African Pygmies: A survey of the health status, serum lipids, and diet of Pygmies in Congo. *Journal of Chronic Diseases* 15:341–71.

McMichael, T. 2001. *Human frontiers, environments and disease: Past patterns, uncertain futures*. Cambridge: Cambridge University Press.

Mercader, J. 2003. *Under the canopy: The archaeology of tropical rain forests*. New Brunswick, NJ: Rutgers University Press.

Morvan, J. M., E. Nakoune, V. Deubel, and M. Colyn. 2000. Écosystèmes forestiers et virus Ebola. *Bulletin de la Société de Pathologie Exotique* 93:172–75.

Myers, N. 1999. What we must do to counter the biotic holocaust. *International Wildlife* 29 (2): 30–39.

Nicolle, C. 1933. *Destin des maladies infectieuses*. Paris: Association des Anciens Élèves de l'Institut Pasteur. Repr. Paris: Éditions France Lafayette, 1993.

Oslisly, R. 2001. The history of human settlement in the middle Ogooué valley (Gabon): Implications for the environment. In *African rain forest ecology and conservation. An interdisciplinary perspective*, ed. W. Weber, L. J. T. White, A. Vedder, and L. Naughton-Treves, 101–18. New Haven: Yale University Press.

Pascual, M., X. Rodó, S. P. Ellner, R. Colwell, and M. J. Bouma. 2000. Cholera dynamics and El Niño–Southern Oscillation. *Science* 289:1766–69.

Peeters, M., V. Courgnaud, B. Abela, P. Auzel, X. Pourrut, F. Bibollet-Ruche, S. Loul, et al. 2002. Risk to human health from a plethora of simian immunodeficiency viruses in primate bushmeat. *Emerging Infectious Diseases* 8:451–57.

Pennington, R. 2001. Hunter-gatherer demography. In *Hunter-gatherers: An interdisciplinary perspective*, ed. C. Panter-Brick, R. Layton, and P. Rowley-Conwy, 170–204. Cambridge: Cambridge University Press.

Rambaud A., D. L. Robertson, O. G. Pybus, M. Peeters, and E. C. Holmes. 2001. Human immunodeficiency virus: Phylogeny and the origin of HIV-1. *Nature* 410:1047–48.

Ratard, R. C., L. E. Kouemeni, M. M. Ekani Bessala, C. N. Ndamkou, M. T. Sama, and B. L. Cline. 1991. Ascariasis and trichuriasis in Cameroon. *Transactions of the Royal Society of Tropical Medicine and Hygiene* 85:84–88.

Rémy, G. 1992. Éléments d'une éco-épidémiologie des maladies transmissibles. In *La santé en société: Regards et remèdes*, ed. C. Blanc-Pamard, 33–56. Paris: Éditions ORSTOM.

Riach, J. R. 2004. Ecosystem approach to rapid health assessments among indigenous cultures in degraded tropical rainforest environments: Case study of unexplained deaths among the Secoya of Ecuador. *EcoHealth* 1:86–100.

Rogers, D. J., A. J. Wilson, S. I. Hay, and A. J. Graham. 2006. The global distribution of yellow fever and dengue. *Advances in Parasitology* 62:181–220.

Saluzzo, J. F., P. Vidal, and J. P. Gonzalez. 2004. *Les virus émergents.* Paris: Éditions IRD.

Soskolne, C. L. 2003. Measuring the impact of ecological disintegrity on human health: A role for epidemiology. In *Managing for healthy ecosystems*, ed. D. J. Rapport, B. L. Lasle, D. E. Rolston, N. O. Nielsen, C. O. Qualset, and A. B. Damania. Boca Raton, FL: Lewis Publishers.

Stearns, S. C., ed. 1998. *The evolution of health and disease.* Oxford: Oxford University Press.

Trevethan, W., J. McKenna, and E. O. Smith. 1998. *Evolutionary medicine.* Oxford: Oxford University Press.

Valleron, A. J., and A. Boumendil. 2004. Epidemiology and heat waves: Analysis of the 2003 episode in France. *Comptes Rendus Biologies* 327:1125–41.

Webb, S. 1995. *Palaeopathology of aboriginal Australians: Health and disease across a hunter-gatherer continent.* Cambridge: Cambridge University Press.

WHO. 2003. Climate change and health: Risks and responses. Geneva, WHO Editions. http://www.who.int/globalchange/climate/en/ccSCREEN.pdf.

WHO. 2006. World health statistics 2006. Geneva: WHO Editions. http://www.who.int/whosis/whostat2006_erratareduce.pdf.

Wolfe, N. D., W. M. Switzer, J. K. Carr, V. B. Bhullar, V. Shanmugam, U. Tamoufe, A. T. Prosser, J. N. Torimiro, et al. 2004. Naturally acquired simian retrovirus infections among Central African hunters. *Lancet* 363:932–37.

13

Biodiversity and Disease

Matthew B. Thomas, Kevin D. Lafferty,
and Carolyn S. Friedman

Human activity is degrading biodiversity in many ecosystems across the earth, with current extinction rates up to four orders of magnitude higher than the background rate in the fossil record (Gaston and Spicer 1998). Concern over the provision of ecosystem services by depauperate ecosystems has led to intense research activity aimed at describing the relationship between diversity and ecosystem function. Our aim in this chapter is to review mechanisms by which biodiversity can affect the risk and spread of infectious disease. We consider biodiversity within an infectious disease system at a range of levels, including pathogen/parasite diversity, host species richness and species composition, and habitat and ecosystem diversity. We do not consider explicitly the range of potential factors that can bring about diversity change, but, in line with the recent heightened awareness of the impact of invasive species, we do pay particular attention to the role of species introductions, examining both disease agents and host/vectors. The links back to human health implications are sometimes direct (e.g., some human infections), sometimes indirect (e.g., changes in biodiversity affecting diseases of crops). In addition, we draw on a broad range of examples of wildlife diseases that, in addition to being important in their own right, serve to illustrate processes and mechanisms that are likely to apply to emerging infectious diseases of humans.

Parasite and Pathogen Diversity

Infectious diseases are an underappreciated component of biodiversity. For example, by some estimates, parasites comprise the majority of animal species (Price 1980). Although the general roles of infectious disease in controlling populations are understood, knowledge of the roles of specific parasites is often lacking or not well defined. Furthermore, the vast majority of host–pathogen/parasite studies consider the direct interaction between one disease agent and one host species. However, evidence from a range of systems indicates that "concomitant" or "mixed" infections involving two or more parasite species or genotypes within a host are not only common, but might be the rule (Cox 2001; Petney and Andrews 1998). These higher-diversity infections have been examined in numerous host systems: mammal (e.g., Behnke et al. 2001; Cox 2001; Nilssen, Haugerud, and Folstad 1998; Petney and Andrews 1998); bird (e.g., Forbes et al. 1999); reptile (Lainson 2002; Schall and Bromwich 1994); fish (Barker, Cone, and Burt 2002; Sousa et al. 1996); and invertebrate (e.g., Tang et al. 2003; Thomas, Watson, and Valverde-Garcia 2003).

Since parasites are a component of biodiversity, we consider the consequences of adding new parasite species to communities already rich with parasites. In mixed infections, complex interactions between parasites and an individual host may arise such that the burden of one or both of the infectious agents may be increased, one or both may be suppressed, or one may be increased and the other suppressed (Cox 2001). For example, Tang et al. (2003) demonstrated that preinfection with infectious hematopoietic and hypodermal necrosis virus (IHHNV) reduced viral load and mortality of shrimps subsequently exposed to white spot syndrome virus. Similarly, preexposure of rainbow trout to a nonpathogenic virus has been shown to reduce impact of a pathogenic virus (Hedrick et al. 1994). The authors suggested that nonspecific or specific humoral factors (host-derived immune response) may play a role in protection, or that one virus is able to reduce the ability of the second virus to infect host cells. In a recent study exploring the competitive interaction between two malaria genotypes within a host, de Roode et al. (2004) showed that the outcome of infection was mediated by host genotype. In a host with an effective immune system, one malaria genotype was able to outcompete a second, less virulent genotype. In an immune-compromised host, however, the less virulent genotype was able to persist and, in fact, do better than it did as a single infection. As such, mixed infection that effectively increases pathogen diversity within the host has the potential to dramatically alter population dynamics and the evolution of a particular disease agent. Such effects apply not only to disease hosts, but also to disease vectors. Vaughan, Trpis, and Turell (1999), for example, showed that infection with microfilariae worms could enhance the infectivity of Venezuelan equine encephalitis virus to *Aedes* mosquitoes, effectively increasing their capacity to transmit this disease. In contrast, the development of malaria parasites can be retarded when mosquito vectors are coinfected with bacteria (Lowenberger et al. 1999), filarial nematodes (Albuquerque and Ham 1995), or fungi (Blanford et al. 2005).

Host Diversity

As indicated earlier, the effects of diversity, in this case of the parasites, can vary depending on the specific nature of the interacting biodiversity elements. The possible effects of host diversity are no less complicated. Species richness and the abundance of individuals in a community tend to be positively associated (Rosenzweig 1995), suggesting that in many natural communities, disease will be higher and more diverse in communities with high host diversity (Hechinger and Lafferty 2005). We will be most concerned, however, when impacts to biodiversity lead to increases in disease. For example, in agricultural systems, the adoption of monocultures has acted to increase the density of a particular species or cultivar while reducing the numbers of crop species, varieties within species, and genetic differences within varieties (Wolfe 2000). The risk this loss of diversity creates is that if a pest or disease is able to exploit the one dominant variety, then it has almost unlimited potential to spread throughout the field and landscape. The conventional approach to deal with this loss of diversity is to breed resistance traits back into the crop and to compensate for loss of resistance with applications of pesticides (i.e., to substitute external inputs for ecosystem services). However, in a recent study in Yunnan Province in China, rice farmers were able to control a key fungal disease (rice blast) through the use of variety mixtures, interplanting one row of a susceptible glutinous rice variety with every four to six rows of a more resistant commercial variety (Zhu et al. 2000). This simple increase in intraspecific diversity led to a substantial reduction in the prevalence of rice blast and an increase in the yield of the susceptible variety. The mechanism appears to be a combination of the disease-resistant variety acting as a physical barrier preventing spread of fungal spores between rows of the resistant variety, coupled with a complex interaction involving induced resistance and multiple pathogen genotypes that prevents the dominance of a single virulent strain of the pathogen.

Similar trends have been observed in monospecies aquaculture systems globally, with often devastating results (Ellis 2001; Vijayan et al. 2004). Tropical shrimp aquaculture has been hampered by catastrophic diseases, most commonly of viral origin (OIE 2003). Shrimp culture is dominated by dense, monospecies culture of *Penaeus monodon*, despite the fact that multiple shrimp species are commonly found in local waters (Vijayan et al. 2004). It has been hypothesized that coculture of a variety of shrimp species would likely lessen the impacts of a pathogen because of the reduced density of each host species, and/or because of differences in innate resistances to pathogens among species. Unlike terrestrial crop systems, few drugs are approved for use in aquaculture in many countries (Schnick 1998; Smith and Pasnik 2002); however, therapeutant availability varies greatly between countries (Jensen 1998; Sano 1998). A lack of approved drugs, or the high cost of applying drugs in developing and rural areas, often leads to the development of alternative management practices such as selective breeding of species free of pathogens (specific-pathogen-free [SPF]; Lotz et al. 1995); breeding for resistance to specific strains (specific-pathogen-resistant [SPR]; Fast and Menasveta 2000); alternate husbandry practices, the application of probiotics (Elston et al. 2004; Sambasivam, Chandran, and Khan 2003); and, for finfish, vaccines (Winton 1998).

With respect to human health, there are relatively few examples where changes in host species diversity have been linked to changes in disease incidence and risk, although as a general rule, the diversity of human infectious diseases is highest in regions with high biodiversity (Guégan, Morand, and Poulin 2005). This relationship is of particular concern when humans encroach on tropical forests. The role of host diversity in the dynamics of zoonotic diseases (those diseases that can be transmitted between animals and humans) has been explored recently by Dobson and Foufopoulos (2001). They used matrix projection models to determine the effects of changes in diversity of hosts on the basic reproductive number (R_0) of the pathogen. R_0 provides a measure of the number of secondary infections produced by the first infective individual to appear in a population (R. Anderson and May 1982). The analysis of Dobson and Foufopoulos (2001) indicated that where increases in species diversity lead to increases in the number of contacts between infected individuals and susceptible hosts, R_0 will increase, creating greater potential for disease outbreaks. Equally, if disease transmission is density dependent, then increasing the number of hosts will increase R_0. On the other hand, where increases in interspecific transmission lead to reductions in intraspecific transmission, or where disease transmission is frequency dependent rather than density dependent, then increased host diversity can reduce R_0. Complementary to this study, Holt et al. (2003) used relatively simple epidemiological models to investigate the role of host community structure on parasite establishment (persistence). Among the various scenarios explored, they confirmed the potential for additional host species to potentially inhibit establishment of infectious diseases. This negative association with diversity is limited to situations where certain host species are more resistant and applies especially to free-living pathogens and those transmitted by vectors. In contrast, they also identified a positive association with diversity whereby parasites with complex life cycles, involving essential passage through alternate hosts, may be particularly vulnerable to extinction in species-depauperate communities (Holt et al. 2003). This latter prediction can be interpreted as support for a decrease in biodiversity being beneficial for human health, especially as landscape changes around human habitations concomitantly can reduce biodiversity and remove, for example, vector breeding sites. However, this viewpoint overlooks other potential benefits of biodiversity and the consequences of reduced biodiversity over the long term.

Further theoretical investigation and empirical support for the hypothesis that diversity can dilute transmission has been provided by Ostfeld and coworkers studying Lyme disease in the United States (LoGiudice et al. 2003; Ostfeld and Keesing 2000; Schmid and Ostfeld 2001). Lyme disease is caused by the spirochete bacterium *Borrelia burgdorferi*, which is transmitted to humans by *Ixodes* spp. ticks. The larvae and nymphs of these ticks are highly polyphagous, feeding on a wide range of vertebrate hosts. Because Lyme disease is vector transmitted, spirochete transmission is more likely to be frequency dependent than density dependent (McCallum, Barlow, and Hone 2001), simplifying predictions of the association between diversity and disease. Vector-transmitted diseases should decrease with host diversity, assuming that vectors are present and that some of the hosts they feed on are not suitable for the vector-transmitted disease.

In the United States, the white-footed mouse (*Peromyscus leucopus*) is the main reservoir for the Lyme disease spirochete. After feeding on these mice, the ticks can transmit the bacterium to other hosts. However, while some hosts are effective reservoirs, many are incompetent reservoirs of Lyme disease (Mather 1993). This variation in reservoir competence creates a situation in which high levels of host diversity (thereby effectively reducing the representation of white-footed mice relative to other hosts) could reduce infection prevalence via a "dilution effect" (LoGiudice et al. 2003; Norman, Begon, and G. Bowers 1994; Ostfeld and Keesing 2000). That is, increasing the frequency of encounters with hosts that are incompetent reservoirs relative to the highly competent white-footed mice reservoir should reduce infection prevalence of ticks, thus leading to a lower risk of exposure to Lyme disease in humans. Empirical support for this dilution effect was provided by Ostfeld and Keesing (2000), who demonstrated significant negative relationships between species richness of small mammals and lizards and the per capita number of Lyme disease cases for each state along the eastern seaboard of the United States from Maine to Florida. Interestingly, a contrasting positive association was identified with species richness of ground-dwelling birds. This pattern suggests that birds may be acting as competent reservoirs contributing to a "rescue effect" whereby increased diversity helps maintain the disease agent at a relatively constant prevalence by buffering against fluctuations of individual host populations (Ostfeld and Keesing 2000). Further empirical studies to test the generality of the dilution effect are urgently needed.

Habitat and Ecosystem Diversity Effects and Diversity of Nonhost Species

Beyond host and pathogen species diversity effects, changes in habitat and ecosystem diversity can also affect risk and incidence of disease. For example, in many cases, large-scale changes in land use associated with agricultural intensification and/or water management have led to increases in human disease. A recent report prepared under the auspices of the World Health Organization and the United Nations Environment Programme (2002) identified a range of causal mechanisms, including the following:

- Changes in habitat diversity, such as clearing of forests, can result in new or increased associations between humans and forest organisms. Increases in yellow fever (Brown 1977), leishmaniasis (Sutherst 1993), malaria (Bradley 1993), and Ebola virus (Walsh et al. 2003) have all been linked to human encroachment on tropical forests and increased contact with key disease vectors.
- Changes in habitat diversity can alter the species diversity of hosts, vectors, reservoirs, and so forth, with implications for disease transmission. For example, changes in grass flora following conversion of pampas habitat in Argentina to maize production, which included use of herbicides to control weeds, favored population growth of the drylands vesper mouse *Calomys musculinus*, the main reservoir of the Junin virus, the cause of Argentine hemorrhagic fever (Daily and Ehrlich 1995).

- Changes in the abundance and distribution of habitats can affect the dynamics and behavior of disease vectors. For example, intensification of rice production in Guyana, which involved an increase in irrigation and reduction in local populations of domestic animals, caused an increase in *Anopheles* mosquito populations and a shift in feeding toward humans, resulting in a malaria epidemic (Desowitz 1991).
- Changes in habitat diversity (often acting in combination with other anthropogenic factors) are also strongly implicated in disease outbreaks in both terrestrial and aquatic wildlife (Daszak, Cunningham, and Hyatt 2001; Dobson and Foufopoulos 2001; Harvell et al. 1999).

Top predators are often the first species to disappear in the process of biodiversity loss. When predators are lost to ecosystems, their prey may increase in abundance, leading to increased disease transmission efficiency (Packer et al. 2003). This trophic effect can have consequences for human health. Many zoonotic reservoir species are rodents (Mills and Childs 1998), whose dynamics can be strongly affected by predators (Ostfeld and Holt 2004 and references therein). While these dynamic effects have rarely been linked to increases in zoonotic disease, in general terms, if rodents are maintained at low density away from humans, disease transmission will be reduced (Ostfeld and Holt 2004). It is proposed that one generic effect of reducing predators could be to increase the equilibrial abundance of infected reservoirs and the fraction of reservoirs infected (Ostfeld and Holt 2004). The loss of vertebrate predators via habitat destruction and environmental degradation could, therefore, lead to a trophic cascade increasing transmission of rodent-borne disease to humans (Ostfeld and Holt 2004). Empirical support that predators can affect disease dynamics is provided by Hudson (1992) and Hudson, Dobson, and Newborn (1998), who studied the interaction between red grouse (*Lagopus scoticus*), predators, and a parasitic nematode. When gamekeepers control predators, grouse populations show cyclic fluctuations driven by the effects of nematodes on grouse fecundity. When predators increase (or the density of gamekeepers falls), grouse populations are less likely to cycle because predators select the infected individuals, reducing the proportion of heavily infected grouse in the population.

Species Introductions

Alien invasive species can have dramatic environmental and economic impacts (Parker et al. 1999; Ruiz et al. 1999). Evidence suggests that invasive species represent the second greatest threat to native biodiversity after habitat loss (Wilcove et al. 1998; but see Gurevitch and Padilla 2004 for an alternative interpretation).

Although invading species leave most of their diseases behind, they do bring a small proportion with them (Mitchell and Power 2003; Torchin et al. 2003). These diseases often encounter naive hosts that have no coevolved resistance, which can lead to relatively high pathogenicity. Many of the most dramatic emerging infectious diseases are attributable to alien invasive (exotic) pathogens. Dobson and Foufopoulos (2001) identified that around 60% of the pathogens responsible for recent wildlife epidemics were of exotic or likely exotic origin. Similarly, an analysis by P. Anderson et al. (2004) indicates that pathogen introductions are the most important driver of emerging infectious diseases in plants. The introduction and spread of viral diseases through the movement of infected stocks is a key impediment to successful penaeid shrimp culture (Flegel 2002; Tang et al. 2003). In fact, of the eight crustacean diseases notifiable by the Office Internationale des Epizooites, seven disease agents are viral (OIE 2003). In addition, many or most of the known molluscan disease epidemics have resulted from the establishment of exotic pathogens in new hosts subsequent to animal transfers (Burreson, Stokes, and Friedman 2000; Friedman and Finley 2003; Naylor, Williams, and Strong 2001). In general, introductions are the primary cause of disease that drive formerly common species to low levels (Lafferty and Gerber 2002).

While most invasive diseases probably go unnoticed, several examples indicate how introduced diseases can impact native communities. Introduction of rinderpest to East Africa led to dramatic declines in buffalo (*Syncerus caffer*) and wildebeest (*Connochaetes* spp.) that then led to large-scale alteration of the Serengeti ecosystem (Plowright 1982). Distemper viruses have presumably crossed from domestic dogs to pinnipeds (Heide-Jorgensen et al. 1992) and African carnivores (Packer et al. 1999), in which they cause mass mortality events. The introduction of *Haplosporidium nelsoni* into native U.S. eastern oyster populations as a result of exposure to infected Pacific oysters (*Crassostrea gigas*) from Japan, where this disease agent is endemic, resulted in catastrophic losses along the mid-Atlantic coast (Burreson, Stokes, and Friedman 2000). *Bonamia ostreae*, a protistan parasite introduced into France from California, resulted in widespread losses of the native European flat oyster, *Ostrea edulis* (Elston, Farley, and Kent 1986). The presumed exotic etiological agent of withering syndrome, *"Candidatus Xenohaliotis californiensis"* (Friedman et al. 2000), has resulted in losses of over 90%–99% of black abalone along the southern and central California coasts (Haaker et al. 1992; Lafferty and Kuris 1993; Raimondi et al. 2002; Richards 2000). In addition, in some cases, an introduced disease can aid an invader's success. In Great Britain, gray squirrels (*Sciurus carolinensis*) from America extirpate native red squirrels (*Sciurus vulgaris*) with the aid of a parapox virus, brought by resistant gray squirrels, that decimates red squirrels (Tompkins, White, and Boots 2003). Similarly, movement of the crayfish plague (*Aphanomyces astaci*)–resistant North American crayfish into Great Britain and the European mainland resulted in extirpation of native crayfish species throughout much of Europe (Alderman 1993; Liley, Cerenius, and Soderhal 1997; Nylund and Westman 1995). Extending these indirect linkages still further, introduction of a mosquito vector to the Hawaiian Islands resulted in transmission of avian malaria and avian pox from resistant introduced bird species to naive native bird species (van Riper and Scott 2001; van Riper, van Riper, and Hansen 2002). This bridging effect resulted in widespread loss of native birds from lowland areas where mosquitoes breed. In general, exotic species introductions are increasing substantially, with changes in species ranges and establishment further influenced by climate change. These changes can result in the ingress of pathogens or invasive species to previously unexposed host populations or habitats (Carlton 2001; Harvell et al. 1999).

The invasion process involves the sequential steps of arrival, establishment, and spread. Abiotic factors are often key parameters driving the establishment of an exotic species and strongly influence the development of clinical disease after pathogen introductions (Gerhardt and Collinge 2003; Moore, Robbins, and Friedman 2000; Occhipinti-Ambrogi and Savini 2003; Thomas and Blanford 2003). Climate-mediated and other environmental stressors can facilitate disease epidemics by impacting host physiology and compromising disease resistance, thereby increasing the frequency of both opportunistic and noninfectious diseases (Carey and Alexander 2003; Harvell et al. 1999; Iguchi et al. 2003; Thomas and Blanford 2003). Emerging diseases often result from host shifts or range expansions of pathogens whose impact or frequency may be influenced by anthropogenic environmental and/or climate change. Water quality and, particularly, environmental fluctuations in temperature, salinity, dissolved gases, and pH are among the primary factors influencing aquatic organismal health, including influences on parasite communities (Calvo, Wetzel, and Burreson 2001; Harvell et al. 1999). Pollutants are most likely to negatively impact parasites, while eutrophication and increased water temperature are more likely to favor parasites (Lafferty 1997). For example, the newly described ostreid herpesvirus causes catastrophic losses of larval and seed Pacific oysters in association with water temperatures of $\geq 20°C \times 24°C$ (Friedman et al. 2005; Le Deuff, Renault, and Gérard 1996). Thermal induction of many other aquatic diseases has been well documented, including such examples as proliferative kidney disease (de Kinkelin and Loriot 2001); withering syndrome of abalone (Moore, Robbins, and Friedman 2000); vibriosis in a variety of finfish and marine invertebrate species (Gay et al. 2003; Karunasagar and Otta 2003); and noninfectious coral bleaching (Harvell, Kim, Burkholder et al. 1999; Harvell, Kim, Quirolo et al. 2001). In terrestrial systems, environmental temperature can also increase or decrease the impact of pathogens and parasites on their ectotherm (see Thomas and Blanford 2003 for a review of invertebrate-based systems) and endotherm (e.g. Hudson et al. 2006) hosts. The key to such effects appears to be a differential in thermal sensitivity between the hosts and parasites, leading to complex "genotype (i.e., host) × genotype (i.e., parasite) × environment" interactions. The influence of environmental change on predator/pathogen efficacy can also have substantial indirect effects on communities through release or suppression of species at different trophic levels (Preisser and Strong 2004). Potential for such trophic complexity suggests considerable value in further research on the role of environmental factors in host × pathogen/parasite interactions. Chapter 14 provides a more detailed discussion on climatic and abiotic influence on organismal health.

The Need to Understand Mechanisms and Complexity

Biodiversity affects disease in complex ways. Increased parasite diversity within hosts (i.e., mixed infections) can increase or decrease the impact of individual disease agents. The greater the diversity of hosts, the greater the diversity of parasites a natural community will tend to support. However, for an individual pathogen or parasite, increased host diversity can actually reduce disease spread. This positive effect of diversity can apply to directly transmitted diseases (see effects of increased rice diversity on the impact of blast disease), but more especially to vector-borne diseases, where an increase in the diversity of hosts for the vector can dilute transmission of the disease if some of the host species are poorer reservoirs. Higher up the ecological hierarchy, then, changes in habitat and ecosystem diversity can also affect disease risk and spread. Loss of habitat and human encroachment on natural areas can increase contact between diseases of wildlife and humans (and their domestic animals). Disease risk is likely to increase further if the pathogens or parasites utilize alternate hosts/vectors, such as rats or mosquitoes, that are adapted to human-impacted environments. An additional mechanism of biodiversity change is alien invasive species. Introduced species can reduce biodiversity through competition and predation and may also impact native species through the introduction of novel pathogens. Many of the emergent infectious diseases currently causing concern result from exotic parasites and pathogens. The spread and establishment of these agents is likely to increase with increased trade and also through alterations in environmental factors linked to global climate change.

References

Albuquerque, C. M. R., and P. J. Ham. 1995. Concomitant malaria (*Plasmodium gallinaceum*) and filaria (*Brugia pahangi*) infections in *Aedes aegypti*: Effect on parasite development. *Parasitology* 110:1–6.

Alderman, D. J. 1993. Crayfish plague in Britain: The first twelve years. *Freshwater crayfish IX*, 266–72.

Anderson, P. K., A. A. Cunningham, N. G. Patel, F. J. Morales, P. R. Epstein, and P. Daszak. 2004. Emerging infectious diseases of plants: Pathogen pollution, climate change, and agrotechnology drivers. *Trends in Ecology and Evolution* 119:535–44.

Anderson, R. M., and R. M. May. 1982. Coevolution of hosts and parasites. *Parasitology* 85:411–26.

Barker, D. E., D. K. Cone, and M. D. B. Burt. 2002. *Trichodina murmanica* (Ciliophora) and *Gyrodactylus pleuronecti* (Monogenea) parasitising hatchery-reared winter flounder, *Pseudopleuronectes americanus* (Walbaum): Effects on host growth and assessment of parasite interaction. *Journal of Fish Diseases* 25:81–89.

Behnke, J. M., C. J. Barnard, A. Bajer, D. Bray, J. Dinmore, K. Frake, J. Osmond, T. Race, and E. Sinski. 2001. Variation in the helminth community structure in bank voles (*Clethrionomys glareolus*) from three comparable localities in the Mazury Lake District region of Poland. *Parasitology* 123:401–14.

Blanford, S., B. H. K. Chan, N. Jenkins, D. Sim, R. J. Turner, A. F. Read, and M. B.

Thomas. Fungal pathogen reduces potential for malaria transmission. *Science* 308:1638–41.

Bradley, D. J. 1993. Environmental and health problems of developing countries. In *Environmental change and human health*, ed. J. V. Lake, G. R. Bock, and K. Ackrill, 234–44. New York: Wiley.

Brown, A. W. A. 1977. Yellow fever, dengue and dengue haemorrhagic fever. In *A world geography of human diseases*, ed. G. M. Howe, 271–317. London: Academic Press.

Burreson, E. M., N. A. Stokes, and C. S. Friedman. 2000. Increased virulence in an introduced pathogen: *Haplosporidium nelsoni* (MSX) in the eastern oyster *Crassostrea virginica*. *Journal of Aquatic Animal Health* 12:1–8.

Calvo, L. M. R., R. L. Wetzel, and E. M. Burreson. 2001. Development and verification of a model for the population dynamics of the protistan parasite, *Perkinsus marinus*, within its host, the eastern oyster, *Crassostrea virginica*, in Chesapeake Bay. *Journal of Shellfish Research* 20:231–41.

Carey, C., and M. A. Alexander. 2003. Climate change and amphibian declines: Is there a link? *Density and Distributions* 9:111–21.

Carlton, J. T. 2001. *Introduced species in coastal waters: Environmental impacts and management priorities*. Arlington, VA: Pew Oceans Commission.

Cox, F. E. G. 2001. Concomitant infections, parasites and immune responses. *Parasitology* 122: S23–S38.

Daily, G. C., and P. R. Ehrlich. 1995. Development, global change, and the epidemiological environment. Working Paper 62, Morrison Institute for Population and Resource Studies, Stanford University, Palo Alto, CA.

Daszak, P., A. A. Cunningham, and A. D. Hyatt. 2001. Anthropogenic environmental change and the emergence of infectious diseases in wildlife. *Acta Tropica* 78:103–16.

deKinkelin, P., and B. Loriot. 2001. A water temperature regime which prevents the occurrence of proliferative kidney disease (PKD) in rainbow trout, *Oncorhynchus mykiss* (Walbaum). *Journal of Fish Diseases* 24:489–93.

De Roode, J. C., R. Culleton, S. J. Cheesman, R. Carter, and A. F. Read. 2004. Host heterogeneity is a determinant of competitive exclusion or coexistence in genetically diverse malaria infections. *Proceedings of the Royal Society of London* B 271:1073–80.

Desowitz, R. S. 1991. *The malaria capers: More tales of parasites and people, research and reality*. New York: Norton.

Dobson, A. P., and J. Foufopoulos. 2001. Emerging infectious pathogens in wildlife. *Philosophical Transactions of the Royal Society of London* B 356:1001–12.

Ellis, A. E. 2001. Mariculture disease and health. In *Encyclopedia of Ocean Sciences*, ed. J. Steele, S. Thorpe, and A. Turekian. Vol. 3, *I–M*, 1555–57. London: Academic Press.

Elston, R. A., C. A. Farley, and M. L. Kent. 1986. Occurrence and significance of bonamiasis in European flat oysters *Ostrea edulis* in North America. *Diseases of Aquatic Organisms* 2:49–54.

Elston, R. A., K. Humphrey, A. Gee, D. Cheney, and J. David. 2004. Progress in the development of effective probiotic bacteria for bivalve shellfish hatcheries and nurseries. *Journal of Shellfish Research* 23:288–89.

Fast, A.W., and P. Menasveta. 2000. Some recent issues and innovations in marine shrimp pond culture. *Reviews in Fisheries Science* 8:151–233.

Flegel, T. W. 2002. Emerging shrimp diseases and innovations to prevent their spread. In

Diseases in Asian aquaculture IV, ed. C. R. Lavilla-Pitogo and E. R. Cruz-Lacierda, 137–50. Manila: Fish Health Section, Asian Fisheries Society.

Forbes, M. R., R. T. Alisauskas, J. D. McLaughlin, and K. M. Cuddington. 1999. Explaining co-occurrence among helminth species of lesser snow geese (*Chen caerulescens*) during their winter and spring migration. *Oecologia* 120: 613–20.

Friedman, C. S., K. B. Andree, K. Beauchamp, J. D. Moore, T. T. Robbins, J. D. Shields, and R. P. Hedrick. 2000. *"Candidatus Xenohaliotis californiensis,"* a newly described pathogen of abalone, *Haliotis* spp., along the west coast of North America. *International Journal of Systematic and Evolutionary Microbiology* 50:847–55.

Friedman, C. S., and C. A. Finley. 2003. Anthropogenic introduction of the etiological agent of withering syndrome into northern California abalone populations via conservation efforts. *Canadian Journal of Fisheries and Aquatic Sciences* 60:1424–31.

Friedman, C. S., N. A. Stokes, E. S. Burreson, B. Barber, R. A. Elston, and K. Reece. 2005. Identification of a herpes-like virus in Pacific oysters, *Crassostrea gigas* Thunberg, in Tomales Bay, California. *Diseases of Aquatic Organisms* 63:33–41.

Gaston, K. J., and J. I. Spicer. 1998. *Biodiversity: An Introduction.* Oxford, UK: Blackwell Science.

Gay, M., D. Saulnier, N. Faury, and F. Le Roux. 2003. Epidemiological study of vibriosis. In *Styli 2003: Thirty years of shrimp farming in New Caledonia,* ed. C. Goarant, Y. Harache, A. Herbland, and C. Mugnier. Proceedings of the symposium held in Nouméa Koné, 2–6 June 2003. Éditions Ifremer, *Actes du Colloque* 38.

Gerhardt, F., and S. K. Collinge. 2003. Exotic plant invasions of vernal pools in the Central Valley of California. *Journal of Biogeography* 30:1043–52.

Guégan, J. F., S. Morand, and R. Poulin. 2005. Are there general laws in parasite community ecology? The emergence of spatial parasitology and epidemiology. In *Parasitism and ecosystems,* ed. F. Thomas, F. Renaud, and J. F. Guégan. Oxford: Oxford University Press.

Gurevitch, J., and D. K. Padilla. 2004. Are invasive species a major cause of extinctions? *Trends in Ecology and Evolution* 19:470–74.

Haaker, P. L., D. V. Richards, C. Friedman, G. Davis, D. O. Parker, and H. Togstad. 1992. Abalone withering syndrome and mass mortality of black abalone, *Haliotis cracherodii* in California. In *Abalone of the world,* ed. S. A. Shephard, M. Tegner, and S. Guzman del Proo, 214–24. Oxford: Blackwell Scientific.

Harvell, C. D., K. Kim, J. M. Burkholder, R. R. Colwell, R. R. Epstein, D. J. Grimes, E. E. Hofmann, et al. 1999. Emerging marine diseases: Climate links and anthropogenic factors. *Science* 285:1505–10.

Harvell, D., K. Kim, C. Quirolo, J. Weir, and G. Smith. 2001. Coral bleaching and disease: Contributors to 1998 mass mortality in *Briareum asbestinum* (Octocorallia, Gorgonacea). *Hydrobiologia* 460:97–104.

Hechinger, R. F., and K. D. Lafferty. 2005. Host diversity begets parasite diversity: Bird final hosts and trematodes in snail intermediate hosts. *Proceedings of the Royal Society of London* B. 272:1059–66.

Hedrick, R. P., S. E. La Patra, S. Yun, K. A. Lauda, G. R. Jones, J. L. Congleton, and P. de Kinkelin. 1994. Induction of protection from infectious hematopoietic necrosis virus in rainbow trout *Oncorhynchus mykiss* by pre-exposure to the avirulent cutthroat trout virus (CTV). *Diseases of Aquatic Organisms* 20:111–18.

Heide-Jorgensen, M. P., T. Harkonen, R. Dietz, and P. M. Thompson. 1992. Retrospective of the 1988 European seal epizootic. *Diseases of Aquatic Organisms* 13:37–62.

Holt, R. D., A. P. Dobson, M. Begon, R. G. Bowers, and E. M. Schauber. 2003. Parasite establishment in host communities. *Ecology Letters* 6:837–42.

Hudson, P. J. 1992. *Grouse in space and time.* Fordingbridge, UK: Game Conservancy.

Hudson, P. J., I. M. Cattadori, B. Boag, and A. P. Dobson. 2006. Climate disruption and parasite-host dynamics: Patterns and processes associated with warming and the frequency of extreme climatic events. *Journal of Helminthology* 80:175–82.

Hudson, P. J., A. P. Dobson, and D. Newborn. 1998. Prevention of population cycles by parasite removal. *Science* 282:2256–58.

Iguchi, K., K. Ogawa, M. Nagae, and F. Ito. 2003. The influence of rearing density on stress response and disease susceptibility of ayu (*Plecoglossus altivelis*). *Aquaculture* 220:515–23.

Jensen, G. L. 1998. Challenges and opportunities with international harmonization of aquaculture therapeutants. *Aquaculture '98 Book of Abstracts*, 271. World Aquaculture Society: Baton Rouge, LA.

Karunasagar, I., and S. K. Otta. 2003. Disease problems affecting fish in tropical environments. *Journal of Applied Aquaculture* 13:231–49.

Lafferty, K. D. 1997. Environmental parasitology: What can parasites tell us about human impacts on the environment? *Parasitology Today* 13: 251–55.

Lafferty, K. D., and L. R. Gerber. 2002. Good medicine for conservation biology: The intersection of epidemiology and conservation theory. *Conservation Biology* 16:593–604.

Lafferty, K. D., and A. M. Kuris. 1993. Mass mortality of abalone *Haliotis cracherodii* on the California Channel Islands: Tests of epidemiologic hypotheses. *Marine Ecology-Progress Series* 96 (3): 239–48.

Lainson, R. 2002. Intestinal coccidia (Apicomplexa: Eimeriidae) of Brazilian lizards. *Emeria carmelinoi* n.sp., from *Kentropyx calcarata* and *Acroeimeria paraensis* n.sp. from *Chemidophorus lemniscatus lemniscatus* (Lacertilia: Teiidae). *Memórias do Instituto Oswaldo Cruz* 97:227–37.

Le Deuff, R., T. Renault, and A. Gérard. (1996). Effects of temperature on herpes-like virus detection among hatchery-reared larval Pacific oysters *Crassostrea gigas*. *Diseases of Aquatic Organisms* 24:149–57.

Liley, J. H., L. Cerenius, and K. Soderhall. 1997. RAPD evidence for the origin of crayfish plague outbreaks in Britain. *Aquaculture* 157:181–85.

LoGiudice, K., R. S. Ostfeld, K. A. Schmidt, and F. Keesing. 2003. The ecology of infectious disease: Effects of host diversity and community composition on Lyme disease risk. *Proceedings of the National Academy of Sciences USA* 100:567–71.

Lotz, J. M., C. L. Browdy, W. H. Carr, P. F. Frelier, and D. V. Lightner. 1995. USMSFP suggested procedures and guidelines for assuring the specific pathogen status of shrimp broodstock and seed. In *Swimming through troubled waters: Proceedings of the special session on shrimp farming*, ed. C. L. Browdy and J. S. Hopkins, 66–75. World Aquaculture Society: Baton Rouge, LA.

Lowenberger, C. A., S. Kamal, S. Paskewitz, P. Bulet, J. A. Hoffman, and B. M. Christensen. 1999. Mosquito-*Plasmodium* interactions in response to immune activation of the vector. *Experimental Parasitology* 91:59–69.

Mather, T. N. 1993. The dynamics of spirochete transmission between ticks and vertebrates. In *Ecology and environmental management of Lyme disease*, ed. H. S. Ginsberg, 43–60. New Brunswick, NJ: Rutgers University Press.

McCallum, H., N. D. Barlow, and J. Hone. 2001. How should transmission be modelled? *Trends in Ecology and Evolution* 16:295–300.

Mills, J. N., and J. E. Childs. 1998. Ecological studies of rodent reservoirs: Their relevance for human health. *Emerging Infectious Diseases* 4:529–37.

Mitchell, C. E., and A. G. Power. 2003. Release of invasive plants from fungal and viral pathogens. *Nature* 421:625–27.

Moore, J. D., T. T. Robbins, and C. S. Friedman. 2000. Withering syndrome in farmed red abalone, *Haliotis rufescens*: Thermal induction and association with a gastrointestinal Rickettsiales-like procaryote. *Journal of Aquatic Animal Health* 12:26–34.

Naylor, R. L., S. L. Williams, and D. R. Strong. 2001. Aquaculture: A gateway for exotic species. *Science* 294:1655–56.

Nilssen, A. C., R. E. Haugerud, and I. Folstad. 1998. No interspecific covariation in intensities of macroparasites of reindeer, *Rangifer tarandus* (L.). *Parasitology* 117:273–81.

Norman, R., M. Begon, and R. G. Bowers. 1994. The population dynamics of microparasites and vertebrate hosts: The importance of recovery and immunity. *Theoretical Population Biology* 46:96–119.

Nylund, V., and K. Westman. 1995. Frequency of visible symptoms of the crayfish plague fungus (*Aphanomyces astaci*) on the signal crayfish (*Pacifastacus leniusculus*) in natural populations in Finland in 1979–1988. *Freshwater Crayfish* 8: 577–88.

Occhipinti-Ambrogi, A., and D. Savini. 2003. Biological invasions as a component of global change in stressed marine ecosystems. *Marine Pollution Bulletin* 46:542–51.

OIE (Office Internationale des Epizooties). 2003. *Manual of diagnostic tests for aquatic animals.* 4th ed., 267–350. Paris: OIE.

Ostfeld, R. S., and R. D. Holt. 2004. Are predators good for your health? Evaluating evidence for top-down regulation of zoonotic disease reservoirs. *Frontiers in Ecology and the Environment* 2:13–20.

Ostfeld, R. S., and F. Keesing. 2000. Biodiversity and disease risk: The case of Lyme disease. *Conservation Biology* 14:722–28.

Packer, C., S. Altizer, M. Appel, E. Brown, J. Martenson, S. J. O'Brien, M. Roelke-Parker, R. Hofmann-Lehmann, and H. Lutz. 1999. Viruses of the Serengeti: Patterns of infection and mortality in African lions. *Journal of Animal Ecology* 68:1161–78.

Packer, C., R. D. Holt, P. J. Hudson, K. D. Lafferty, and A. P. Dobson. 2003. Keeping the herds healthy and alert: Implications of predator control for infectious disease. *Ecology Letters* 6:797–802.

Parker, I. M., D. Simberloff, W. M. Lonsdale, K. Goodell, M. Wonham, P. M. Kareiva, M. H. Williamson, et al. 1999. Impact: Toward a framework for understanding the ecological effects of invaders. *Biological Invasions* 1:3–19.

Petney, T. N., and R. H. Andrews. 1998. Multiparasite communities in humans and animals: Frequency, structure and pathogenic significance. *International Journal of Parasitology* 28: 377–93.

Plowright, W. 1982. The effects of rinderpest and rinderpest control on wildlife in Africa. *Symposia of the Zoological Society of London* 50:1–28.

Preisser, E. L., and D. R. Strong. 2004. Climate affects predator control of an herbivore outbreak. *American Naturalist* 163:754–62.

Price, P. W. 1980. *Evolutionary biology of parasites.* Princeton, NJ: Princeton University Press.

Raimondi, P. T., C. M. Wilson, R. F. Ambrose, J. M. Engle, and T. E. Minchinton. 2002. Continued declines of black abalone along the coast of California: Are mass mortalities related to El Niño events? *Marine Ecological Progress Series* 242:143–52.

Richards, D. V. 2000. The status of rocky intertidal communities in Channel Islands National Park. In *Proceedings of the Fifth California Islands Symposium, 29 March–1 April 1999, Santa Barbara, CA,* ed. D. R. Browne, K. L. Mitchell, and H. W. Chaney, 356–58. Camarillo, CA: U.S. Minerals Management Service.

Rosenzweig, M. L. 1995. *Species diversity in space and time.* Cambridge: Cambridge University Press.

Ruiz, G. M., P. Fofonoff, A. H. Hines, and E. D. Grosholz. 1999. Nonindigenous species as stressors in estuarine and marine communities: Assessing invasion impacts and interactions. *Limnology and Oceanography* 44:950–72.

Sambasivam, S., R. Chandran, and S. A. Khan. 2003. Role of probiotics on the environment of shrimp pond. *Journal of Environmental Biology* 24:103–6.

Sano, T. 1998. Control of fish disease, and the use of drugs and vaccines in Japan. *Journal of Applied Ichthyology* 14:131–37.

Schall, J. J., and C. R. Bromwich. 1994. Interspecific interactions tested: Two species of malarial parasite in a West African lizard. *Oecologia* 97:326–32.

Schmid, K. A., and R. S. Ostfeld. 2001. Biodiversity and the dilution effect in disease ecology. *Ecology* 82:609–19.

Schnick, R. A. 1998. Approval of drugs and chemicals for use by the aquaculture industry. *Veterinary and Human Toxicology* 40:9–17.

Smith, S. A., and D. J. Pasnik. 2002. Drugs approved for aquaculture use in the United States. In *Proceedings of the 4th International Conference on Recirculating Aquaculture,* (np). *Virginia Polytechnic Institute and State University, Roanoke, VA, July 18–21, 2002,* ed. T. T. Rakestraw, L. S. Douglas, and G. Flick. U.S. Department of Agriculture, Cooperative Extension Service, Virginia Polytechnic Institute and State University.

Sousa, J. A., J. L. Romalde, A. Ledo, J. C. Eiras, J. L. Barja, and A. E. Toranzo. 1996. Health status of two salmonid aquaculture facilities in North Portugal: Characterisation of the bacterial and viral pathogens causing notifiable diseases. *Journal of Fish Diseases* 19: 83–89.

Sutherst, R. W. 1993. Arthropods as disease vectors in a changing environment. In *Environmental Change and Human Health,* ed. J. V. Lake, G. R. Bock, and K. Ackrill, 124–39. New York: Wiley.

Tang, K. F. J., S. V. Durand, B. L. White, R. M. Redman, L. L. Mohney, and D. V. Lightner. 2003. Induced resistance to white spot syndrome virus infection in *Penaeus stylirostris* through pre-infection with infectious hypodermal and hematopoietic necrosis virus: A preliminary study. *Aquaculture* 216:19–29.

Thomas, M. B., and S. Blanford. 2003. Thermal biology in insect-pathogen interactions. *Trends in Ecology and Evolution* 18:344–50.

Thomas, M. B., E. L. Watson, and P. Valverde-Garcia. 2003. Mixed infections and insect-pathogen interactions. *Ecology Letters* 6:183–88.

Tompkins, D. M., A. R. White, and M. Boots. 2003. Ecological replacement of native red squirrels by invasive greys driven by disease. *Ecology Letters* 6:189–96.

Torchin, M. E., K. D. Lafferty, A. P. Dobson, V. J. McKenzie, and A. M. Kuris. 2003. Introduced species and their missing parasites. *Nature* 421:628–30.

van Riper, C., and J. M. Scott. 2001. Limiting factors affecting Hawaiian native birds. In *Evolution, ecology, conservation, and management of Hawaiian birds: A vanishing avifauna,* ed. J. M. Scott, S. Conant, and C. van Riper, 221–33. Studies in Avian Biology No. 22. Lawrence, KS: Allen Press.

van Riper, C., S. G. van Riper, and W. R. Hansen. 2002. Epizootiology and effect of avian pox on Hawaiian forest birds. *Auk* 119:929–42.

Vaughan, J. A., M. Trpis, and M. J. Turell. 1999. *Brugia malayi* microfilaria (Nematoda: Filaridae) enhance the infectivity of Venezuelan equine encephalitis virus to *Aedes* mosquitoes (Diptera: Culicidae). *Journal of Medical Entomology* 36:758–63.

Vijayan, K. K., C. P. Balasubramanian, S. V. Alavandi, and T. C. Santiago. 2004. Introduction of exotic penaeids to India and its implications to shrimp farming and biodiversity. In *Proceedings of the National Seminar on New Frontiers in Marine Bioscience Research, National Institute of Ocean Technology, Chennai, India, January 22–23, 2004,* ed. S. A. H. Abidi, M. Ravindran, R. Venkatesan, and M. Vijayakumaran, 179–86. New Delhi: Allied Publishers.

Walsh, P. D., K. A. Abernethy, M. Bermejo, R. Beyers, P. De Wachter, M. E. Akou, B. Huijbregts, et. al. 2003. Catastrophic ape decline in western equatorial Africa. *Nature* 422 (6932): 611–14.

Wilcove, D. S., D. Rothstein, J. Dubow, A. Phillips, and E. Losos. 1998. Quantifying threats to imperiled species in the United States. *BioScience* 48:607–615.

Winton, J. R. 1998. Molecular approaches to fish vaccines. *Journal of Applied Ichthyology* 14:153–58.

Wolfe, M. S. 2000. Crop strength through diversity. *Nature* 406:681–82.

World Health Organisation and the United Nations Environment Programme. 2002. *Biodiversity: Its importance to human health,* ed. E. Chivian. Boston: Center for Health and the Global Environment, Harvard Medical School.

Zhu, Y., H. Chen, J. Fan, Y. Wang, Y. Li, J. Chen, J.-X. Fan, et al. 2000. Genetic diversity and disease control in rice. *Nature* 406:718–22.

14

Climate Change, Wildlife, and Human Health

Camille Parmesan and Pim Martens

Human-driven climate change is already affecting human health (Kuhn et al. 2004; WHO 2002, 2003). For the year 2000, the World Health Organization estimated that 6% of malaria infections, 7% of dengue fever cases, and 2.4% of diarrhea incidence could be attributed to climate change (Campbell-Lendrum, Pruss-Ustun, and Corvalán 2003). Increases in these diseases, which are facilitated by standing water and/or contamination of drinking water, were associated with increases in intensity and frequency of flood events, which in turn have been linked to greenhouse gas–driven climate change (Easterling, Evans, et al. 2000; Easterling, Meehl, et al. 2000; IPCC 2007a). Floods directly promote transmission of water-borne diseases by causing mingling of untreated or partially treated sewage with freshwater sources, as well as indirectly from the breakdown of normal infrastructure causing postflood loss of sanitation and freshwater supplies (Atherholt et al. 1998; Curriero et al. 2001; Patz et al. 2003; Rose et al. 2000). Further, high precipitation can cause increased abundances of insect vectors (particularly mosquitoes) whose populations benefit from excess standing water. In looking over all climate-related impacts (particularly death from floods and heat waves and associated increases in diarrhea, but also increases in malnutrition due to drought and flood-related crop failure), WHO estimated that for a single year, 150,000 deaths were due to climate change (WHO 2002).

The links between climate change and human health are strongest in terms of the direct impacts of extreme weather and climate events on mortality. Extreme events are generally defined as having a low probability of occurrence (e.g., less than 10%) during a base period. Extreme temperature and precipitation events have been increasing over the past 100 years, and this increase has been linked to greenhouse gas–driven climate change (Easterling, Evans, et al. 2000; Easterling, Meehl, et al. 2000; IPCC 2007a; Karl, Knight, and Baker 2000). The heat wave of 2003 resulted in 35,000 deaths across western Europe.

Recent analyses of observational climate data coupled with results of modeling studies indicate that western Europe has a particularly great risk of extreme high temperatures from ongoing anthropogenic climate change (Beniston 2005; Frich et al. 2002; Meehl and Tebaldi 2004; Schär et al. 2004; Stott, Stone, and Allen 2004). While attributing any single climate event to long-term climate change is inappropriate, these results do suggest that climate change may have already increased deaths from heat stress.

These numbers may seem small compared with other major health risk factors (e.g., AIDS-related deaths), but given that health-related deaths result from a very complex set of interacting factors, it is surprising that a climate signal could be detected at all. There is general agreement that the health sectors are strongly buffered against responses to climate change, and that a suite of more traditional factors is more often responsible for both chronic and epidemic health problems. These factors include quality and accessibility of health care, sanitation infrastructure and practices, land use change (particularly practices that alter timing and extent of standing water), pollution, population age structure, presence and effectiveness of vector control programs, and general socioeconomic status (Campbell-Lendrum et al. 2003; Gubler 1998; Gubler et al. 2001; IPCC 2001b, 2007b; Kuhn et al. 2004; Patz et al. 2001; Wilkinson, Campbell-Lendrum, and Bartlett 2003).

Methods used to estimate deaths due to climate change come partly from empirical studies of the relationships between temperature and hospital admissions. This design is a simple but effective way to control for the effects of confounding factors, as all data are gathered from the same base human population and geographic area. For example, in Lima, Peru, Checkley et al. (2000) analyzed long-term variations in both climate and daily hospital admissions for diarrhea. After accounting for seasonal and long-term trends, temperature (but not rainfall or humidity) was significantly correlated with diarrhea. They estimated an 8% increase in cases per 1°C temperature rise, with a doubling of admissions during the 1997/1998 El Niño year (figure 14.1). In a similarly designed study in Fiji, Singh et al. (2001) found significant correlations between both temperature and rainfall and total monthly doctor reports of infant diarrhea. Here, an increase of 1°C increased diarrhea cases by only 3%, perhaps because of the additional effects of rainfall. Interestingly, neither effect was totally simple: effects of temperature had a one-month lag, and both extreme low and extreme high rainfall were associated with increased diarrheal cases.

In contrast, it is generally assumed that diarrhea incidence in developed countries, which have better sanitation infrastructure, has little or no association with climate (Kuhn et al. 2004; WHO 2003). Studies for the United States indicate that this assumption—that developed countries have low vulnerability—may be premature. Independent studies have repeatedly concluded that water- and food-borne pathogens (that cause diarrhea) will likely increase in the United States and Europe with projected increases in regional flooding events, primarily by contamination of main waterways (Ebi et al. 2006; Rose et al. 2001).

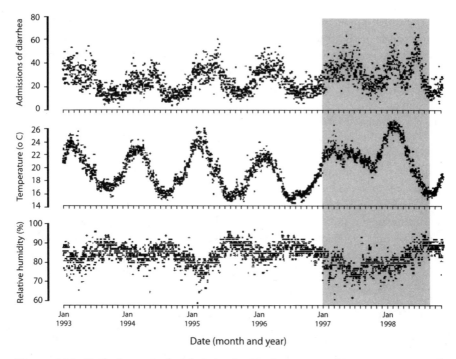

Figure 14.1. Daily time series for admission for diarrhea, mean ambient temperature, and relative humidity in Lima, Peru. Time series spans from January 1, 1993 to November 15, 1998. The shaded area to the right of the graph indicates the 1997/1998 El Niño event. Reprinted with permission from Checkley et al. 2000.

The contrasting conclusions on diarrhea and climate change (noted above) highlight the lack of hard evidence for health impacts of climate change. Existing data are often inadequate, and many changes in disease patterns have been attributed to climate change without conclusively ruling out other explanations. The term *attribution* has been coined by the Intergovernmental Panel on Climate Change (IPCC) to refer to the process of pinpointing and quantifying the specific impacts of climate change on a system of interest (in this case, human health) above and beyond the simultaneous impacts of many other confounding forces (Ahmad et al. 2001). There has been some recognition among health scientists that many studies are weak on rigorous "attribution" (Kovats et al. 2001; Wilkinson, Campbell-Lendrum, and Bartlett 2003).

Issues of detection and attribution methodologies have been much more fully explored for study of climate change impacts on wild species (Ahmad et al. 2001; Parmesan 2005; Parmesan and Yohe 2003), but these methods are not easy to apply to human health given the manner in which most disease data are collected. Rather than coming from long-term, systematic monitoring of disease organisms and their vectors in the wild, or even from systematic monitoring of actual disease incidences in human populations, changes in disease dynamics most often come much more indirectly from changes in numbers of people diagnosed and recorded with a disease by a doctor, hospital, or clinic. This method is subject to underrepresentation of disease in isolated, rural, or poor regions that don't have easy access to medical facilities, as well as to misdiagnosis of uncommon diseases, such as is likely to happen in the early stages of shifts in the geographic distributions of diseases.

Further, effects of climate change have tended to be studied independently by researchers focusing either on natural systems or on humans. Yet, because human diseases and their vectors are inherently part of nature, this division is more historical than real. Cross-links among these processes can be inferred from individual studies, but rarely is the entire pathway directly studied. Ultimately, however, effective mitigation of climate change impacts on human health rely on mechanistic understanding of complex links operating through natural communities. In this chapter, we review some of the studies for which the linkages from climate change to alterations in distributions, behaviors, or life cycles of wild species to impacts on human health are explicitly explored and documented.

The Big and the Small: Plants, Animals, and Microbes

Too often we forget that the organisms that cause human diseases—protozoa, bacteria, and viruses—are all part of life's rich pageant of biodiversity. Though tiny in size, microbes make up a huge portion of Earth's biomass (see this volume, chapter 11). In numbers, relatively few cause harm; but occasional microbial population booms in humans have caused massive deaths, economic woes, and social unrest. The bubonic plague in medieval Europe was catastrophic in affected societies, and the influenza epidemic of North America and Europe in 1918 effectively shut down industries in severely affected areas. We tend to focus on their human impacts, but microbes exist as a part of nature; are subject to the same climatic, and therefore geographic, constraints on their global distributions (see this volume, chapter 11); and are just as susceptible to alterations of global climate as much more visible plants and animals.

The direct impacts of temperature and precipitation on disease survival, reproduction, and transmission have been moderately studied for a few serious, common diseases. A WHO world health report in 2004 stated explicitly that "most viruses, bacteria and parasites do not replicate below a certain temperature threshold (e.g., 18°C for the malaria parasite *Plasmodium falciparum* and 20°C for the Japanese encephalitis virus; MacDonald 1957; Mellor and Leake 2000)" (Kuhn et al. 2004). When temperatures rise above this threshold, WHO notes that "increases in temperature reduce the time taken for vector populations to breed. Increases in temperature also decrease the incubation period of the pathogen (e.g., malaria parasite, dengue or yellow fever virus), meaning that vectors become infectious more quickly (MacDonald 1957)" (WHO 2003).

While documentation of climate change impacts on disease microbes and their vectors is still sparse, it's very clear that human-driven climate change has already had substantial and sometimes irreversible impacts on natural biodiversity (IPCC 2007b; Parmesan 2006; Parmesan and Yohe 2003; Walther et al. 2002). An analysis of more than 1,500 species across the globe estimated that half (50%) of all wild species have shown a significant response to local or regional warming trends (Parmesan and Yohe 2003). Observed responses include earlier spring phenologies, poleward and upward range shifts, as well as extinctions of individual populations and entire species. Impacts have been detected in all studied taxonomic groups (plant, vertebrate, and invertebrate) and in terrestrial, freshwater, and marine systems (Parmesan 2006; Parmesan and Yohe 2003).

Of particular interest with respect to human health issues is the advance into the temperate zone of historically tropical species (Parmesan 2006; Parmesan and Galbraith 2004). A recent study by Guernier, Hochberg, and Guégan (2004) showed that the global distribution of key human pathogens (bacteria, viruses, protozoa, fungi, and other parasites) follows the same latitudinal diversity gradient as most animals and plants (see this volume, chapter 11). Higher pathogen species richness in the tropics was related to climatic variables. Laboratory experiments with pathogens show that, as with large-bodied organisms, many can only develop within particular temperature limits, and often have a quite narrow temperature range for optimal growth (Kuhn et al. 2004; MacDonald 1957; Patz et al. 2001). Therefore, even though data on wild populations are sparse, it is likely that pathogens are mirroring responses observed in large plants and animals, and so similar advances of "tropical" microbial species into temperate zones would be expected.

Sixty-two percent of human pathogens also use wild animals as hosts (Cleveland, Laurenson, and Taylor 2001). Therefore, changes in disease dynamics in wild animals have strong implications for human disease threat. There is emerging evidence that climate change has caused increases in diseases among many wild populations of plants and animals (Daszak, Cunningham, and Hyatt 2000; Harvell et al. 1999). For example, in the Rocky Mountain range of the United States, the mountain pine beetle (*Dendroctonus ponderosae*) has responded to warmer temperatures by altering its life cycle. The mountain pine beetle life cycle now only takes one year per generation rather than the previous two years, allowing large increases in population abundances, which, in turn, have increased incidences of a fungus the beetle transmits (pine blister rust, *Cronartium ribicola*) (Logan, Régnière, and Powell 2003). Increased abundances of a nematode parasite have also occurred as its life cycle shortened in response to warming trends. This has had associated negative impacts on its wild musk oxen host (*Ovibos moschatus*), causing decreased survival and fecundity (Kutz et al. 2005).

When disease dynamics are altered in wild populations, humans may be impacted when affected plants and animals are part of our diet, allowing for transmission of the pathogen to humans. For example, in the Gulf of Mexico, incidence of the *Vibrio vulnificus* bacterium in oysters rises dramatically with sea temperatures. In one study, 60% of the variation in *V. vulnificus* concentrations in the Gulf of Mexico was explained by temperature (Motes et al. 1998). In another study, 89% of human contraction of *V. vulnificus* came from eating oysters harvested in waters with mean annual temperatures >22°C, and disease incidence rose sharply during summer months (Shapiro, Altekruse, and Griffin 1998). The human impact is severe, as 30% to 48% of people die after contracting this bacterium (Shapiro, Altekruse, and Griffin 1998).

Diseases of the animals we eat may not infect humans, but these diseases can still impact human well-being by reducing abundances of human food resources (see this volume, chapter 2). In a single year (1991), the oyster parasite *Perkinsus marinus* extended its range northward from Chesapeake Bay to Maine—a 500 km shift. Censuses from 1949 to 1990 showed a stable distribution of the parasite from the Gulf of Mexico to its northern boundary at Chesapeake Bay. The rapid expansion in 1991 has been linked to above-average winter temperatures rather than human-driven introduction or genetic change (Ford 1996). A kidney disease has been implicated in low-elevation trout (*Salmo trutta*) declines in Switzerland. High mortality from infection occurs above 15°C–16°C, and water temperatures have risen in recent decades. High infection rates (27% of fish at 73% of sites) at the hottest sites below 400 m have been associated with a 67% decline in catch; midelevation sites had lower disease incidence and only moderate declines in catch; and the highest (coldest) sites (800–3029 m) had no disease present and relatively stable catch rates (Hari et al. 2006).

Evidence for Human Health Impacts via Climate Change Alterations of Biodiversity

While hard evidence for three-way linkages between climate change, biodiversity change, and human health impacts is rare, the examples that do exist offer useful, if sobering, insights.

Alterations of Wild Species' Ecologies and Human Health

Few studies combine basic ecology of wild pathogens and their associated reservoirs with human disease epidemics. A rare example of such a link was made by Colwell (1996) in a broad global synthesis of cholera outbreaks and associated basic research over the past 200 years. References for the following can be found in Colwell's 1996 review. The bacterium that causes human cholera, *Vibrio cholerae*, has wide salt tolerances and can live in anything from seawater to brackish river water. When environmental conditions are not conducive to reproduction, *V. cholerae* can live in a viable but quiescent state (suspected up to several months) and become pathogenic when conditions become more favorable. A diversity of marine life can serve as reservoirs, with *V. cholerae* having been found living in the mucosa of many shellfish and in crustacean zooplankton, and subsequently found in fish that have fed on plankton hosts. *V. cholerae* has also been found in aquatic vegetation, such as water hyacinth, and in blue-green bacterium.

Correlational analyses show strong repeated patterns in which extreme warm water temperatures cause algae blooms, which then promote rapid increases in zooplankton abundances and hence also in their associated *V. cholerae* bacteria. A single copepod has been found to contain 10,000 *V. cholerae* individuals. Analyses of long-term data sets from Peru and Bangladesh (from 18 years up to 70 years) show that cholera has recently become associated with El Niño events, suggesting that a threshold for high transmission has only recently been commonly surpassed as El Niño events have become stronger and more frequent in the past three decades (Pascual et al. 2000; Rodó et al. 2002). Even when known epidemiological dynamics are taken into account (such as cycling of immunity in human populations), a strong El Niño signal in cholera dynamics is maintained (Koelle et al. 2005). In summary, there is compelling evidence for links between climate variability, climate change (via increases in strength of El Niños), native plankton dynamics, bacterial dynamics in the wild, and cholera disease epidemics.

Though with less dramatic consequences than those caused by *V. cholerae*, movements of species poleward and upward have brought some old problems into new areas. Over the past 32 years, the pine processionary moth (*Thaumetopoea pityocampa*) has expanded 87 km at its northern range boundary in France and 110–230 m at its upper altitudinal boundary in Italy (Battisti et al. 2005). Laboratory and field experiments have linked the feeding behavior and survival of this moth to minimum nighttime temperatures, and its expansion has been associated with warmer winters. The hairs of this caterpillar are irritating to lung linings, with asthmatics and those with an allergic reaction being particularly sensitive. Expansion of this insect into new territory is already a concern to the health community.

On a different note, increases in atmospheric carbon dioxide are directly affecting the physiology of wild plants, with some unexpected effects on human health. Experimental research has shown that doubling CO_2 levels from about 300 to 600 ppm induces a fourfold increase in the production of ragweed pollen (Ziska and Caulfield 2000a, 2000b). Carbon dioxide is already up to 380 ppm—higher than was ambient at the time the experiment was done.

Extinctions, Declines, and Human Well-Being

Extinctions of wild species in and of themselves may have direct consequences on human health and well-being. Extinctions have been occurring at a higher rate in recent centuries because of a variety of anthropogenic causes, but climate change is likely to exacerbate population declines sufficiently for some species to completely disappear. To date, 74 known species have been driven extinct by recent climate change (Pounds, Fogden, and Campbell 1999; Pounds et al. 2006). All of these were montane neotropical frogs. However, many more species may be being lost than are documented. Some 30% of tropical coral reefs have been killed off because of multiple recent high sea-surface temperature events, and the diversity of coral reefs is poorly documented in many regions (Hoegh-

Guldberg 1999; Wilkinson 2000). Further, drastic regional declines of particular species have already been directly attributed to climate change. Populations of many species of butterfly have gone extinct along their southern range margins, in association with 20th-century warming trends (Parmesan 1996; Parmesan et al. 1999). The abundance of zoo-plankton has declined by 80% off the California coast. This decline has been related to the gradual warming of sea-surface temperatures (Roemmich and McGowan 1995).

A recent study estimated risk of species extinctions by synthesizing the results of many different climate envelope models applied to different data sets from around the world (Thomas et al. 2004). The results provided a rough estimate of extinction risks by 2050. Even if one assumes perfect dispersal capabilities, an absolute reduction in climati-cally suitable space suggests that about 9% to 13% of species could be ultimately driven extinct even with minimal warming of another 2°C. Under midrange warming projec-tions (2°C–3°C), extinction estimates varied from 21% to 32% of species.

The loss of species diversity could have a direct negative health effect due to losses in genetic resources (see this volume, chapter 15). For example, tropical frogs have been the source of numerous medicinal drugs commonly prescribed in industrial countries, with diverse uses ranging from analgesics to muscle relaxants to treatment of cancer. Science does not know what potential drugs have been lost to the world by the climate change–driven extinctions of harlequin frogs (*Lysapsus* spp.) and the Golden toad (*Bufo peri-glenes*) in Central America (Pounds, Fogden, and Campbell 1999; Pounds et al. 2006). Wild species, insects in particular, are a vast reservoir for the ever-rising demands for new biological control agents, loss of which will exacerbate the spread of (new) diseases or pests. Climate change–driven extinctions could also result in negative impacts on food supplies. Declines in Arctic sea ice have already resulted in reduced catches of seals by local Inuit peoples, partly because of the increased difficulty of hunting on thinner ice with shorter winter duration, but also because of real declines of numbers of seals as their ice habitat dwindles (ACIA 2004; see also chapter 2 in this volume).

Climate Change and the Spread of Vector-Borne Disease

It has been suggested that observed warming has already affected vector-borne disease by altering vector distributions, increasing the intensity of disease transmission, and increas-ing the probability of outbreaks (Epstein 2000; Gubler et al. 2001; Kovats et al. 2003; Kuhn et al. 2004; Loevinsohn 1994). In 2004, the World Health Organization assessed the evidence for climatic controls on human diseases and for epidemics in particular (Kuhn et al. 2004). Links between disease outbreaks and climate vary strongly by region (Kovats et al. 2003), but overall global patterns are striking: out of 17 common com-municable human diseases, 71% were weakly to strongly affected by changes in rainfall and temperature. Of the common diseases for which epidemics were linked to climatic conditions, 67% are transmitted by a wild animal vector, or are dependent on wild ani-mals as reservoirs for some life history stage. Table 14.1 shows the WHO assessment for

Table 14.1. World Health Organization assessment of observed links between climate and human diseases.

Disease (organism that causes disease)	Transmission	Distribution	Evidence for interannual variability[1]	Strength of climate sensitivity[2]
Malaria (protozoa)	Bite of female mosquitoes, genus *Anopheles*	Endemic in >100 countries in tropics and subtropics	*****	*****
	CLIMATE-EPIDEMIC LINK: Changes in temperature and rainfall associated with epidemics. Abundances and distributions of *Anopheles* mosquitoes are highly related to climate. Many other locally relevant factors include vector characteristics, immunity, population movements, drug resistance, etc.			
Dengue fever (virus)	Bite of female mosquitoes, genus *Aedes*	Africa, Europe, South and Central America, Southeast Asia, west Pacific	****	***
	CLIMATE-EPIDEMIC LINK: High temperature, humidity, and heavy rain associated with epidemic. Nonclimatic factors may have more important impact on epidemic outbreaks.			
Japanese & St. Louis encephalitis (virus)	Bite of female mosquitoes, genus *Culex* & *Aedes*	North and South America	***	***
	CLIMATE-EPIDEMIC LINK: High temperature and heavy rains associated with epidemic. Status of animal reservoirs important.			
Rift Valley fever	Bite of female mosquitoes, genus *Culex* & *Aedes*	Sub-Saharan Africa	***	***
	CLIMATE-EPIDEMIC LINK: Heavy rains associated with onset of epidemic. Cold weather often associated with end of epidemic. Status of animal reservoirs important.			
Leishmaniasis (protozoa)	Bite of female sandflies	Africa, central Asia, Europe, India, South America	**	***
	CLIMATE-EPIDEMIC LINK: Increases in temperature and rainfall associated with epidemics.			

(continued)

Table 14.1. *(continued)*

Disease (organism that causes disease)	Transmission	Distribution	Evidence for interannual variability[1]	Strength of climate sensitivity[2]
Ross River virus (virus)	Bite of female mosquitoes, genus *Culex* & *Aedes*	Australia, Pacific islands	**	***
CLIMATE-EPIDEMIC LINK: High temperature and heavy precipitation associated with onset of epidemic. Host immune factors and status of animal reservoirs are also important.				
Murray Valley fever	Bite of female mosquito, genus *Culex*	Australia	**	***
CLIMATE-EPIDEMIC LINK: Heavy rains and below-average atmospheric pressure associated with epidemic.				
Yellow fever (virus)	Bite of female mosquitoes, genus *Aedes* & *Haemagogus*	Africa, South and Central America	****	**
CLIMATE-EPIDEMIC LINK: High temperature and heavy rain associated with epidemic. Intrinsic population factors are also important.				
West Nile virus (virus)	Bite of female *Culex* mosquito	Africa, central & southwest Asia, Europe	***	**
CLIMATE-EPIDEMIC LINK: High temperature and heavy precipitation associated with onset of epidemic. Nonclimatic factors may have more important impact on epidemic outbreaks.				
Lymphatic filariasis	Bite of female mosquitoes in genus *Culex* & *Anopheles*	Africa, India, South America, south Asia	—	**
CLIMATE-EPIDEMIC LINK: Temperature and rainfall determine the geographical distribution of disease.				

Disease (organism that causes disease)	Transmission	Distribution	Evidence for interannual variability[1]	Strength of climate sensitivity[2]
African trypanosomiasis	Bite of male and female tsetse flies	Sub-Saharan Africa	***	**
	CLIMATE-EPIDEMIC LINK: Changes in temperature and rainfall may be linked to epidemics. Tsetse fly's distribution is related to climate. Cattle density and vegetation patterns are also relevant factors.			
Schistosomiasis	Water-borne transmission involving intermediate snail host	Africa, east Asia, South America	*	*
	CLIMATE-EPIDEMIC LINK: Increases in temperature and rainfall can affect seasonal transmission and geographical distributions.			
Chagas' disease (American trypano- somiasis)	Bite of blood- feeding Reduviid bugs	South and Central America	*	*
	CLIMATE-EPIDEMIC LINK: Presence of bugs associated with high temperatures, low humidity, and specific vegetation types.			
Lyme disease	Bite of ticks in genus Ixodid	North America, Europe, Asia	*	*
	CLIMATE-EPIDEMIC LINK: Temperature and vegetation patterns associated with distributions of tick vectors and of disease organism.			

[1] — no interannual variability, * very weak variability, ** some variability, *** moderate variability, **** strong variability, ***** very strong variability.

[2] * climate link is weak, ** climate plays a moderate role, *** climate plays a significant role, **** climate is an important factor, ***** climate is the primary factor in determining at least some epidemics, and the strength of the association between climate and disease outbreaks has been assessed on the basis of published quantitative (statistical) rather than anecdotal evidence.

This table is modified from table 2 in Kuhn et al. (2004) and lists only those diseases that use an animal vector and for which a link was made between climate and disease incidence or trans- mission in a scientific publication. The table has been slightly modified by the authors to include some additional information in the "Climate-epidemic link" column.

this latter category. A notable feature of the WHO assessment is that although climate is reported to have a measurable or even "important" role in epidemics of several diseases, only one disease—malaria—is reported as having rigorous, quantitative, statistical evidence that climate variability plays the primary role in disease outbreaks (table 14.1).

Aside from the cholera paradigm, the clearest examples of climate change–biodiversity-disease links come from the southwestern desert of the United States. This area experienced extreme levels of rainfall during the intense 1992/1993 El Niño. High precipitation promoted lush vegetative growth, which in turn was correlated with population booms of deer mice (*Peromyscus maniculatus*). This wild rodent carries the hantavirus that is transmissible, and frequently lethal, to humans. The virus is normally present at moderate levels in wild mouse populations. While researchers who routinely handle desert rodents must exercise caution when working with these wild populations, humans in nearby settlements have little exposure through normal activities. However, during 1993, local overcrowding resulting from the wet-year population boom caused spillover rodent activity in nearby human settlements. Subsequent increased human contact and transmission rates led to a major regional epidemic of hantavirus (Engelthaler et al. 1999; Glass et al. 2000; Hjelle and Glass 2000).

Similar dynamics have been shown for plague in the western United States, with increased incidence of plague following extreme high-precipitation events (Gubler et al. 2001; Parmenter et al. 1999). The plague bacterium (*Yersinia pestis*) is carried by fleas, which in turn are carried by rodents and hence are also sensitive to El Niño events. El Niño has increased in frequency and intensity during the 20th century, and at least one global model projects that greenhouse-gas warming will create a permanent "El Niño–like" state in the eastern Pacific (Meehl et al. 2000). Should such a state occur, the associated increased rodent and flea densities could carry a higher risk of hantavirus and plague outbreaks in nearby human settlements.

The geographic distributions of vector-borne diseases are limited by the climatic tolerance of their vectors and by biological restrictions that limit the survival and incubation of the infective agent in the vector population. Moderate (10–50 km) to substantial (>1000 km) poleward range shifts have been detected in thousands of insects, birds, and mammals over the past 30 to 130 years (IPCC 2007b; Parmesan 2006; Parmesan and Yohe 2003). Many species in each of these groups act as vectors of human diseases— either directly as the host of the disease, or indirectly by harboring the host. It is likely, then, that climate change has already had a significant effect on the geographical range of many vector species.

Evidence for range shifts in known disease vectors is rare, but studies are beginning to emerge. Marked increases in abundances of the disease-transmitting tick *Ixodes ricinus* have been documented along its northernmost range limit in Sweden (Lindgren, Tälleklint, and T. Polfeldt 2000). Between the early 1980s and 1994, numbers of ticks found on domestic cats and dogs increased by 22%–44% along the tick's northern range boundary across central Sweden. In the same time period, this region has had a marked decrease in numbers of extremely cold days (< –12°C) in winter, and a marked increase in warm days (> 10°C) during the spring, summer, and fall. Previous studies on temperature developmental and activity thresholds of this tick species indicated that the observed warmer temperatures would be associated with decreased mortality and longer growing seasons (Lindgren, Tälleklint, and T. Polfeldt 2000).

The linkages between climate change impacts on natural systems and disease outbreaks among humans can be complex. In the United States, the same genus of tick, *Ixodes*, carries Lyme disease, caused by the spirochete *Borrelia burgdorferi*. A 32-fold increase between 1982 and 1998 in incidence of Lyme disease (from about 500 cases to nearly 16,000 cases per year) has largely been attributed to land use change, primarily reforestation leading to exploding white-tailed deer (*Odocoileus virginia*), and hence tick, populations (Gubler et al. 2001). However, recent studies have shown a strong positive correlation between acorn abundances, white-footed mice (*Peromyscus leucopus*) and deer abundances (mammals that act as reservoirs for *Ixodes* ticks), and tick abundance in northeastern oak (*Quercus* spp.) forests (Jones et al. 1998; Ostfeld, Hazler, and Cepeda 1997; Ostfeld, Jones, and Wolff 1996). Jones et al. (1998) documented that 67% of the variation in tick abundances among years was related to acorn abundance. Mast years in oaks have some genetic basis, but environmental controls are moderately strong: seasonal patterns of temperature and rainfall explained some 30% of yearly variation in masting patterns in an intensive eight-year study (Sork, Bramble, and Sexton 1993). The association between yearly climate and masting suggests that climate change—in particular, warmer springs and dryer summers—could also be a factor in the observed increases in incidence of Lyme disease via increased acorn production allowing population increases of tick hosts.

Projected Impacts of Climate Change

The World Health Organization conducted an ambitious study to estimate the relative contributions of 26 risk factors to incidence of disease in the global human population (WHO 2003). Although quantitative estimates for the relative impact of climate change have large uncertainty, there are some interesting trends in the numbers. For example, evidence from long-term hospital records indicates that there is an observed 5% increase in diarrhea for each 1°C increase in mean temperature in developing countries, but no relationship with temperature in developed countries. Differences in sanitation likely play a key role, but this result is also attributed to differences in the ecologies of the microbes, with intestinal disease organisms being mostly tropical (e.g., warm-adapted amoebae and bacteria) in the developing world, and mostly cold-adapted ones in developed regions (e.g., the *Giardia* protozoa). It is difficult to assess to what extent these results can be extrapolated into the future to provide projections of disease incidences under different levels of projected warming, but they suggest a systematic increase in diarrhea in tropical climates, and potential increased sensitivity to temperature of diarrhea outbreaks in higher-latitude countries as historically subtropical and temperate zones acquire tropical parasites.

One of the frequently used approaches for making health projections is to model the effects of climate change on vector-borne diseases. Several types of models have been developed to forecast the impact of climate change on vector-borne disease transmission, most models focusing on the best-understood disease—malaria, caused by species of protozoa in the genus *Plasmodium*. The current distributions of some mosquito species have been successfully mapped in Africa using meteorological data (Lindsay, Parson, and Thomas 1998). Projected distribution shifts can then be modeled. For example, David Rogers has mapped the projected changes of three important disease vectors (ticks, tsetse flies, and mosquitoes) in southern Africa under three climate change scenarios (Hulme 1996). The results indicate significant changes in areas suitable for each vector species, with a net increase for malaria mosquitoes (*Anopheles gambiae*).

Martin and Lefebvre (1995) developed a malaria potential occurrence zone (MOZ) model. This model was combined with five general circulation models (GCMs) to estimate the changes in malaria risk based on moisture and minimum and maximum temperatures required for parasite development. An important conclusion of this modeling exercise was that all simulation runs showed an increase in seasonal (unstable) malaria transmission under climate change, at the expense of perennial (stable) transmission.

Rogers and Randolph (2000), using a multivariate empirical-statistical model, estimated that, for the IPCC IS92a climate change scenario (more or less assuming a business as usual future), there would be no significant net change by 2050 in the estimated portion of world population living in malaria transmission zones: malaria increased in some areas and decreased in others.

An integrated, process-based model to estimate climate change impacts on malaria that is part of the MIASMA modeling framework (*m*odeling framework for the health *i*mpacts *a*ssessment of *m*an-induced *a*tmospheric changes) has been developed by Martens and colleagues (Martens 1999; Martens et al. 1995, 1999). This model differs from the others in that it takes a broad approach in linking GCM-based climate change scenarios with a module that uses the formula for the basic reproduction rate (R_0) to calculate the "transmission or epidemic potential" of a malaria mosquito population. The results of this model show a projected increase of the population at risk of potential malaria transmission due to climatic changes.

Tol and Dowlatabadi (1999) integrated the results of MIASMA within an economic framework to estimate the trade between climate change and economic growth on malaria risk. The first results of this exercise show the importance of economic variables in estimating changes in future malaria risk.

All of the examples discussed above have their specific disadvantages and advantages. For example, the model developed by Rogers and Randolph (2000) incorporates information about the current social, economic, and technological modulation of malaria transmission. It assumes that those contextual factors will apply in the future in unchanged fashion. This adds an important, though speculative, element of multivariate realism to the modeling—but the model thereby addresses a qualitatively different question from the biological model. The biological model of Martens and colleagues assumes that there are known and generalizable biologically mediated relationships. Also, this biological modeling, in its early stages, did not include the horizontal integration of social, economic, and technical change.

Looking at the current knowledge on the relationships between climate and vector-borne diseases, we see that substantial evidence already exists that most vector-borne diseases are sensitive to climate variations either in the field or in the laboratory. Consideration of the vector biology and epidemiology of each of these diseases suggests that malaria, African trypanosomiasis, tick-borne diseases, and to a lesser extent leishmaniasis, may be among the diseases whose distributions will change most quickly. Dengue, lymphatic filariasis, and onchocerciasis are also likely to be affected, but it will be particularly difficult to differentiate the effects of climate change from those of control programs and other influences.

On balance, hard evidence that observed climate change has affected many vector-borne diseases remains elusive. Probably the greatest current limitation on direct demonstration of the effects of gradual climate change is that the climate changes that have so far occurred have been relatively minor. As with other diseases, outbreaks of vector-borne diseases are strongly buffered by societal infrastructures (sanitation, medical care, and adaptive behaviors), and thus effects of climate change on human incidences of these diseases are expected to be less dramatic than direct responses of wild species (Gubler et al. 2001).

A recent clear example of the influence of social structure on disease outbreak comes from two towns situated on either side of the Rio Grande, which separates the United States from Mexico. In 1999, this area had a dengue virus outbreak. However, blood samples from inhabitants on the U.S. side of the river had only 1/20 to 1/2 the levels of dengue antibodies of individuals on the Mexican side, reflecting significantly less exposure to, and transmission of, the disease (Reiter et al. 2003). The *Aedes aegypti* mosquito vector is common on both sides of the river, but lowered incidence of dengue in humans on the U.S. side is likely due to better-sealed buildings and much higher prevalence of air-conditioning in the U.S. town compared with the Mexican town.

Research Directions

The differences in disease incidence in climatically similar regions, often with similar arrays of wild vectors (e.g., the dengue fever example above), make it clear that models that consider climate alone without incorporating societal aspects of disease growth and transmission are likely to overestimate disease expansion with climate change (Reiter 2001). The rate of climate change, however, is expected to increase rapidly in coming decades (IPCC 2007a). Careful consideration should therefore be given to maximizing the chances of detection of the effects of these changes on disease distributions and dynamics in the wild, as well as on incidences in human populations.

It is difficult to detect and attribute climate change effects using data collected for other purposes. Consideration should be given to setting up long-term surveillance programs to specifically monitor sensitive aspects of climate variability and long-term climatic trends on the dynamics of pathogens and their vectors and reservoirs in the wild. While it may take 20 years before such monitoring schemes bear fruit, projections of climate change impacts on human health are severely limited in the absence of this information.

There is a need to identify vulnerable areas—that is, where diseases or their vectors are at the climatological limits of their distribution, and vulnerable human populations—those that are at most risk of emerging diseases. In the meantime, investigations of the effects of shorter-term climate variations will increase our understanding of climate-disease relationships, and allow us to make more confident assessments and predictions of the effects of long-term climate change.

There is a clear dearth of studies that provide rigorous links all the way from changes in wild populations of microbes and their hosts to actual disease outbreaks in humans. While much responsibility for this dearth surely lies in data limitations, there is also a lack of cross-collaborations among disciplines, in this case between basic ecologists and disease epidemiologists. The few good studies we have come from collaborations between researchers in these two groups. To make the fullest use of what data we have requires a concerted effort to build interdisciplinary teams whose research spans from microbial ecology in natural systems to wild life ecology to sociology to epidemiology. Effective mitigation of climate change impacts on human health relies on mechanistic understanding of complex links from wild pathogens to human populations operating through natural communities.

References

Ahmad, Q. K., R. A. Warrick, T. E. Downing, S. Nishioka, K. S. Parikh, C. Parmesan, S. H. Schneider, F. Toth, and G. Yohe. 2001. Methods and tools. In *Climate change 2001: Impacts, adaptation, and vulnerability. Contribution of Working Group II to the Third Assessment Report of the Intergovernmental Panel on Climate Change,* ed. J. J. McCarthy, O. F. Canziani, N. A. Leary, D. J. Dokken, and K. S. White, 105–43. Cambridge: Cambridge University Press.

Arctic Climate Impact Assessment. 2004. *Impacts of a warming Arctic.* Cambridge: Cambridge University Press.

Atherholt, T. B., M. W. LeChevallier, W. D. Norton, and J..S. Rosen. 1998. Effect of rainfall on giardia and crypto. *Journal of the American Water Works Association* 90 (9): 66–80.

Battisti, A., M. Stastny, S. Netherer, C. Robinet, A. Schopf, A. Roques, and S. Larsson. 2005. Expansion of geographic range in the pine processionary moth caused by increased winter temperatures. *Ecological Applications* 15:2084–96.

Beniston, M. 2005. Warm winter spells in the Swiss Alps: Strong heat waves in a cold season? A study focusing on climate observations at the Saentis high mountain site. *Geophysical Research Letters* 32, L01812, doi:10.1029/2004GL021478.

Campbell-Lendrum, D., A. Pruss-Ustun, and C. Corvalán. 2003. How much disease could climate change cause? In *Climate change and human health: Risks and responses*, ed. A. J. McMichael, D. H Campbell-Lendrum, C. F. Corvalán, K. L. Ebi, A. K. Githeko, J. D. Scheraga, and A. Woodward, 133–58. Geneva: World Health Organisation, World Meteorological Organization, and United Nations Environment Programme.

Checkley, W., L. D. Epstein, R. H. Gliman, D. Figueroa, R. I. Cama, J. A. Patz, and R. E. Black. 2000. Effects of El Niño and ambient temperature on hospital admissions for diarrhoeal diseases in Peruvian children. *Lancet* 355 (9202): 442–50.

Cleveland, S., M. K. Laurenson, and L. H. Taylor. 2001. Diseases of humans and their domestic mammals: Pathogen characteristics, host range and the risk of emergence. *Philosophical Transactions of the Royal Society, London* B 356:991–99.

Colwell, R. R. 1996. Global climate and infectious disease: The cholera paradigm. *Science* 274 (5295): 2025–31.

Curriero, F., J. A. Patz, J. B. Rose, and S. Lele. 2001. The association between extreme precipitation and waterborne disease outbreaks in the United States, 1948–1994. *American Journal of Public Health* 91 (8): 1194–99.

Daszak P., A. A. Cunningham, and A. D. Hyatt. 2000. Emerging infectious diseases of wildlife: Threats to biodiversity and human health. *Science* 287:443–49.

Easterling, D. R., J. L. Evans, P. Y. Groisman, T. R. Karl, K. E. Kunkel, and P. Ambenje. 2000. Observed variability and trends in extreme climate events: A brief review. *Bulletin of the American Meteorological Society* 81:417–25.

Easterling, D. R., G. A. Meehl, C. Parmesan, S. Chagnon, T. Karl, and L. Mearns. 2000. Climate extremes: Observations, modeling, and impacts. *Science* 289:2068–74.

Ebi, K. L., D. M. Mills, J. B. Smith, and A. Grambsch. 2006. Climate change and human health impacts in the United States: An update on the results of the U.S. National Assessment. *Environmental Health Perspectives* 114 (9): 1318–24.

Engelthaler, D. M., D. G. Mosley, J. E. Cheek, C. E. Levy, K. K. Komatsu, P. Ettestad, T. Davis, et al. 1999. Climatic and environmental patterns associated with hantavirus pulmonary syndrome, Four Corners region, United States. *Emerging Infectious Diseases* 5:87–94.

Epstein, P. R. 2000. Is global warming harmful to health? *Scientific American* 283 (2): 50–57.

Ford, S. E. 1996. Range extension by the oyster parasite *Perkinsus marinus* into the northeastern United States: Response to climate change? *Journal of Shellfish Research* 15:45–56.

Frich, P., L. V. Alexander, P. Della-Marta, B. Gleason, M. Haylock, A. M. G. K. Tank, and T. Peterson. 2002. Observed coherent changes in climatic extremes during the second half of the twentieth century. *Climatic Research* 19:193–212.

Glass, G. E., J. E. Cheek, J. A. Patz, T. M. Shields, T. J. Doyle, D. A. Thoroughman, D. K. Hunt, et al. 2000. Using remotely sensed data to identify areas of risk for hantavirus pulmonary syndrome. *Emerging Infectious Diseases* 6:238–47.

Gubler, D. J. 1998. Resurgent vector-borne diseases as a global health problem. *Emerging Infectious Diseases* 4 (3): 442–50.

Gubler, D. J., P. Reiter, K. L. Ebi, W. Yap, R. Nasci, and J. A. Patz. 2001. Climate variability and change in the United States: Potential impacts on vector- and rodent-borne diseases. *Environmental Health Perspectives* 109 (Suppl. no. 2): 223–33.

Guernier, V., M. E. Hochberg, and J. F. Guégan. 2004. Ecology drives the worldwide distribution of human diseases. *PLoS Biology* 2:740–46.

Hari, R. E., D. M. Livingstone, R. Siber, P. Burkhardt-Holm, and H. Güttinger. 2006. Consequences of climatic change for water temperature and brown trout population in Alpine rivers and streams. *Global Change Biology* 12:10–26.

Harvell, C. D., K. Kim, J. M. Burkholder, R. R. Colwell, P. R. Epstein, D. J. Grimes, E. E. Hofmann, E. K. Lipp, et al. 1999. Emerging marine diseases: Climate links and anthropogenic factors. *Science* 285 (5433): 1505–10.

Hjelle, B., and G. E. Glass. 2000. Outbreak of hantavirus infection in the Four Corners region of the United States in the wake of the 1997–1998 El Niño–Southern Oscillation. *Journal of Infectious Diseases* 181 (5): 1569–73.

Hoegh-Guldberg, O. 1999. Climate change, coral bleaching and the future of the world's coral reefs. *Marine and Freshwater Research* 50:839–66

Hulme, M. 1996. *Climate change and Southern Africa: An exploration of some potential impacts and implications in the SADC region.* Norwich, UK: WWF International Climate Research Unit.

IPCC. 2001a. *Climate change 2001: Impacts, adaptation, and vulnerability. Contribution of working group II to the third assessment report of the intergovernmental panel on climate change,* ed. J. J. McCarthy, O. F. Canziani, N. A. Leary, D. J. Dokken, and K. S. White. Cambridge: Cambridge University Press.

———. 2001b. *Climate change 2001: The science of climate change. Contribution of working group I to the third assessment report of the intergovernmental panel on climate change,* ed. J. T. Houghton, Y. Ding, D. J. Griggs, M. Noguer, P. J. van der Linden, X. Dai, K. Maskell, and C. A. Johnson. Cambridge: Cambridge University Press.

IPCC. 2007a. Summary for Policymakers. In *Climate change 2007: The physical science basis. Contribution of working group I to the fourth assessment report of the intergovernmental panel on climate change,* ed. S. Solomon, D. Qin, M. Manning, Z. Chen, M. Marquis, K.B. Averyt, M.Tignor and H.L. Miller. Cambridge, UK, and New York: Cambridge University Press.

———. 2007b. Summary for Policymakers. In *Climate change 2007: Climate change impacts, adaptation and vulnerability. Contribution of working group II to the fourth assessment report of the intergovernmental panel on climate change,* ed. M. L. Parry, O. F. Canziani, J. P. Palutikof, P. J. van der Linden and C. E. Hanson. Cambridge, UK: Cambridge University Press.

Jones, C. G., R. S. Ostfeld, M. P. Richard, E. M. Schauber, and J. O. Wolff. 1998. Chain reactions linking acorns to gypsy moth outbreaks and Lyme disease risk. *Science* 279:1023–26.

Karl, T. R., R. W. Knight, and B. Baker. 2000. The record-breaking global temperatures of 1997 and 1998: Evidence for an increase in global warming? *Geophysical Research Letters* 27 (5): 719–22.

Koelle, K., X. Rodó, M. Pascual, M. Yunus, and G. Mostafa. 2005. Refractory periods and climate forcing in cholera dynamics. *Nature* 436:696–700.

Kovats, R. S., D. H. Campbell-Lendrum, A. J. McMichael, A. Woodward, and J. St. H. Cox. 2001. Early effects of climate change: Do they include changes in vector-borne disease? *Philosophical Transactions of the Royal Society, London* B 356:1057–68.

Kovats, R. S., B. Menne, M. J. Ahern, and J. A. Patz. 2003. National assessments of health impacts of climate change: A review. In *Climate change and human health: Risks and responses*, ed. A. J. McMichael, D. H. Campbell-Lendrum, C. F. Corvalán, K. L. Ebi, A. K. Githeko, J. D. Scheraga, and A. Woodward, 181–203. Geneva: World Health Organisation, World Meteorological Organization, and United Nations Environment Programme.

Kuhn, K., D. Campbell-Lendrum, A. Haines, and J. Cox. 2004. *Using climate to predict infectious disease outbreaks: A review*. Ed. M. Anker, C. Corvalán. Geneva: World Health Organisation.

Kutz S. J., E. P. Hoberg, L. Polley, and E. J. Jenkins. 2005. Global warming in changing the dynamics of Arctic host-parasite systems. *Proceedings of the Royal Society of London* B 272:2571–76.

Lindgren, E., L. Tälleklint, and T. Polfeldt. 2000. Impact of climatic change on the northern latitude limit and population density of the disease-transmitting European tick *Ixodes ricinus*. *Environmental Health Perspectives* 108:119–23.

Lindsay, S. W., L. Parson, and C. J. Thomas 1998. Mapping the ranges and relative abundance of the two principal African malaria vectors, *Anopheles gambiae sensu stricto* and *A. arabiensis*, using climate data. *Proceedings of the Royal Society of London* B 265:847–54.

Loevinsohn, M. 1994. Climatic warming and increased malaria incidence in Rwanda. *Lancet* 343:714–18.

Logan, J. A., J. Régnière, and J. A. Powell. 2003. Assessing the impacts of global warming on forest pest dynamics. *Frontiers in Ecology and the Environment* 1:130–37.

MacDonald, G. 1957. The epidemiology and control of malaria. London: Oxford University Press.

Martens, P. 1999. *MIASMA: Modelling framework for the health impact assessment of man-induced atmospheric changes*. Electronic Series on Integrated Assessment Modeling No. 2. Modeling software. CD-ROM.

Martens, P., R. S. Kovats, S. Nijhof, P. de Vries, M. T. J. Livermore, D. J. Bradley, J. Cox, and A. J. McMichael. 1999. Climate change and future populations at risk of malaria. *Global Environmental Change* 9 (Suppl. no. 1): 89–107.

Martens, W. J., L. W. Niessen, J. Rotmans, T. H. Jetten, and A. J. McMichael. 1995. Potential impacts of global climate change on malaria risk. *Environmental Health Perspectives* 103 (5): 458–64.

Martin, P., and M. Lefebvre. 1995. Malaria and climate: Sensitivity of malaria potential transmission to climate. *Ambio* 24 (4): 200–207.

Meehl, G., and C. Tebaldi. 2004. More intense, more frequent and longer lasting heat waves in the 21st century. *Science* 305:994–97.

Meehl, G. A., F. Zwiers, J. Evans, T. Knutson, L. Mearns, and P. Whetton. 2000. Trends in extreme weather and climate events: Issues related to modeling extremes in projections of future climate change. *Bulletin of the American Meteorological Society* 81:427–36.

Mellor, P. S., and C. J. Leake. 2000. Climatic and geographic influences on arboviral infections and vectors. *Revue Scientifique et Technique de l'Office International des Epizooties* 19 (1): 41–54.

Motes, M. L., A. DePaola, D. W. Cook, J. E.Veazey, J. C. Hunsucker, W. E. Garthright, R. J. Blodgett, and S. J. Chirtel. 1998. Influence of water temperature and salinity on

Vibrio vulnificus in Northern Gulf and Atlantic Coast oysters (*Crassostrea virginica*). *Applied Environmental Microbiology* 64 (4): 1459–65.

Ostfeld, R. S., K. R. Hazler, and O. M. Cepeda. 1997. The ecology of Lyme-disease risk. *American Scientist* 85:338–46.

Ostfeld, R. S., C. G. Jones, and J. O. Wolff. 1996. Of mice and mast: Ecological connections in eastern deciduous forests. *BioScience* 46 (5): 323–30.

Parmenter, R. R., E. P. Yadav, C. A. Parmenter, P. Ettestad, and K. I. Gage. 1999. Incidence of plague associated with increased winter-spring precipitation in New Mexico. *American Journal of Tropical Medicine and Hygiene* 61:814–21.

Parmesan, C. 1996. Climate and species' range. *Nature* 382:765–66.

———. 2005. Range and abundance changes. Chap. 12 in *Climate change and biodiversity*, ed. T. Lovejoy and L. Hannah. New Haven, CT: Yale University Press.

———. 2006. Observed ecological and evolutionary impacts of contemporary climate change. *Annual Reviews of Ecology and Systematics* 37:637–69.

Parmesan, C., and H. Galbraith. 2004. *Observed impacts of global climate change in the U.S.* Arlington, VA: Pew Center on Global Climate Change.

Parmesan, C., N. Ryrholm, C. Stefanescu, J. K. Hill, C. D. Thomas, H. Descimon, B. Huntley, et al. 1999. Poleward shifts in geographical ranges of butterfly species associated with regional warming. *Nature* 399:579–83.

Parmesan, C., and G. Yohe. 2003. A globally coherent fingerprint of climate change impacts across natural systems. *Nature* 421:37–42.

Pascual, M., X. Rodó, S. P. Ellner, R. Colwell, and M. J. Bouma. 2000. Cholera dynamics and El Niño–Southern Oscillation. *Science* 289:1766–69.

Patz, J. A., A. K. Githeko, J. P. McCarty, S. Hussein, U. Confalonieri, and N. de Wet. 2003. Climate change and infectious diseases. Chap. 6 in *Climate change and human health: Risks and responses*, ed. A. J. McMichael, D. H. Campbell-Lendrum, C. F. Corvalán, K. L. Ebi, A. K. Githeko, J. D. Scheraga, and A. Woodward. Geneva: World Health Organisation, World Meteorological Organization, and United Nations Environment Programme.

Patz, J. A., M. A. McGeehin, S. M Bernard, K. L. Ebi, P. R. Epstein, A. Grambsch, D. J. Gubler, et al. 2001. Potential consequences of climate variability and change for human health in the United Status. In *The U.S. national assessment on potential consequences of climate variability and climate change*, 437–58. Executive Office of the President of the United States of America, United States Global Change Research Program, National Assessment Synthesis Team. Cambridge: Cambridge University Press.

Pounds, J. A., M. R. Bustamente, L. A. Coloma, J. A. Consuegra, M. P. L. Fogden, P. N. Foster, E. L. Marca, et al. 2006. Widespread amphibian extinctions from epidemic disease driven by global warming. *Nature* 439:161–67.

Pounds, J. A., M. P. L. Fogden, and J. H. Campbell. 1999. Biological response to climate change on a tropical mountain. *Nature* 398:611–15.

Reiter, P. 2001. Climate change and mosquito-borne disease. *Environmental Health Perspectives* 109 (Suppl. no. 1): 141–61.

Reiter, P., S. Lathrop, M. Bunning, B. Biggerstaff, D. Singer, T. Tiwari, L. Baber, et al. 2003. Texas lifestyle limits transmission of dengue virus. *Emerging Infectious Diseases* 9 (1): 86–89.

Rodó, X., M. Pascual, G. Fuchs, and A. S. G. Faruque. 2002. ENSO and cholera: A nonstationary link related to climate change? *Proceedings of the National Academy of Sciences* 99 (20): 12901–6.

Roemmich, D., and J. McGowan. 1995. Climatic warming and the decline of zooplankton in the California Current. *Science* 267:1324–26.

Rogers, D. J., and S. E. Randolph. 2000. The global spread of malaria in a future, warmer world. *Science* 289:1763–66.

Rose, J. B., S. Daeschner, D. R. Easterling, F. C. Curriero, S. Lele, and J. A. Patz. 2000. Climate and waterborne outbreaks in the U.S.: A preliminary descriptive analysis. *Journal of the American Water Works Association* 92: 77–86.

Rose, J. B., P. R. Epstein, E. K. Lipp, B. H. Sherman, S. M. Bernard, and J. A Patz. 2001. Climate variability and change in the United States: Potential impacts on water- and foodborne diseases caused by microbiologic agents. *Environmental Health Perspectives* 109 (Suppl. no 2): 211–221.

Schär, C., P. L. Vidale, D. Lüthi, C. Frei, C. Häberli, M. A. Liniger, and C. Appenzeller. 2004. The role of increasing temperature variability in European summer heatwaves *Nature* 427:332–36.

Shapiro, R. L., S. Altekruse, and P. M. Griffin. 1998. The role of Gulf Coast oysters harvested in warmer months in *Vibrio vulnificus* infections in the United States, 1988–1996. *Journal of Infectious Diseases* 178 (3): 752–59.

Singh, R. B. K., S. Hales, N. de Wet, R. Raj, M. Hearnden, and P. Weinstein. 2001. The influence of climate variation and change on diarrhoeal disease in the Pacific Islands. *Environmental Health Perspectives* 109 (2): 155–59.

Sork, V., J. Bramble, and O. Sexton. 1993. Ecology of mast-fruiting in three species of North American oaks. *Ecology* 74:528–41.

Stott P. A., D. A. Stone, and M. R. Allen. 2004. Human contribution to the European heatwave of 2003. *Nature* 432:610–14.

Thomas, C. D., A. Cameron, R. E. Green, M. Bakkenes, L. J. Beaumont, Y. C. Collingham, B. F. N. Erasmus, et al. 2004. Extinction risk from climate change. *Nature* 427:145–48.

Tol, R. S. J., and H. Dowlatabadi. 1999. Vector-borne diseases, economic growth and climate change. Paper presented at the EMF Workshop on Climate Change Impacts and Integrated Assessment, July 26–August 5, 1999, Snowmass, Colorado.

Walther, G.-R., E. Post, A. Menzel, P. Convey, C. Parmesan, F. Bairlen, T. Beebee, J. M. Fromont, and O. Hoegh-Guldberg. 2002. Ecological responses to recent climate change. *Nature* 416:389–95.

WHO (World Health Organisation). 2002. *World health report 2002: Reducing risks, promoting healthy life.* Geneva: World Health Organisation.

———. 2003. *Climate change and human health: Risks and responses,* ed. A. J. McMichael, D. H. Campbell-Lendrum, C. F. Corvalán, K. L. Ebi, A. K. Githeko, J. D. Scheraga, and A. Woodward. Geneva: World Health Organisation, World Meteorological Organization, and United Nations Environment Programme.

Wilkinson C. R., ed. 2000. *Global coral reef monitoring network: Status of coral reefs of the world in 2000.* Townsville, Queensland: Australian Institute of Marine Science.

Wilkinson, P., D. H. Campbell-Lendrum, and C. L. Bartlett. 2003. Monitoring the health effects of climate change. Chap. 10 in *Climate change impacts on the United States: The potential consequences of climate variability and change.* National Assessment Synthesis Team Report for the United States Global Change Research Program. Cambridge: Cambridge University Press.

Ziska, L. H., and F. A. Caulfield. 2000a. The potential influence of raising atmospheric carbon dioxide (CO_2) on public health: Pollen production of common ragweed as a test case. *World Resources Review* 12:449–57.

———. 2000b. Pollen production of common ragweed (*Ambrosia artemisiifolia*), a known allergy-inducing species: Implications for public health. *Australian Journal of Plant Physiology* 27:893–98.

PART V
Biodiversity as a Resource for Medicine

Part V (chapter 15) of this book discusses native biodiversity as a resource for medicine. Developed nations have increasingly embraced alternative medicine and healing practices as part of health and well-being. Paul Alan Cox, in chapter 15, focuses on the challenges and rewards of searching for new medicines through native biodiversity. Cox briefly reviews the history and various strategies used for collecting and identifying biological diversity in species-rich countries for potential medical uses and describes both the positive and negative effects of international agreements to conserve biodiversity. A clear message from this chapter is that the benefits to human health from the discovery of novel compounds have yet to be supplanted by purely laboratory approaches.

15

Biodiversity and the Search for New Medicines

Paul Alan Cox

The molecular diversity of the natural world is profound. During the 3.5-billion-year history of life on Earth, evolution has produced a staggering array of bioactive molecules. Some, such as enzymes in mosquitoes that serve to detoxify DDT (dichloro-diphenyl-trichloroethane) are of very recent date, while others, such as highly specific motor neuron toxins in cycads like beta-N-methylamino-L-alanine (BMAA), likely evolved in response to dinosaur herbivory (Cox, Banack, and Murch 2003). Despite the laboratory prowess of modern chemists, a variety of molecular structures present in plants, animals, fungi, bacteria, and viruses are difficult, if not impossible, to synthesize. Only 60% of the 11,500 different structural types of natural products are represented by synthetic compounds (Henkel et al. 1999). Despite remarkable progress in synthetic and organic chemistry over the last four centuries, human efforts have scarcely been able to document, let alone duplicate, the plethora of bioactive molecules produced by life on earth.

It is therefore not surprising that human cultures throughout the world rely on biodiversity as a source of medicines. The World Health Organization (WHO) estimates that 85% of the people on our planet depend on plants for their primary health care. Over 50% of Western pharmaceuticals are derived from biodiversity (Grifo and Rosenthal 1997). Useful bioactive molecules, however, do not appear to be equally derived from all life forms. In a convergence between strategies employed by modern pharmaceutical researchers and indigenous healers, the search for new drugs has not focused on highly mobile organisms. Rare indeed is the poison derived from birds such as the hooded pitohui (*Pitohui dichrous*) in New Guinea (Dumbacher et al. 1992), or the elixir derived from mammals, such as from the gall bladder of the Asian bear species (*Ursus* spp.) (Servheen, Herrero, and Peyton 1999). Instead, both scientists and indigenous peoples direct the majority of their attention to sessile organisms, particularly plants and marine invertebrates. While perhaps the immobility of such organisms facilitates ease of mapping and subsequent recollection, it appears that sessile organisms also produce the most potent bioactive molecules. In some cases, the interests of indigenous peoples and scientists converge on the same toxic molecules from sessile organisms.

A few kilometers from the manicured grounds of the National Tropical Botanical Garden in Hana, Hawaii, exists a small, nondescript tidal pool that does not appear from superficial inspection to be different in any way from the hundreds of other pools etched into the black lava of the east Maui coast. Local Hawaiians, however, not only avoid this pool, but seldom speak of it to visitors, because it contains *limu make o Hana*—the death algae of Hana. It is claimed that ancient Hawaiian warriors, under extreme threat, could dip their spears in the pool to gain lethal power over their enemies. The pool, however, was essentially under *kapu*—taboo—and was totally avoided by the local population. Research by Paul Scheuer at the University of Hawaii led to the discovery of an unusual coelenterate, *Palythoa toxica*, which produces one of the most poisonous substances known (Moore and Scheuer 1971). Palytoxin is a low-molecular-weight toxin (2,678.5 daltons) with an LD_{50} of 0.15 µg/g in mice (Moore 1985; Moore and Scheuer 1971). Palytoxin has since been found in other species of the genus *Palythoa*. Even a scientist such as Scheuer could not completely avoid the ill effects of violating a native *kapu*, however—it is claimed that on his return to Honolulu, he found his laboratory burned to the ground (Lars Bohlin, pers. comm.).

Why should marine invertebrates and plants produce such potent molecules? Photosynthesis, energy storage, and hormonal integration of reproduction require a variety of complex molecular entities. Plants, which are not carbon limited, generate copious secondary metabolites and other organic discards. Some of these are pharmacologically active, but most bioactive molecules likely arose as chemical defenses. Sessile organisms must mediate their interaction with the world—including parasites, predators, and competitors—primarily with chemicals. Evolutionary pressures have selected for toxins that fulfill this protective role. Certainly the avoidance of the pool by native Hawaiians benefited *Palythoa toxica*; fish that sought to forage on *Palythoa* almost certainly did not survive the attempt.

Different Search Strategies for Bioactive Molecules

Bioprospecting has gained much recent popular attention, particularly spurred by the apocryphal image of adventurous ethnobotanists penetrating at considerable peril the darkest recesses of the jungle at the behest of rich pharmaceutical firms. However, there is no large pharmaceutical firm that currently bases a majority or even a significant component of its research program on searching for new molecules from rainforests. Although the pendulum is slowly swinging back toward biodiversity, there is unfortunately an industry-wide aversion to natural products. "Rational approaches" such as combinatorial chemistry and computer-aided design are much more in vogue.

Even the term *bioprospecting* poorly describes the activities of the few remaining programs, mostly headquartered in universities, that seek to evaluate the pharmaceutical potential of biodiversity. The mining industry, to which the prospecting metaphor refers, is extractive, consumptive, and, too often, exploitive. In contrast, biodiversity research, like other academic disciplines, seeks the creation of new knowledge. Like other public goods—poetry, historical monographs, or descriptions of new galaxies—studies of biodiversity can be consumed without being depleted. And although some progressive mining companies have taken great efforts to minimize environmental damage and social upheaval from their activities, the often negative interactions of prospectors and miners with local indigenous communities generally do not apply to biodiversity research. Ethnobotanists who obtain government, village, and healer approval to study traditional medicines have emerged as protectors of biological conservation and advocates of indigenous peoples (Cox 2001).

At least five different approaches are involved in selecting organisms for study in biodiversity research.

Random collections, particularly of plants and marine invertebrates, have been used by the National Cancer Institute (NCI) in the search for molecules active against various forms of human cancer. The collection program, initiated in 1960 and discontinued in 1982, resulted in 114,000 plant extracts and 18,000 extracts of marine organisms. However, only seven anticancer drugs from plants have been approved by the U.S. Food and Drug Administration (FDA) since 1960. The most prominent success of the NCI program was Taxol (Paclitaxel), a chemotherapeutic agent for the treatment of ovarian and breast cancer (Cragg et al. 1993). NCI is currently conducting trials of bryostatin, a natural product from marine invertebrates, for the treatment of leukemia, lymphoma, and skin cancers, while Calanolide A and Prostratin from rainforest trees are being investigated as anti-HIV/AIDS therapies.

Ecological screening of biodiversity allows specific interactions between organisms to be examined for clues to the presence of novel bioactive molecules. For example, Kurt Hostettmann and his collaborators at the University of Lausanne have been seeking antimolluscidal compounds to help control schistosomiasis in Africa (Hostettmann and Marston 2002). Clues can be found in plants that live in environments exposed to mollusk herbivory, particularly those that show little evidence of snail damage. Another example of ecological screening is the study of symbioses between bacteria and nematodes that allow nematodes to kill their insect hosts within 24 to 48 hours. Such studies resulted in the discovery of molecules used as quorum signals by prokaryotes as well as the characterization of genes responsible for their production (Derzelle et al. 2002).

Phylogenetic approaches are becoming increasingly sophisticated due to the advent of cladistics and the increased availability of molecular phylogenies. In a search for new anti-inflammatory molecules, Lars Bohlin at the University of Uppsala and his collaborators are plotting inhibition of cyclooxygenase-mediated biosynthesis of prostaglandin by both COX-2 enzymes and COX-2 gene expression on angiosperm phylogenies (Perera et al. 2001). The goal is to identify lineages of higher plants that are rich in COX-2 inhibitory compounds such as those derived from the toi tree of Samoa, *Atuna racemosa* (Chrysobalanaceae) (Dunstan et al. 1997). Certain families of angiosperms are known to contain bioactive molecules of the same structural family—for example, the Euphorbiacae and phorbol esters (Evans 1986) —so investigators can focus their efforts on families and lineages of particular interest.

Genetic approaches have been pioneered by researchers such as Craig Venter. In the spring of 2004, Venter and his team sampled microbial diversity in the Sargasso Sea, looking not only for novel microbes, but more importantly, for novel DNA sequences. Venter reported finding 1.2 million genes, 70,000 of which were novel, as well as 1,800 new microbial species. Venter's "environmental genome shotgun sequencing" approach does not rely on isolating individual microbes, but instead on merely breaking up their DNA and using it to probe novel genetic targets. This approach resulted in the discovery of 782 new rhodopsin-like photo receptor genes (Venter et al. 2004).

Ethnobotanical approaches are the oldest, but perhaps most successful, techniques in discovering new pharmaceuticals from biodiversity. Pioneered as early as 1732 by Carl Linnaeus, who interviewed Sami healers in northern Lappland, the ethnobotanical approach to drug discovery relies on using the knowledge of indigenous peoples, particularly healers, to guide pharmaceutical research efforts (Cox and Balick 1994). Notable compounds discovered through ethnobotany include vincristine and vinblastine, used for the treatment of pediatric leukemia, and quinine, used for malaria prophylaxis, discovered through observations of indigenous peoples in Ecuador in the 16th century (Balick and Cox 1996). A current compound of interest discovered through ethnobotany is the anti-AIDS drug candidate prostratin, which is being developed by the AIDS Research Alliance as part of a combination therapy for AIDS (Brown 2001). Prostratin was isolated by an NCI team from *Homalanthus nutans* (Euphorbiaceae), a small rainforest tree used by traditional healers in Samoa for the treatment of hepatitis (Cox 2001; Gustafson et al. 1992).

Bioassay-Guided Fractionation and Structural Determination

In order to elucidate the molecules responsible for bioactivity in a plant or animal, an iterative process known as bioassay-guided fractionation is employed. While in some pharmaceutical firms, bioassay-guided fractionation utilizes computer-controlled robots capable of screening thousands of extracts in a single microwell plate for precise enzyme binding assays, the origins of this approach are much more prosaic. "In the year 1775, my opinion was asked concerning a family receipt for the cure of the dropsy," wrote physician William Withering. "I was told that it had long been kept a secret by an old woman in Shropshire, who had sometimes made cures after the more regular practitioners had failed. ... This medicine was composed of twenty or more different herbs; but it was not very difficult for one conversant in these subjects, to perceive, that the active herb could

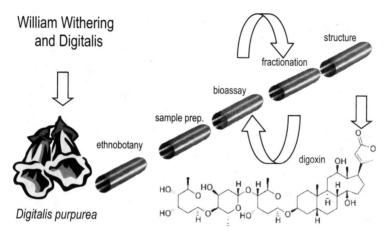

Figure 15.1. Discovery pipeline beginning with ethnobotanical research and leading to digitoxin.

be no other than Foxglove" (Withering 1785, 2). Withering used the only bioassay he had access to—his own patients. A recent analysis shows that between 65% and 80% of his patients recovered from dropsy, systematic edema due to congestive heart failure (Aronson 1985). Withering soon discovered the secret of the efficacy of foxglove leaves (*Digitalis purpurea*): "It has a power over the motion of the heart, to a degree yet unobserved in any other medicine" (Withering 1785, 192). In the early 20th century, chemists were able to isolate more than 30 distinct cardiac glycosides from dried foxglove leaves, including digitoxin and digoxin, the latter of which is marketed as Lanoxin by Burroughs Wellcome (Samuelsson 1992). Each year over 1,500 kg of digoxin and 200 kg of digitoxin are prescribed to patients throughout the world. The discovery and isolation of the cardiac glycosides from *Digitalis* can be illustrated as a pipeline beginning with Withering's ethnobotanical interviews, his preparation of dried foxglove leaves, and then bioassay-guided fractionation and ultimately structural elucidation of the active compounds (figure 15.1).

The discovery of the antihypertensive drug reserpine followed a similar process. People in the Bihar region of India use roots of snakeroot (*Rauvolfia serpentina,* Apocynacae) to treat insomnia and anxiety, and in Ayurvedic medicine as a remedy for snakebite (Sinha 1996, 25). In 1931, a report published in the *Indian Medical Record* indicating that the powdered roots dramatically lower blood pressure obtained little attention until a chemist at CIBA Pharmaceuticals in Basel, Switzerland, named Emil Schlittler, read a clinical study of *Rauvolfia* by R. J. Vakil in the *British Heart Journal*. Together with his colleague Hans Schwarz, Schlittler isolated an alkaloid that in remarkably low doses—for example, 0.1 mg/kg—lowers blood pressure. Named reserpine, this new alka-

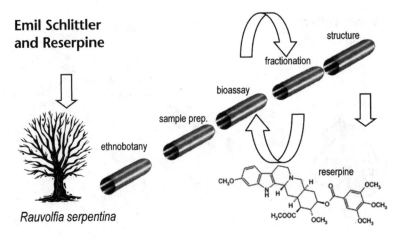

Figure 15.2. Discovery pipeline beginning with ethnobotanical research and leading to reserpine.

loid was the first drug that controlled hypertension without vasodilation (CIBA 1954). Again, the discovery process that ultimately led to the isolation of reserpine can be characterized as a discovery pipeline beginning with ethnobotany (figure 15.2).

Modern techniques for fractionation include using solvents of different polarity to produce an increasingly refined sample. Each fraction is then tested against the original bioassay to determine bioactivity. Chromatographic techniques, including high-pressure liquid chromatography, thin-layer chromatography, and paper chromatography can be used to separate crude extracts into their components.

Once a relatively pure fraction is obtained, it can be screened using mass spectroscopy, and then checked against a computer library of molecular weights to see if it is a known compound. If the molecule is novel, nuclear magnetic resonance (NMR) spectroscopy can be used to determine the molecular structure. In difficult cases, x-ray crystallography can be used.

Biodiversity-Derived Pharmaceuticals and Global Needs

The disease targets that are currently being pursued in biodiversity research often focus on patterns of illness prevalent in northern, industrialized countries. As a result, remedies for heart disease, cancer, complications associated with obesity, clinical depression, loss of libido, and male erectile dysfunction tend to be emphasized. For example, red yeast (*Monascus pureus*) from rice has been shown to lower serum cholesterol levels. Synthetic analogues of the responsible statin molecules marketed as Zocor and Lipitor have generated hundreds of millions of dollars in revenues. Licensing of Taxol, derived from *Taxus brevifolia* (Taxaceae), for treatment of ovarian and breast cancer led to a small resurgence in natural product pharmacology in the 1990s. However, Zocor, Taxol, and other Western pharmaceuticals are often so expensive that they are beyond the economic reach of entire populations in developing countries.

The state of pharmacological research in biodiversity unfortunately too often mirrors research into tropical crops: only species valued as cash commodities in the Western world, including coffee, sugar, cocoa, and rubber, are carefully studied, while subsistence crops such as breadfruit, taro, and cocoyams receive scant attention, despite the hundreds of millions of people who depend on them as staples. Given that 85 percent of the world's population depends directly on plants for primary health care, it seems important that studies of safety and efficacy of traditional remedies should receive high priority. Equally important is translating innovative research in biodiversity-derived pharmaceuticals into products that can be marketed at a low cost in developing countries for diseases that may not be common in the industrialized world, but which are devastating in the developing world.

Such a disease that presents little threat in the industrialized world, but which causes millions of deaths and untold suffering throughout the developing world, is malaria. Although the first major antimalarial drug—quinine—was itself derived from biodiversity, indiscriminate use rapidly led to the evolution of resistance in the various *Plasmodium* parasites. Synthetic antimalarial drugs, including Maloprim and Fansidar, which interfere with the folic acid cycle of the parasite, are expensive and soon proved ineffective in parts of the world where *Plasmodium vivax* had gained resistance. Recently, artemisinin, a sesquiterpene derivative of a traditional Chinese remedy made from the wormwood plant, *Artemisia annua* (Compositae), has emerged as an effective drug against different resistant *Plasmodium* strains, including *P. falciparum,* which causes cerebral malaria. Careful plant breeding in China, as well as in Vietnam, has increased yields from 0.5% to 2% dry weight (Samuelsson 2004). Oral capsules, as well as suppositories, are being mass-marketed with a complete course of treatment for malaria currently costing US$2.50. However, the success of artemisinin threatens to outstrip both natural and cultivated supplies of the plant, while the cost of treatment—modest by Western standards—still excludes millions of the poor. An innovative approach by Jay Keasling and his colleagues at the University of California, Berkeley, to clone the artemisinin gene and insert it into a bacterium was generously funded in December 2004 through a US$42.6 million Gates Foundation grant administered by the Institute for OneWorld Health. Keasling's goal (pers. comm.) is to produce a microbial product that can be marketed at US$0.25 for a three-day course of treatment.

One of the biggest killers of children in the developing world is diarrhea. While so easily treated in the industrialized world that pediatric diarrhea is an extremely rare cause of infant mortality, diarrhea represents an ongoing and lethal threat to indigenous and marginalized populations in developing countries. Michael Heinrich and his colleagues at the London School of Pharmacy have performed an extraordinary analysis of plants used to treat gastrointestinal disorders in Oaxaca, Mexico. This groundbreaking work, coupled with Heinrich's careful analysis of the physiological impact of rice water on the colon, provides hope that simple, easily obtainable plants and dietary items can be used to provide effective remedies throughout the world (Heinrich, Rimpler, and Barrera 1992).

A final example of rigorous biodiversity research leading to pharmaceuticals appropriate for the developing world can be found in studies of *Ipomoea pes-caprae* (Convolvulaceae), a beach morning glory used in Asia by fishermen to treat jellyfish stings. Ubonwan Pongprayoon and her collaborators in Thailand and Sweden analyzed the potent effect of leaves in neutralizing jellyfish venoms (Pongprayoon, Bohlin, and Wasuwat 1991). Using bioassay-guided fractionation, they were able to isolate two antispasmodic norisoprenoids (Pongprayoon et al. 1992). Pongprayoon's research led to clinical trials within Thailand of a crude extract of *Ipomoea pes-caprae* for topical treatment of inflammation. In a carefully controlled double-blind test, the crude extract was found to be safe and effective. It currently is marketed at pharmacies throughout Thailand, where it sells for a fraction of the cost of other anti-inflammatory drugs.

Similar studies of safety and efficacy are needed of traditional medicines throughout the world. Innovative research programs in Mexico, Nigeria, Thailand, and China to document traditional medicines need to be more widely adopted, and incentives need to be developed for pharmaceutical firms to help fill the "gray" pharmaceutical market for inexpensive crude extracts and tinctures from biodiversity.

Even such low-cost, locally produced remedies, however, are often beyond the means of the poorest of the poor. It would be useful if the World Health Organization (WHO) assembled an international team of scientists to develop a kit of 10 to 12 medicinal plants that could be easily cultivated in warm regions of the world. Such a kit might focus on guava (*Psidium guajava* Myrtaceae) leaves for diarrhea, beach morning glory (*Ipomoea pes-caprae*) leaves for inflammation, and *Mimosa tenuiflora* Leguminosae bark for burns (Camargo-Ricalde 2000). WHO could request pharmaceutical firms to test the proposed species for safety and efficacy using the same rigorous standards as those for pharmaceuticals in the industrialized world. Ecologists could be asked to ensure that the proposed species would not become invasive weeds, and soil scientists and horticulturalists could screen the proposed species for ease of cultivation in a variety of soils and rainfall regimes. Such a grouping of effective medicinal plants for common ailments, could, with the imprimatur of WHO, be distributed throughout the world, assisting millions of people to have improved health.

The Convention on Biological Diversity

Prior to the Convention on Biological Diversity in 1992 (CBD), asymmetries between biodiversity-rich developing countries and technologically rich industrialized countries too often left little opportunity for developing nations to analyze and economically benefit from their own biodiversity. The intent of the CBD was to encourage sustainable use of biodiversity by establishing an international mechanism for its commercial development (Glowka, Burhenne-Guilmin, and Synge 1994). By recognizing the sovereignty of each nation over its own biological resources, and by establishing national clearinghouses for scientific studies of biodiversity, it was hoped that commercial returns from biodiversity development could provide a powerful impetus for biological conservation.

Although the CBD by diplomatic standards is remarkable—more nations have signed the CBD than nearly any other treaty in the world's history—its success in establishing a level playing field between wealthy technology-rich countries and poorer biodiversity-rich countries has been somewhat slower. The uneven distribution of taxonomic resources, including well-curated museums, herbaria, and botanical gardens as well as a cohort of trained taxonomists, has emerged as an impediment for many nations in documenting and developing their biodiversity. However, the Global Environment Facility (GEF) has attempted to remedy this paucity of biosystematics in developing countries by funding a variety of national stock-taking efforts.

A positive example of international cooperation in sustainable development of biodiversity is the 2004 agreement between the Independent State of Samoa, an island nation in the South Pacific, and the University of California, Berkeley, to clone the gene sequence from the rainforest tree *Homalanthus nutans,* responsible for the anti-AIDS drug prostratin. Samoa, which under the CBD asserted sovereignty over the gene sequence, will enjoy a 50% share in any revenues from the gene product, and is actively collaborating with U.C. Berkeley scientists in their research. This agreement follows a previous arrangement between Samoa and the AIDS Research Alliance, which gives a 20% share of prostratin's profits to Samoa. In both cases, Samoa's share of the revenues will be divided between the Samoan government, villages hosting the research, and families of traditional healers (Cox 2001).

These Samoan agreements highlight the importance of article 8(j) of the CBD, in which the parties to the convention agree to (1) respect, preserve, and perpetuate indigenous biological knowledge, (2) promote wider applications of traditional knowledge, and (3) encourage equitable sharing of the benefits arising from such knowledge. Samoa argues that prostratin would not have been discovered as an anti-AIDS remedy without active participation of Samoan healers in ethnobotanical studies. Although Samoa and U.C. Berkeley have agreed to provide prostratin at low or no cost to developing countries, and although Samoa asserts no restriction concerning traditional use of *Homalanthus nutans* elsewhere, the nation does claim significant shares of any income resulting from the commercialization of prostratin as an antiviral drug. Equally important, both the AIDS Research Alliance and the University of California, Berkeley, accept these claims of Samoan sovereignty, and have negotiated equitable sharing of benefits with the Samoan government, the affected villages, and the families of the healers who participated in the original research.

The Future of Biodiveristy Research

Historically, biodiversity has been the major source of pharmaceuticals, and today is relied on by 85% of the world's population for primary health care. A large proportion of pharmaceuticals used by the remaining 15% of the world's population can be traced to biodiversity-derived compounds. Innovative programs of biodiversity research, such as the search for novel small molecules from marine and soil microorganisms by Jon Clardy and his colleagues at Harvard (Thomas 2005) have been greatly facilitated by innovative new technologies. High-throughput screens and bioassay-guided fractionation have led to remarkable new techniques in which thousands of plant or animal extracts can be rapidly screened for hundreds of new disease targets.

Such research, however, has been hampered by habitat loss and species extirpations, which have particularly impacted sessile organisms in tropical rainforests and coral reefs. The Convention on Biological Diversity aims to provide incentives for the sustainable use of biodiversity, but unforeseen by the architects of the CBD was the resultant bureaucratization of biodiversity research. Although designed to facilitate and encourage international collaborations between biodiversity-rich countries and technologically rich institutions, many smaller nations, lacking the financial and scientific resources to assess their own biodiversity, have essentially closed their doors to foreign researchers. In a world in which biodiversity conservation was a priority and in which anthropogenic extinctions were rare, such a strategy could have political saliency—why not hold on to biodiversity and protect it from foreign exploitation, particularly if its value is rising through time? However, biodiversity is not static. By closing borders to foreign researchers, but allowing natural habitats to be destroyed by loggers and miners, some small nations have in effect doomed their biodiversity to extinction before it can be assessed for pharmaceutical potential. Species extinctions and habitat destruction have accelerated rather than declined since the signing of the CBD. Indigenous knowledge of biodiversity is also rapidly disappearing from the world (Cox 2000). Unfortunately, it is now far easier in many parts of the world to clear-cut a rainforest than to obtain permission to take 50 mg samples of the selected plant species for pharmacological analysis. As long as plant and animal species are easier to destroy rather than to study, biodiversity will continue to decline.

Acknowledgments

I thank M. K. Asay for her assistance in preparing this manuscript.

References

Aronson, J. K. 1985. *An account of the foxglove and its medical uses 1785–1985*. Oxford: Oxford University Press.

Balick, M. J., and P. A. Cox. 1996. *Plants, people and culture: The science of ethnobotany*. New York: Scientific American Library.

Brown, S. J. 2001. Prostratin: A potential adjuvant therapy for HAART against HIV-1 reservoirs. *Searchlight* (Spring): 3–8.

Camargo-Ricalde, S. L. 2000. Descripción, distribución, anatomía, composición química y usos de *Mimosa tenuiflora* (Fabaceae-Mimosoideae) en México. *Revista deBiología Tropical* 48 (4): 939–54.

CIBA. 1954. *The Rauwolfia story: From primitive medicine to alkaloidal therapy.* Summit, NJ: CIBA Pharmaceutical Products.

Cox, P. A. 2000. Will tribal knowledge survive the millennium? *Science* 287 (5450): 44–45.

———. 2001. Ensuring equitable benefits: The Falealupo covenant and the isolation of antiviral drug prostratin from a Samoan medicinal plant. *Pharmaceutical Biology* 39 (Supplement): 33–40.

Cox, P. A., and M. J. Balick. 1994. The ethnobotanical approach to drug discovery. *Scientific American* 270 (6): 82–87.

Cox, P. A., S. A. Banack, and S. J. Murch. 2003. Biomagnification of cyanobacterial neurotoxins and neurodegenerative disease among the Chamorro people of Guam. *Proceedings of the National Academy of Sciences* 100 (23): 13380–83.

Cragg, G. M., M. R. Boyd, J. H. Cardellina II, M. R. Grever, S. Schepartz, K. M. Snader, and M. Suffness. 1993. The search for new pharmaceutical crops: Drug discovery and development at the National Cancer Institute. In *New crops*, ed. J. Janick and J. E. Simon, 161–67. New York: Wiley.

Derzelle, S., E. Duchaud, F. Kunst, A. Danchin, and P. Bertin. 2002. Identification, characterization and regulation of a cluster of genes involved in carbapenem biosynthesis in *Photohabdus luminescens. Applied and Environmental Microbiology* 68 (8): 3780–89.

Dumbacher, J. P., B. M. Beehler, T. F. Spande, H. M. Garraffo, and J. W. Daly. 1992. Homobatrachotoxin in the genus *Pitohui*: Chemical defense in birds? *Science* 258:799–801.

Dunstan, C. A., Y. Noreen, G. Serrano, P. A. Cox, P. Perera, and L. Bohlins. 1997. Evaluation of some Samoan and Peruvian medicinal plants by prostaglandin biosynthesis and rat ear oedema assays. *Journal of Ethnopharmacology* 57:35–56.

Evans, F. J. 1986. *Naturally occurring phorbol esters.* Boca Raton, FL: CRC Press.

Glowka, L., F. Burhenne-Guilmin, and H. Synge. 1994. *A guide to the convention on biological diversity.* Gland, Switzerland: IUCN.

Grifo, F., and J. Rosenthal, eds. 1997. *Biodiversity and human health.* Washington, DC: Island Press.

Gustafson, K. R., J. H. Cardellina, J. B. McMahon, R. J. Gulakowski, J. Ishitoya, Z. Szallasi, N. E. Lewin, et al. 1992. A non-promoting phorbol from the Samoan medicinal plant *Homalanthus nutans* inhibits cell killing by HIV-1. *Journal of Medicinal Chemistry* 35: 1978–86.

Heinrich, M., H. Rimpler, and N. A. Barrera. 1992. Indigenous phytotherapy of gastrointestinal disorders in a lowland Mixe community (Oaxaca, Mexico): Ethnopharmacologic evaluation. *Journal of Ethnopharmacology* 36 (1): 63–80.

Henkel, T., R. M. Brunn, H. Müller, and F. Reichel. 1999. Statistical investigation into the structural complementarity of natural products and synthetic compounds. *Angewandte Chemie International Edition* 38:643–47.

Hostettmann, K., and A. Marson. 2002. Twenty years of research into medicinal plants: Results and perspectives. *Phytochemistry Reviews* 1 (3): 275–85.

Moore, R. E. 1985. Structure of palytoxin. *Progress in the Chemistry of Organic Natural Products* 48:82.

Moore, R. E., and P. J. Scheuer. 1971. Palytoxin: A new marine toxin from a coelenterate. *Science* 172:495–98.

Perera, P., T. Ringbom, U. Huss, M. Vasänge, and L. Bohlin. 2001. Search for natural products which affect cyclooxygenase-2. In *Bioactive compounds from natural sources: Isolation, characterisation and biological properties*, ed. C. Tringali. New York: Taylor & Francis.

Pongprayoon, U., P. Baeckstrom, U. Jacobsson, M. Lindstrom, and L. Bohlin. 1992. Antispasmodic activity of beta-damascenone and E-phytol isolated from *Ipomoea pes-caprae*. *Planta Medica* 58 (1): 19–21.

Pongprayoon, U., L. Bohlin, and S. Wasuwat. 1991. Neutralization of toxic effects of different crude jellyfish venoms by an extract of *Ipomoea pes-caprae* (L.) R. Br. *Journal of Ethnopharmacology* 35 (1): 65–69.

Samuelsson, G. 1992. *Drugs of natural origin*. Stockholm: Swedish Pharmaceutical Press.

———. 2004. *Drugs of natural origin*. 5th ed. Stockholm: Swedish Pharmaceutical Press.

Servheen, C., S. Herrero, and B. Peyton, eds. 1999. *Status survey of the bears of the world and global conservation action plan*. Gland, Switzerland: IUCN.

Sinha, R. K. 1996. *Ethnobotany: The renaissance of traditional herbal medicine*. Jaipur, India: Ina Shree Publishers.

Thomas, P. 2005. The chemical biologists. *Harvard Magazine* 107 (4): 38–47.

Venter, J. C., K. Remington, J. F. Heidelberg, A. L. Halpern, D. Rusch, J. A. Eisen, D. Wu, et al. 2004. Environmental genome shotgun sequencing of the Sargasso Sea. *Science* 304 (5667): 58–60.

Withering, W. 1785. *An account of the foxglove and some of its medical uses: With practical remarks on dropsy and other diseases*. Birmingham, AL: M. Swinney.

16

Human Health, Biodiversity, and Ecosystem Services: The Intertwined Challenging Future

Frederick A.B. Meyerson, Laura A. Meyerson, Camille Parmesan, and Osvaldo E. Sala

In a world where scarcity of food, water, and fuel is an immediate challenge and not just a future specter, the maintenance of biodiversity and ecosystem services is critical to human health and welfare. The climate change wild card adds insecurity as to what biotic resources and tools may be needed for human adaptation to food supply disruptions, ecosystem alterations, shifting disease patterns, and other health threats. There is great uncertainty about the future interaction between human health and changes in biodiversity. In many ways, we are collectively experimenting blindly with the ecology of the entire planet. However, we know enough to describe several major challenges and important research needs.

State of the Science: Strengths and Shortcomings

This book has explored the effects of biodiversity on four major determinants of human health and well-being: ecosystem services, constraint of infectious disease, medicinal resources, and quality of life (see this volume, chapter 1). We have described what is known about those relationships, as well as the significant gaps in knowledge, particularly about human drivers and biodiversity. The movement of species to new areas and other anthropogenic alterations in biological diversity have both positive and negative effects, many of which are unanticipated. We need to increase our understanding of the optimal relationship between people and other species—one that maximizes human health and welfare as well as biological diversity in both the short and long term.

For some anthropogenic drivers, there are very clear, well-documented links that lead all the way from biodiversity alteration or degradation to impacts on human health. For example, the shift to irrigated agriculture, which increases food production and nutrition, also increases available habitat for mosquitoes, raising the incidence of mosquito-borne diseases such as malaria and dengue fever (see this volume, chapter 2). A similar case can be made for Lyme disease, for which human-caused habitat fragmentation, suburbanization, and the resultant growing deer and deer mice populations with little predation have boosted both the number of ticks and the potential for interactions with people.

For other drivers, the links are inferred, but rigorous documentation is lacking. For example, many studies have demonstrated cause-and-effect relationships among anthropogenic land conversion, degradation of watersheds, and human health threats (e.g., eutrophication, increased contaminants, and pathogen introduction) (see this volume, chapters 3, 8, 9, and 10). However, few data exist to link and quantify the relationships between alteration of biodiversity and water quality, or to explain the role of aquatic biodiversity in human health (see this volume, chapter 8).

For certain drivers, our understanding of the links is still very weak. Climate change is an important example, in part because the scope of the projected changes in temperature, precipitation, and other factors are outside our recent human experience, particularly during the short period in which we have been conducting ecological research. The studies that do exist are almost all correlational, with essentially no experimental work. However, logical extrapolation and inference all lead to an expectation of increases in tropical diseases in North America and Europe. Yet lack of rigorous studies (even correlational ones) and lack of experimental work to nail down mechanisms (especially geographic variation in the thermal ecology of diseases) make projections very vague, with high uncertainty for any given region (see this volume, chapter 14).

Future Research Needs

Science has been successful in unearthing a vast amount of information on human physical and mental health, biodiversity, and ecosystem functions and processes. Less clear, however, are the relationships between and among these diverse areas, such as the links between microbial diversity, human health, and ecosystem services. In large part, this may be due to the paucity of collaboration across disciplines—particularly the medical, natural, and social sciences. The recent call for the expansion of interdisciplinary education has been answered by many universities around the world, but this is still a new and complicated endeavor. It is still uncertain whether those educated in this manner will be more effective at addressing current and future challenges than strictly disciplinary scientists and practitioners.

While interdisciplinary research teams have made headway into unraveling some complex global phenomena such as the biogeochemical processes of climate change, other relationships, such as the consequences of biodiversity loss for human well-being and health, remain opaque. Identifying and mediating the trade-offs among global human population growth, agriculture, and maintenance of native biodiversity is becoming more urgent. Which land use practices will best serve human health (e.g., degrees of agricultural intensification) while maintaining current levels of biodiversity (e.g., location of agricultural land)? Can we apply biological and ecological knowledge to improve sustainability and avoid, or at least manage, irreversible changes in biodiversity, ecosystem functioning, and related factors affecting human health and well-being?

A few additional research needs stand out. The increasing scarcity of freshwater around the world, the concomitant loss of aquatic biological diversity, and the ongoing introduction of invasive plants, animals, and pathogens highlight the need for expanding investigations into the effects of altering biodiversity on ecosystem function, water quality, and disease transmission. Our knowledge of the functional effects of altered microbial communities on ecosystems, and ultimately on human health, lacks both depth and breadth and will require systematic investigation to fill this gap. Perhaps most challenging is gaining a better understanding of the ecological role of pathogens and the relative trade-offs between human health and well-being and conservation of biological diversity.

We also cannot overstress the importance of new, rigorous interdisciplinary studies that combine study of the autecology of diseases and their vectors in the wild, research on the ecology of disease transmission from the wild into human populations, and the study of societal susceptibility to potential new epidemics. This is a complex set of relationships, mediated in part by the health system in place, including the ability to recognize and treat diseases, to respond and adapt quickly, and to anticipate the potential spread of diseases into surrounding regions.

Global-Scale Challenges

One critical factor that may be very difficult to predict is human reaction to future scarcity and stress in response to these interlocking challenges and trade-offs. In the short time that it has taken to write and edit this volume, biofuels have first been hailed and widely adopted as the solution to both fossil fuel shortages and greenhouse gas emissions, then broadly rejected as contributing to high food prices, scarcity, and starvation, and having a negative, or at best neutral, effect on net carbon emissions. Biofuel decisions, which may be distorted by political or economic crises, could have global-scale consequences for land use and the distribution of species.

In the background of the relationship between human health and biodiversity churns the continuing rapid growth of humanity. For at least the next 20 years, the net addition to population is projected to average 75 million people per year globally (United Nations 2007). Much of the population growth will occur in areas where human health is at greatest risk because of poverty and natural resource scarcity, and where there is the significant potential for disease transmission from the wild into human populations. The trade-offs between human well-being and the competing uses of solar energy, resources, and physical space for food supply, energy production, ecosystem services, and biodiversity maintenance will become more difficult as the global population rises toward 9 billion over the next few decades. Those decisions may well be mediated more by political events and immediate crises rather than science or long-term ecological and social planning (see this volume, chapter 5).

As countries with large populations and rapid economic growth such as China and India move to the foreground of world leadership, the balance of trade and its pathways are shifting. With this change both the center of power for environmental policy and the philosophy behind it may also shift. It could also be the case that global corporations more than national governments control the future spread of species, particularly genetically modified organisms. The ecological and human health ramifications for the planet in the 21st century may be just as dramatic as those that occurred when the New World began to exchange species and diseases with Europe, Africa, and Asia several centuries ago (Mann 2005).

In the context of these changes in the distribution of species and the human forces shaping it, further development of the concept and practice of biosecurity is necessary (Meyerson, Meyerson, and Reaser, in press). Biosecurity should comprehensively address not only chronic threats such as invasive alien species that erode the security of natural resources, food resources, human health, and economy stability, but also acute situations such as sudden offers of aid in the form of seed in a famine or natural disaster.

Well-intentioned policy solutions may also have unforeseen effects. For instance, the Convention on Biological Diversity (CBD) has had unintended consequences not only for species, but also for the science needed to catalog and understand them. Some smaller nations, without the resources to assess their own biodiversity or the capacity to enter into international scientific collaborations under the CBD, have essentially closed their doors to foreign researchers. Meanwhile, logging and other practices that cause habitat destruction and species loss are much less constrained in practical terms (see this volume, chapter 15).

Similarly, long-term policy choices may be overtaken by events. In response to recent high food prices and food supply uncertainty, there appears to have been a significant decline in resistance to the use of genetically modified crops and organisms (GMOs). Faced with life- and health-threatening disruptions such as hurricanes or droughts and the possibility of starvation, concerns over the potential ecological effects of GMOs may be quickly put aside in favor of immediate food supply (Zerbe 2004). In a crisis, the factor that determines the introduction of new species or genotypes could well be the willingness of particular donors at that moment rather than consideration of long-term effects. In human society as well as in ecosystems, major disturbances can provide opportunities for the introduction of new species and diseases.

Our global-scale unplanned ecological experiment has been under way for centuries, since the development of worldwide trade routes (Mann 2005). It has accelerated with human advances such as the green revolution, technological discoveries and improvements, and ubiquitous and constant air travel. In the long run, the ability to continue improvement in human well-being and to extend its benefits to the largest number of people may depend on our ability to identify and maintain the equilibrium between conservation and production. In other words, humans need to understand and maintain biodiversity in a way that maximizes the multiple immediate and long-term benefits for humans. Identifying that optimal condition depends on our understanding of the effects of biodiversity on human long-term well-being, the global heterogeneity of needs and resources over time, and the varying circumstances among regions. This understanding is still in its early stages.

Planetary processes such as climate change, population growth, and the introduction and spread of invasive species (including emerging infectious diseases) further complicate the situation and the challenges. Recent heightened concern about the limits of the global agricultural system to supply both the food and fuel needs of the human population may have the welcome side effect of broadening public and political understanding about the interconnectedness and fragility of the biological assets of the planet and their direct link to human health and well-being. In response to these serious challenges, this book adds to our understanding of the complex links between biodiversity and human health, with the hope that this knowledge will lead to decisions that maintain the best possible balance.

References:

Mann, C. R. 2005. *1491: New revelations of the Americas before Columbus.* New York: Knopf.

Meyerson, F. A. B., L. A. Meyerson, and J. K. Reaser. In press. Biosecurity from the ecologist's perspective: Developing a broader view. *International Journal of Risk Assessment and Management.*

United Nations Department for Economic and Social Information and Policy Analysis Population Division. 2007. *World population prospects: The 2006 revision.* New York: United Nations.

Zerbe, N. 2004. Feeding the famine? American food aid and the GMO debate in Southern Africa. *Food Policy* 29 (6): 593–674.

About the Editors

DR. OSVALDO SALA is the Sloan Lindemann Professor of Biology at Brown University. As president of the Scientific Committee on Problems of the Environment and a coordinating lead author of the Millennium Ecosystem Assessment, Sala is an international leader in ecological science and global environmental policy. Osvaldo Sala has explored several topics throughout his career from water controls on carbon and nitrogen dynamics in arid and semi-arid ecosystems to the consequences of changes in biodiversity on the functioning of ecosystems, including the development of biodiversity scenarios for the next 50 years. Osvaldo Sala is an elected member of the American Academy of Arts and Sciences, the Argentinean National Academy of Sciences, and the Argentinean National Academy of Physical and Natural Sciences.

DR. LAURA A. MEYERSON is an Assistant Professor in the Department of Natural Resources Science at the University of Rhode Island. Her research addresses the mechanisms underpinning plant invasions and the consequences of this process for ecosystem function and the provision of ecosystem services to humans. She has also made significant contributions in the area of environmental policy at national and international levels working with the U.S. Environmental Protection Agency, the U.S. National Invasive Species Council, the Global Invasive Species Programme, and the Convention on Biological Diversity. Laura Meyerson started her work in policy as an Environmental Fellow with the American Association for the Advancement of Science and later as a staff scientist at the H. John Heinz III Center for Science, Policy and the Environment. She seeks to generate new scientific knowledge and clearly communicate this science to the public, policy makers and the scientific community to bringing invasive species and biosecurity issues to the center stage of policy.

DR. CAMILLE PARMESAN is an Associate Professor in Integrative Biology at the University of Texas at Austin. She was Lead Author for the Third Assessment Report of the Intergovernmental Panel on Climate Change, which won the Nobel Peace Prize in 2007. She has been active in climate change programs for international conservation organizations, as well as being an expert advisor to the U.S. and other governments. Her field work has concentrated on impacts of climate change on butterfly ecology and distributions, while synthetic work has embraced global-scale syntheses of biological responses to anthropogenic global warming across broad taxonomic groups in aquatic and terrestrial systems.

List of Contributors

Dana Blumenthal
USDA-ARS, Rangeland Resources
  Research Unit
Crops Research Laboratory
1701 Center Avenue
Fort Collins, CO 80526
USA
E-mail: dana.blumenthal@ars.usda.gov

Hélène Broutin
Fogarty International Center
National Institutes of Health
9000 Rockville Pike
Building 16, Room 303
Bethesda, MD 20892
USA
E-mail: broutinh@mail.nih.gov

Paul Cox
Executive Director
Institute for Ethnomedicine
3214 North University Avenue, #316
Provo, Utah 84604
USA
E-mail: paul@ethnomedicine.org

Peter Daszak
Consortium for Conservation
  Medicine
460 West 34th Street, 17th Floor
New York, NY 10001
USA
E-mail: daszak@conservationmedicine
  .org

Marie-Laure Desprez-Loustau
UMR Biodiversité Gènes et
  Ecosystèmes
Equipe de Pathologie Forestière
INRA Bordeaux
71 avenue Edouard Bourlaux
33883 Villenave d'Ornon cedex
France
E-mail: loustau@bordeaux.inra.fr

Carolyn Friedman
School of Aquatic and Fishery Sciences
University of Washington
Box 355020
Seattle, WA 98195
USA
E-mail: carolynf@u.washington.edu

Alain Froment
Eco-anthropologie et ethnobiologie
IRD UR 092
Musée de l'Homme
75116 Paris
France
E-mail: afroment@anth.umd.edu
E-mail: froment@mnhn.fr

Jean-François Guégan
Centre IRD de Montpellier
Génétique & Evolution des Maladies
  Infectieuses
UMR 2724 IRD-CNRS
911 avenue Agropolis
BP 64501
34394 Montpellier cedex 5
France
E-mail: guegan@mpl.ird.fr

Robert W. Howarth
Department of Ecology and
  Evolutionary Biology
Cornell University
E311 Corson Hall
Ithaca, NY 14853
USA
E-mail: rwh2@cornell.edu

Kerstin Hund-Rinke
Fraunhofer-Institut für
  Molekularbiologie und Angewandte
  Oekologie IME
Postfach 12 60
57377 Schmallenberg
Germany
E-mail: kerstin.hund-rinke@ime.
  fraunhofer.de

Peter Jutro
National Homeland Security Research
  Center
Office of Research and Development
U.S. Environmental Protection Agency
  (EPA 8801 R)
1200 Pennsylvania Avenue, NW
Washington, DC 20460
USA
E-mail: jutro.peter@epa.gov

Stephen R. Kellert
Yale University
School of Forestry and Environmental
  Studies
205 Prospect Street
New Haven, CT 06511
USA
E-mail: stephen.kellert@yale.edu

Kevin D. Lafferty
University of California Santa Barbara
Marine Science Institute
Santa Barbara, CA 93106
USA
E-mail: lafferty@lifesci.ucsb.edu

Anne Larigauderie
DIVERSITAS Executive Director
DIVERSITAS Secretariat
Muséum National d'Histoire Naturelle
57 Rue Cuvier
CP41
75231 Paris Cedex 05
France
E-mail: anne@diversitas-international.org

Prof. Dr. Pim Martens
Professor of Sustainable Development
Director ICIS
International Centre for Integrated
    Assessment & Sustainable
    Development
Maastricht University
PO Box 616
6200 MD Maastricht
The Netherlands
E-mail: p.martens@icis.unimaas.nl

Luiz A. Martinelli
CENA, Av. Centenario 303
13416-000, Piracicaba SP
Brazil
E-mail: martinelli@cena.usp.br

Asit Mazumder
Department of Biology
University of Victoria
3800 Finnerty Road
Victoria, BC V8W 3N5
Canada
E-mail: mazumder@uvic.ca

Frederick A. B. Meyerson
Natural Resources Science
University of Rhode Island
1 Greenhouse Road
Kingston, RI 02881 USA
E-mail: fmeyerson@uri.edu

Laura A. Meyerson
Natural Resources Science
University of Rhode Island
1 Greenhouse Road
Kingston, RI 02881
USA
E-mail: Laura_Meyerson@uri.edu

Charles Mitchell
Department of Biology
University of North Carolina
Chapel Hill, NC 27599
USA
E-mail: mitchell@bio.unc.edu

Camille Parmesan
Integrative Biology C0930
1 University Station
University of Texas
Austin, Texas 78712
USA
E-mail: parmesan@mail.utexas.edu

Anne-Hélène Prieur-Richard
DIVERSITAS Science Officer
DIVERSITAS Secretariat
Muséum National d'Histoire Naturelle
57 Rue Cuvier
CP41
75231 Paris Cedex 05
France
E-mail:anne-helene@diversitas-
    international.org

Aditya Purohit
Indian National Science Academy
181/1 Dobhalwala
DEHRADUN - 248001
Uttarakhand
India
E-mail: purohit_aditya@hotmail.com

David J. Rapport
Principal
EcoHealth Consulting
217 Baker Road
Salt Spring Island
British Columbia V8K 2N6
Canada
E-mail: eco_health@hotmail.com

Marie Roué
CNRS-MNHN
Départment Hommes Natures Sociétés
57 rue Cuvier
75231 Paris
France
E-mail: roue@mnhn.fr

Osvaldo E. Sala
Department of Ecology and
    Evolutionary Biology
Brown University
Box 1951
Providence, RI 02912
USA
E-mail: Osvaldo_Sala@Brown.edu

Manju Sharma
C5/10 (GF)
Vasant Kunj
New Delhi-110070
India
E-mail: manjuvps@gmail.com

Suzanne Skevington
World Health Organization Centre for
    the Study of Quality of Life
Department of Psychology
University of Bath
Bath BA2 7AY
United Kingdom
E-mail: s.m.skevington@bath.ac.uk

Matthew B. Thomas
Center for Infectious Disease Dynamics
    and Department of Entomology,
    Chemical Ecology Lab, Penn State
University Park 16802, PA
USA
Email: mbt13@psu.edu

Alan Townsend
Institute of Arctic and Alpine Research
    (INSTAAR) & Dept. of Ecology and
    Evolutionary Biology
Campus Box 450
Boulder, CO 80309
USA
E-mail: Alan.Townsend@colorado.edu

Andrew Wilby
Department of Biological Sciences
University of Lancaster
Bailrigg, Lancaster
LA1 4YQ
United Kingdom
E-mail: a.wilby@lancaster.ac.uk

Anne Winding
National Environmental Research
    Institute, Aarhus University
Box 358
Frederiksborgvej 399
4000 Roskilde
Denmark
E-mail: aw@dmu.dk

SCOPE Series List

SCOPE 1–59 are now out of print. Selected titles from this series can be downloaded free of charge from the SCOPE Web site (http://www.icsu-scope.org).

SCOPE 1:   *Global Environment Monitoring*, 1971, 68 pp
SCOPE 2:   *Man-made Lakes as Modified Ecosystems*, 1972, 76 pp
SCOPE 3:   *Global Environmental Monitoring Systems (GEMS): Action Plan for Phase I*, 1973, 132 pp
SCOPE 4:   *Environmental Sciences in Developing Countries*, 1974, 72 pp
SCOPE 5:   *Environmental Impact Assessment: Principles and Procedures*, Second Edition, 1979, 208 pp
SCOPE 6:   *Environmental Pollutants: Selected Analytical Methods*, 1975, 277 pp
SCOPE 7:   *Nitrogen, Phosphorus and Sulphur: Global Cycles*, 1975, 129 pp
SCOPE 8:   *Risk Assessment of Environmental Hazard*, 1978, 132 pp
SCOPE 9:   *Simulation Modelling of Environmental Problems*, 1978, 128 pp
SCOPE 10: *Environmental Issues*, 1977, 242 pp
SCOPE 11: *Shelter Provision in Developing Countries*, 1978, 112 pp
SCOPE 12: *Principles of Ecotoxicology*, 1978, 372 pp
SCOPE 13: *The Global Carbon Cycle*, 1979, 491 pp
SCOPE 14: *Saharan Dust: Mobilization, Transport, Deposition*, 1979, 320 pp
SCOPE 15: *Environmental Risk Assessment*, 1980, 176 pp
SCOPE 16: *Carbon Cycle Modelling*, 1981, 404 pp
SCOPE 17: *Some Perspectives of the Major Biogeochemical Cycles*, 1981, 175 pp
SCOPE 18: *The Role of Fire in Northern Circumpolar Ecosystems*, 1983, 344 pp
SCOPE 19: *The Global Biogeochemical Sulphur Cycle*, 1983, 495 pp
SCOPE 20: *Methods for Assessing the Effects of Chemicals on Reproductive Functions*, SGOMSEC 1, 1983, 568 pp
SCOPE 21: *The Major Biogeochemical Cycles and Their Interactions*, 1983, 554 pp
SCOPE 22: *Effects of Pollutants at the Ecosystem Level*, 1984, 460 pp

SCOPE 47: *Long-Term Ecological Research. An International Perspective,* 1991, 312 pp

SCOPE 48: *Sulphur Cycling on the Continents: Wetlands, Terrestrial Ecosystems and Associated Water Bodies,* 1992, 345 pp

SCOPE 49: *Methods to Assess Adverse Effects of Pesticides on Non-target Organisms, SGOMSEC 7,* 1992, 264 pp

SCOPE 50: *Radioecology after Chernobyl,* 1993, 367 pp

SCOPE 51: *Biogeochemistry of Small Catchments: A Tool for Environmental Research,* 1993, 432 pp

SCOPE 52: *Methods to Assess DNA Damage and Repair: Interspecies Comparisons, SGOMSEC 8,* 1994, 257 pp

SCOPE 53: *Methods to Assess the Effects of Chemicals on Ecosystems, SGOMSEC 10,* 1995, 440 pp

SCOPE 54: *Phosphorus in the Global Environment: Transfers, Cycles and Management,* 1995, 480 pp

SCOPE 55: *Functional Roles of Biodiversity: A Global Perspective,* 1996, 496 pp

SCOPE 56: *Global Change, Effects on Coniferous Forests and Grasslands,* 1996, 480 pp

SCOPE 57: *Particle Flux in the Ocean,* 1996, 396 pp

SCOPE 58: *Sustainability Indicators: A Report on the Project on Indicators of Sustainable Development,* 1997, 440 pp

SCOPE 59: *Nuclear Test Explosions: Environmental and Human Impacts,* 1999, 304 pp

SCOPE 60: *Resilience and the Behavior of Large-Scale Systems,* 2002, 287 pp

SCOPE 61: *Interactions of the Major Biogeochemical Cycles: Global Change and Human Impacts,* 2003, 384 pp

SCOPE 62: *The Global Carbon Cycle: Integrating Humans, Climate, and the Natural World,* 2004, 526 pp

SCOPE 63: *Alien Invasive Species: A New Synthesis,* 2004, 352 pp.

SCOPE 64: *Sustaining Biodiversity and Ecosystem Services in Soils and Sediments,* 2003, 308 pp

SCOPE 65: *Agriculture and the Nitrogen Cycle,* 2004, 320 pp

SCOPE 66: *The Silicon Cycle: Human Perturbations and Impacts on Aquatic Systems,* 2006, 296 pp

SCOPE 67: *Sustainability Indicators: A Scientific Assessment,* 2007, 448 pp

SCOPE 68: *Communicating Global Change Science to Society: An Assessment and Case Studies,* 2007, 240 pp

SCOPE 69: *Biodiversity Change and Human Health: From Ecosystem Services to Spread of Disease,* 2008, in press

SCOPE 70: *Watersheds, Bays, and Bounded Seas: The Science and Management of Semi-Enclosed Marine Systems,* 2008, in press

SCOPE Executive Committee 2005–2008

President
Prof. O. E. Sala (Argentina)

Vice President
Prof. Wang Rusong (China-CAST)

Past President
Dr. J. M. Melillo (USA)

Treasurer
Prof. I. Douglas (UK)

Secretary-General
Prof. M. C. Scholes (South Africa)

Members
Prof. W. Ogana (Kenya-IGBP)
Prof. Annelies Pierrot-Bults (The Netherlands-IUBS)
Prof. V. P. Sharma (India)
Prof. H. Tiessen (Germany)
Prof. R. Victoria (Brazil)

Index